PRECIPICE

DANIEL POLLOCK

This book is a work of fiction. Names, characters, places, and incidents are either products of the author's imagination or are used fictitiously. Any resemblance to actual events or locales or persons, living or dead, is entirely coincidental.

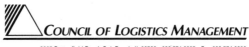

2805 Butterfield Road, Oak Brook, IL 60523 • 630.574.0985 • Fax 630.574.0989
E-Mail: clmadmin@clm1.org • Web Site: www.clm1.org

For information, address: Council of Logistics Management, 2805 Butterfield Road, Suite 200, Oak Brook, IL 60523.

ISBN: 0-9658653-0-4

Council of Logistics Management hard cover printing: September 1997

Front cover design by Evolution Communications Design

Printed in the U.S.A.

ACKNOWLEDGMENTS

From concept to final manuscript, this novel has been a collaborative project. The primary triad consisted of Elaine M. Winter, director of communications and research for the Council of Logistics Management; Joel L. Sutherland, vice president of logistics, Con Agra/Monfort and chairperson of the Logistics Novel Project Committee for the Council of Logistics Management; and Toby Stein, project and book editor. I owe a debt to all three for their guidance and forbearance. In the book's creation, Toby Stein's involvement was passionate and unstinting, her contributions immeasurable. Many of my favorite parts of the book are hers.

I wish to thank the Council of Logistics Management's 1994-97 Executive Committee members; Mark Garvey, editor of *Writer's Market*, who helped determine the feasibility of the logistics novel project; and the Logistics Novel Project Committee: Kenneth B. Ackerman, President, K. B. Ackerman Company; Cheryl S. Byrne, President/Owner, Marketpower; Robert Lorin Cook, Professor of Marketing and Logistics, Central Michigan University; Craig M. Gustin, Principal, CGR Management Consultants; and Donna K. Richmond, President, Richmond Group.

And special thanks to Council members who responded to repeated queries and solicitations for "war stories" and intriguing supply chain problems and solutions, as well as to my experts and guides: Joseph E. Angle, Manager of Warehousing and Transportation, Joico Laboratories, Inc.; Rosemary Coates, Senior Manager, KPMG Peat Marwick LLP; John S. Dillon, Facility Manager, Montgomery Ward West Coast Combined Distribution Facility; James J. Gill, Vice President, California Cartage Company; Christopher Hamm, Assistant Manager,

Ralph's Grocery Company; Franklyn C. Hathaway, Vice President Customer Operations, Avery-Dennison Corporation-Office Products; Andrea L. Manning, Communications Coordinator, Council of Logistics Management; Charles E. Marr, Group Supply Chain Manager, Hewlett-Packard Company; C. Roy Martin, Director, Customer Partnership and Service, Avery-Dennison Corporation; Ken Mason, Director Technology Group, United Parcel Service-Worldwide Logistics; William C. McCaughey, Director of Logistics Services, Joico Laboratories, Inc.; Robert Michel, Technical Coordinator, Newscom, *Los Angeles Times;* Masao Nishi, Vice President, Sabre Decision Technologies; Larry Pryor, Professor of Journalism, University of Southern California and Director of the Annenberg Center for Online Publishing; Mike Self, Corporate Director of Scheduling, Con Agra/Monfort; Richard C. S. Scott, Vice President, International Transportation Service; Colonel Harry G. Summers, Distinguished Fellow, Army War College; Joe Tomasone, Digital Radio Officer, Suffolk County, NY; Steve Whitney, President, Friends of Meigs Field (Chicago); Jack R. Wiley, Manager of Transportation, Hunt Wesson, Inc.; Guy L. Wilks, Community Relations-Southeast California District, United Parcel Service; and Philip Worwa, San Diego Regional Sales Manager, Club Demonstration Services.

P R O L O G U E

DHAHRAN, SAUDI ARABIA
FEBRUARY 25, 1991

With air raid sirens wailing all over Dhahran, Captain Jane Akers gunned her Humvee toward the harbor district, gripping the wheel tightly enough to turn her knuckles white. The real threat of the nightly Scud attacks, she decided, was not the missiles themselves. They either broke up in-flight or were knocked down by Patriot antimissiles. Far more dangerous was the ground war being fought by Saudi motorists. In their flowing robes and gas masks, they looked like giant hooded insects— and drove like madmen, totally ignoring oncoming vehicles as they scanned the skies for ballistic fireworks. Of course, if one of the vehicles she was battling for the road decided to hit her head-on, her own helmet might prove useful. Her gas mask, on the other hand, would probably be imbedded in her face. A death mask.

No. If she was going to die in the service of her country, it wouldn't be three hundred miles from the real action. Besides, if her life was to be aborted by a berserk motorist—Jane swerved out of the path of a careening truck—it ought to happen in Paris or Rome, where she would at least have eaten a decent last meal.

The truth was, on this Day Three of the ground offensive against Saddam Hussein's forces, Dhahranian drivers were not the only reason she wanted to be closer to Kuwait or southern Iraq. There she might be of some use.

"Don't get left in the dust," General Schwarzkopf had instructed his logistics chief, Lieutenant General William G.

"Gus" Pagonis, in the early days of Desert Shield. And ever since Jane's arrival in Saudi Arabia six months ago, she and her colleagues in Pagonis's 22nd Support Command had worked eighteen-hour days to ensure that their supply lines could keep pace with *any* assault. In some cases they were actually able to *outpace* an assault by setting up huge supply depots near the Iraqi border *before* the ground war was launched.

Right now, in her mind's eye, she could see an armada of Chinook transport helicopters crossing hundreds of miles of battlefront, to provide fuel, water, food, and ammunition for the fighting advance. Backing up the Chinooks would be truck convoys crisscrossing the rocky desert with ammunition crates, boxed rations, tankloads of water, medical supplies, pallets of Pepsi, portable showers, latrines—even Abrams battle tanks.

In recent weeks Jane's "log team" had spent time working on the logistical problems posed by POWs. Now that Saddam's soldiers were surrendering by the battalion, hundreds of emptied trucks and buses would be filled for their return from the front with live cargo—captured Iraqis. Each of them would have to be cleaned and clothed, housed and fed, and given urgent medical attention.

Urgent—the word that accurately described all her assignments up to now. Every single day she'd felt useful in a way life had denied her previously. No doubt what she was in Dhahran to accomplish would also prove useful, but the assignment to mesh timetables with the logistics officer of the 14th Quartermaster Detachment, whose job would be to enter Kuwait as soon as the fighting stopped and set up large water-purification units . . . well, it felt bland compared to what she'd been doing.

But this was the army, and you did what they gave you to do.

Right, Captain. Jane saluted herself mentally, and refocused. Her meeting, just up ahead in a converted-warehouse barracks in Khobar City, might not seem urgent right now, but Jane knew that if any segment of the war's logistics was out of sync with another, it would eventually create a problem.

Right now, just getting to her meeting, not three miles from the military air base, was becoming a problem Jane saw as she was yanked back to the street scene by several cars lane jumping to get to the side of the road. She slowed the Hummer to watch as their drivers got out, all eyes raised toward an incoming Scud. The air raid sirens kept up their frenzied whoop, as though there were actual danger. The next sounds would be the mighty kettledrum roar of a Patriot antimissile battery, then the sonic boom, followed by the midair interception of the Scud.

No roar. No boom. What in hell—

It wasn't possible! Jane peered up through the windshield, blinked at a fire streak racing across the desert night. Damn! The Scud was coming right toward her!

Behind the Hummer, tires screeched, metal crumpled. Jane veered sharply and pulled off the road. There was a lot of shouting in Arabic by robed figures running into the night. With an unfamiliar sense of helplessness, Jane followed the flaming trajectory as it slammed to earth beyond the other side of the road in a blinding orange flash. Her wide-wheelbased vehicle rocked in the blast wave, so it was impossible to tell if her body was shaking of its own accord. Then came a roar and a rain of sparks and shrapnel across the Humvee's roof and hood. She opened her eyes to see fiery debris falling onto a panel truck, which skidded across the road, then slammed into a parked taxi.

You can't just sit here, soldier! Jane wrenched open the door, scrambled out. People were scattering in all directions. A young man dodged across six lanes of highway to a chain-link fence already fringed with spectators. Beyond were red-and-yellow flames—in precisely the area of Khobar City where Jane was supposed to report. Darting through gaps in the chaotic traffic, she, too, raced across the road and joined the crowd at the fence. Where the Al Khobar warehouse-barracks had stood, she beheld an inferno.

At this hour a hundred people could easily have been inside. Suddenly she was in the belly of the war, and it was terrifying. Jane backed away from the fence, her body moving of its own accord, her mind having shut down rather than take in the incredible tragedy.

Think! Her assignment voided, plainly what she had to do was get back in the Hummer and get the hell out of there. Make room for others—*qualified* others—who would be converging on the disaster scene. They'd be arriving soon—emergency crews, MPs and Saudi police, firefighters, medics. People trained for nightmare duty. She wasn't! She'd only be in the way.

She was starting back toward the Hummer when shouting off to her right commanded her attention. Several men were crowding through a gap in the chain-link fence, forcing it wider, beckoning others to follow them.

They didn't mean her; they weren't even looking her way. She turned and kept walking. Stopped. *You're going through that hole and do what you can to help, soldier.* Her face clenched, Jane hurried to join the group squeezing through the opening in the fence. Her gas mask hood caught and tore, rendering it useless. She yanked it off and was assaulted by a sulfurous stench. Conventional explosive. Hope to God the warhead was, too. Without the mask goggles she could see better. She wasn't at all sure that qualified as a blessing.

She made her way through the dark field to a narrow road—and stopped. Beyond the road, the blaze held dominion. Through coiling flames and showering sparks was visible a twisted skeleton of steel girders. The warehouse roof and walls were gone, evaporated in the explosion.

In a moment her fear likewise evaporated, and Jane advanced into the searing heat, through a maze of parked trucks and paralyzed onlookers. A muscular black man, wearing only his dog tag and underwear, stumbled against her, swiping at blood on his chest. Jane shouted at him, asking how she could help. The man pointed to his ear, shook his head, lurched on. Two more men staggered past her, one naked, both wounded and screaming. Other dazed survivors, men and women, sagged against rescuers now streaming out of the barracks beside the leveled warehouse.

A gas-masked soldier dashed toward the blaze, stopped short, backpedaled—yelling, "Gunfire!" Jane heard it, too, over the crackle—a sound like M-16 rifle bursts. But more likely stored ammunition, she told herself, exploding in the flames. Which didn't make it any less deadly.

She looked in vain for ambulances, helicopters, paramedics. It didn't make sense just to be part of a swarming crowd. But what could she do? The answer came hopping toward her through swirling smoke—a fire-grimed GI, propelled on one bare foot, nothing showing below the bloodied cuff fringe of his other pants leg. He fell just out of Jane's reach. Instinctively she stepped forward and levered him up onto his good leg, bracing him with her shoulder and arm, then screamed for help.

A tall, long-jawed man in T-shirt and blue jeans came out of nowhere. "You hang on," he told the GI. "We're gonna get you help. I can carry him, Captain. You just tell me where."

But Jane had no idea where. And there was no time to plan this one out. The badly wounded youth, now cradled in the big stranger's arms, could bleed to death in minutes. She scanned the rubble-strewn, vehicle-clogged street. Beyond an overturned pickup was a school bus, dented by shrapnel, its yellow sides charred, the windows blown out—but at least it was resting on its wheels. If they could just get it going, she thought, it could bulldoze all the other wreckage out of its way.

"The bus!" she yelled. "The nearest hospital is just north of the airport—five minutes from here!"

The bus's rear exit door was off its hinges. Jane jumped in

and swept shattered glass from the nearest bench seat, not even feeling the sliver that ended up in the side of her hand. Her helper eased the young man down, then charged forward.

"No keys, Captain!" he called back. "But I think I can hot-wire it."

"Do it!" Jane yelled. "But first give me your T-shirt so I can bandage this boy's foot." She had no time to entertain her panic as she wrapped a pressure bandage around the young soldier's ankle stump.

He'd be okay for a little while. She glanced outside at the teeming street. A young woman in a bloody bathrobe, arm pressed against her chest, was staring blankly straight ahead of her. "Be right back!" Jane shouted.

She scrambled off the bus. "You hit?"

"Flying glass," the woman said, eyes returning to that awful place. "Couldn't pull it out to stop the bleeding."

"Come on—inside—we'll get you to the hospital."

Settling her beside the injured soldier, Jane said, "Keep his leg elevated and keep talking to him. We'll get you both to the hospital as fast as we can."

She rushed forward. Her helper was on his back, wedged under the steering wheel, probing with a flashlight and swearing nonstop.

Never waste transportation space. Logistics 101. "How much time till you get this thing going?" she shouted. "There's a lot of people out there who could use a ride to the hospital."

"It'll take a few more minutes to get the son of a bitch rolling," he growled back. "You round up some of those folks."

Instituting a one-quick-glance triage system, Jane moved quickly and, with more strength than she had ever been called on to test, herded sixteen more injured people on board. Those who seemed less injured she enlisted to help her get others aboard. Several were in worse shape than the young man who'd lost his foot, but all looked as if getting to a hospital soon might save them.

There were still no emergency crews. Across the street others were hoisting survivors onto a flatbed truck.

The tall, shirtless man was in the driver's seat now, bent forward, peering beneath the dash.

Jane glanced around. One soldier was weeping from his excruciating pain; a luckier one had passed out. "We gotta go *now*," she yelled in the ersatz mechanic's ear. "Or move everyone to that flatbed."

The reply was a sizzling spark and a long spasm of grinding while he tromped on the gas pedal. Then the engine rumbled to life, the entire bus frame shuddering as they pulled into traffic. "She's raring to go, Captain."

"Get us through that." She pointed at the abandoned vehicles jamming the street ahead.

"Right. Tell everybody to hang on."

A few minutes later, as they careened back onto Ahd Dhahran Street, nearing the hospital, Jane heard the distant thrash of choppers. Finally.

It was over weak coffee in the canteen, only after they'd turned over their busload of injured to the hospital staff, that Jane found out the name of her volunteer bus thief—Cheval Johnson: "Chevy since the day I found out that cheval is horse in French and decided I'd rather be a car."

"And what did you decide to be after that?"

Chevy laughed. "A pilot. Getting around, one way or the other, that's my motto."

He filled in a few details. A commercial pilot and major in the air force reserve, he was currently flying C-141s for the Military Airlift Command and had been billeted next door to the destroyed barracks.

"Nice job of hot-wiring," she said, sipping from the steaming cup of coffee.

"My misspent youth finally paid off."

"Sorry about commandeering your T-shirt," she said, laughing as she pointed at the hospital gown he had put on in its place. Despite his skewed, youthful smile, Jane guessed that Chevy Johnson had ten or twelve years on her twenty-five. But this was no day to bank on a guess. Hadn't her own face, caught in a hospital rest room mirror, shocked her? What she'd just witnessed seemed scrawled all over it.

Bodies were still arriving at the hospital, more going to the morgue now than Emergency. All those she and Chevy had helped evacuate were still alive on arrival, although one young man was burned so badly, the doctor upstairs told them his chances of pulling through were not good.

"What are you gonna do when all this is over?" Chevy asked. "Make a career of the service?"

"Don't think so. But I'm going to stick with logistics—what I'm doing here—definitely. That's my thing, like getting around is yours." She smiled. "Working here has been an extraordinary experience, but I don't want to spend the next ten years waiting for another that comes close. So I'll probably be looking for a job with a private company."

"Like?"

"It almost doesn't matter, at least not right away. Every business has logistical parts to their operation—and plenty of them are fouled up." She grinned. "I like finding out how things work, then figuring out ways to make them work better. Later on, if I can prove my stuff in a company, or a big consulting firm, I may want to go out on my own. How about you, flyboy?"

"I'll go back to flying commercial. I like the life. And, heck, I've invested a lot of time perfecting my drawl."

"Well, if I ever hear you over the intercom at thirty-seven thousand feet, I'll knock on your flight deck door."

"You do that, Captain Akers."

That sounded like good-bye to her. She stood. "I'm going to check on some of our passengers."

"I'll come with you," he said. "You always have so many thoughtful ideas in one day?"

The badly burned young soldier was dead. "Damn," Jane said. "Damn damn *damn.*"

He held her until she got her tears in check.

"Sorry," she said, moving away. She glanced at her watch. Chevy had said he was hitching a ride with a reporter back to the air base. "Your ride must be waiting."

"He's probably still trying to get through to his editor."

Jane definitely wasn't in any hurry. She was waiting to be picked up by an NCO in her group and find out whether the logistics officer of the 14th had made it. Someone else would retrieve her Humvee. Even if she could get back to the vehicle, she felt way too shaky to drive.

They walked together toward the nearest bank of elevators. Halfway there, Chevy stopped her and kissed her on the cheek.

"What's that for?"

"For you," he said. "You're quite a soldier, Jane. Just wanted to say that."

"Back at you," she said.

"And quite a lady," he said, a slight hoarseness in his voice.

She had to tilt her face up to look him in the eye. The corridor was dimly lighted, but that wasn't why their eyes fastened hard. Jane waited, surprised but not shocked by the shudder of desire that ran through her or by her sudden certitude that they would spend what was left of this night together. Somewhere. If there was a better way to obliterate a little of what they'd seen and heard, she didn't know it.

She reached up to touch his rough-stubbled jaw. Her hand was captured by his in midair, her knuckles kissed. Their eyes remained locked.

"Oh, boy," Jane breathed huskily, dropping her head onto his shoulder.

"I'll get you back to base tomorrow."

"Now who's full of good—" Her eyes registered a wedding band on Chevy's left hand, just inches from her face. She stiffened slightly.

Tracking her glance, Chevy grimaced. "Shoulda taken that off. No dice, right?"

She shook her head. "I just can't, Chevy. I'm not making any judgments, it's just that—"

"Hey, no explanations called for."

And hers would certainly not have erased the awkwardness as they said their good-byes. At the last moment, he kissed her. On the forehead. Not quite what her body had had in mind. In her last glimpse of him, Chevy still wore a rueful smile.

A quarter of an hour later, she was being driven away by the wrong man, angry for the thousandth time at her father. If he hadn't . . .

It was all such a long time ago, but she could still remember how stupidly the sun was shining as he drove off to work after the horrendous fight with her mother over *that* woman. Jane was not—ever—going to be the "other woman" and risk tearing some other family apart.

She felt a sting and mindlessly rubbed the place on her palm. *Ouch.* She looked, found a sliver of glass. She had no idea where she'd acquired it. But it felt right. A *little* pain never hurt anyone.

P A R T   O N E

I

She was en route to the taxi stand at Honolulu International when there he was, coming straight at her along the central concourse, a man she knew. Past tense—had known. But who? Dark blue flight blazer slung over one shoulder. Tall, lanky body, moving with the nonchalance of a man used to being admired. Joking with his Orient Air-uniformed crew mates. Then memory dealt its face card, that lopsided grin—topping another uniform, eight years and half a world away.

Just before they would have passed each other, she stepped in front of him.

"Hello, Chevy," she said.

His grin turned quizzical. *Why not? Last time he saw me, I had on a grimy uniform and a grimier face.* Now here she was in a navy pinstriped pant suit, with a good haircut and a not naked face.

He'd changed a bit, too. Hair saltier than she recalled, the face just a little fleshier. His ring finger was bare.

"At least give me a hint," he said.

"Ever driven a school bus, Major?"

"I'll be damned! Hey, guys, you're looking at a genuine Gulf War hero." Chevy's grin now split his face. Jane had the distinct impression that she was about to be vacuumed into a bear hug. Then she saw his focus flick to a willowy, redheaded flight attendant flanking him pretty closely. His open-armed gesture stopped just short of an embrace. "How could I ever forget?"

"So what's my name, Major?"

"Hold on!" He furrowed his forehead. Then, triumphantly,

"Captain Jane Akers, U.S. Army. How's that?"

"Catch you later, Chevy." The redhead wheeled on by.

He tossed her a wave. The rest of the flight crew continued down the concourse in her wake.

"I didn't mean to break up your—"

"Halt right there, Cap'n. I've been cooped up with those mates for the last five hours. God, it's great to see you!" Here came that hug. Red must have turned the corner.

"Jane, look, I'm off duty. Are you? Whatever counts as duty for you these days. Do you have time for a drink?"

"That would be lovely."

"To pure blind luck!" Chevy said, hoisting his glass of ginger ale and tapping hers.

"Once, I thought I recognized your voice, and asked a flight attendant the pilot's name. It was Czernowski or something."

He laughed. "A long-lost cousin, no doubt. We do all sound alike, don't we? But here I am, doing just about what I told you I'd be, flying a big old bus back and forth across the big water. And you"—he indicated her attire—"whatever branch that uniform represents, you've obviously moved up a rank. Is it still logistics? See, I do remember." The smile was warm and obviously genuine.

"Still doing it, still loving it. Worked for other folks until last year, now I have my own little consulting business in San Francisco. Malcolm and Associates."

"If it's your shop, who's Malcolm?"

"I am."

"Oh."

"Don't look so crestfallen, it's my mother's maiden name."

"I don't follow."

"Long story. Not especially interesting."

"But you're not married?"

Jane glanced at his hand. *Given the fella's globe-trotting job and charming ways, Jane-girl, that bare hand doesn't mean a thing.*

Chevy, following her eyes, laughed. "You're right. Fair's fair. Me first. Took me a while, but, yes, I *am* divorced."

"I'm not." Jane managed not to smile—for about a second. "I'm not married, Chevy. Been way too busy."

"Being busy certainly agrees with you. You look . . . "

"Happy? Let you in on a secret. I'm about to land my first blockbuster contract—if tomorrow goes right."

"I'd love to hear about it, Jane. Tell me all about it?"

"I'd really like that, but I'm meeting my associate at the Hilton for a strategy session in"—Jane glanced at her watch—"thirty-five minutes. Wish me luck tomorrow?"

"I doubt you'll need it, but sure. Here's to lavish luck for Cap'n Jane." They touched glasses again. "How about dinner after your strategy powwow? Get your mind off the big game?"

"I thought you wanted to hear all about it?"

"I do, I do." Chevy grinned. "Okay, I'll help you prep. What do you say? That too much for my boyish heart to hope for?"

"You're incorrigible."

"Is that a yes?"

"You're lucky I'm in the kind of mood I'm in."

"*That's* a yes."

She laughed, and stood. "Call me at the Hilton Hawaiian Village, the Alii Tower, with where and when—and if I need a hibiscus in my hair."

"You're going out—tonight!—with some guy you met in the airport?" Mark Gibian, Jane's associate, tried to keep his voice calm and failed.

"I *met* him in Dhahran eight years ago, if that's pertinent. Look, we're ready for any questions that might crop up tomorrow. If we weren't, we'd keep at it."

"You need a good night's sleep."

"Mark! Would you get out of here so I can change?"

From anywhere in the Hanohano Room, the sunset sweep of Waikiki and Diamond Head was breathtaking. Like other diners in the Sheraton's rooftop restaurant, they had been seated side by side, facing the windows. But Chevy had promptly switched so he faced Jane.

"You're missing the dazzling view," she said.

"Don't think so."

Jane had changed into a black silk sheath. You never knew where a client might want to have dinner. "Thank you," she said.

"To heck with mai tais or rum punches, let's celebrate your deal right now." Chevy ordered a bottle of Cristal.

"My word, Chevy. A world-class view, first-class champagne. The job's not in my pocket yet."

"Whatever else I may have blocked out, I haven't forgotten your terrific efficiency. That job's yours."

"I'll drink to that," she said, slipping into his mood. It wasn't hard. The man's charm quotient went off the chart.

She studied the bubbles in her champagne. "People ever ask you at parties—you know, the CNN warriors—if *being* there was even more exciting than watching the war on TV?"

"Maybe once or twice. I'm not much of a party-goer."

"Oh, right."

"I skipped tonight's party, Captain. And I'm all ears. Tell me about your meeting tomorrow."

"You really want to hear this?"

"I'd *like* to hear about *everything* you've been doing since I had to drive off with that reporter, but—"

"You'll settle for hearing about my deal." She laughed, and then she began to tell him about TranSonic Airex, which had been muscling FedEx and UPS among others for the U.S. package distribution market. They wanted to extend the challenge into the Asian air parcel business.

After a brief interruption while they gave their order, Jane said, "What's the chuckle for?"

"Just that I've been flying around the Pacific long enough to remember when cargo was nothing more than an add-on to the passenger service. Now here we are, with Asia exploding as the brand-new frontier—the wild, wild East—and all the big boys are fighting over who gets to move the most brown boxes."

"You've got the picture. Well, over the last few months, nearly a dozen contract logistics consulting outfits devised Asian strategies for TranSonic Airex. My firm was among the least-known players."

"But you outfoxed them all?"

"Won't know for sure till tomor—"

"I thought we settled that."

"Right. If I did, it was equal parts creative thinking and sheer mental drudgery. Those marathon skull sessions we endured in the Gulf are second nature now."

"So what did you come up with?"

"A good idea. You obviously know Asia. TSA—that's what their people call it—wanted to enter the Asian market as a big integrator like FedEx and UPS. Meaning they would offer door-to-door package service across the region, with both airplanes and a ground network—and try to do it better than their rivals."

"A tall order," Chevy said.

"They were gung ho. By the time I entered the picture, they had about decided to build their own air cargo hub at the old Clark Air Force Base in the Philippines."

Chevy whistled. "That's practically on top of where FedEx has its hub. Subic Bay isn't more than fifteen miles from Clark Field. In your face, FedEx, huh?"

"Exactly."

"I'm no logistician," Chevy said, "but I'd vote against that pick. I had my doubts when I heard FedEx was going in there. Manila is—what?—more than an hour from any landmass.

Seven hundred miles from Hong Kong, forget Seoul or Singapore. How'm I doing so far? Maybe you should hire me."

"Go on, make the case."

The arrival of their food gave Chevy a moment to think. He chewed a bite of his steak, then said, "Well, if you were shipping overnight from Bangkok to Hong Kong, say, you'd have to travel the hell out of your way to reach a Philippines sortation hub. You'd end up flying twice as far. Which means a lot of extra time and a lot of fuel."

"Bravo! So why did FedEx pick it? Great curry!"

"Good. Hey, don't ask me. They've made gaffes before. That ZapMail idea? And didn't they make heavy investments in the European package market, then end up pulling back?"

"Yes, but they do a lot of things very well. And there are actually a lot of advantages to the Philippines as a sortation center. Infrastructure's already in place. Subic's a designated economic development zone. There's cheap, nonunion yet technically skilled labor with fluent English. Throw in twenty-four-hour customs clearance, good weath—"

"Enough already!" He topped off her champagne flute. "So how'd *you* talk 'em out of it? Fair warning, when you smile that way you destroy my concentration."

"You want to hear about my little triumph or not?"

"Stop smiling, and I'll listen with both ears. Deal?"

"Deal." Only it suddenly wasn't easy not to smile. When was the last time a man had made her feel so pretty?

For someone who's getting used to taking control of a situation . . .

Jane stopped smiling and picked up her story.

"I used all the nifty arguments you mentioned, but my main point was that they shouldn't go it alone. They need a partner, a company with an existing Asian infrastructure—air and ground network, landing rights, ocean connections."

"Makes sense to me," Chevy said. "So where'd you pick?

Taipei might do nicely, but UPS is already there, right?"

"Right. As for in your face, if you want to do business with China, locating your hub in Taiwan is not exactly the smartest strategic move. I advised TSA against it."

"What about Hong Kong?" Chevy asked.

"Pretty good. Another consultant explored a Hong Kong airline, but couldn't reach an agreement. One more guess."

Chevy was clearly concentrating now. "Let's see. Japan and Korea, too far north; Bangkok and Singapore, too far west—except for overnight deliveries to India." He sighed. "I give up."

"Hey, if you'd have come up with the answer just like that, how would I feel? This one wasn't a snap, believe me."

Jane told him her idea. Chevy chewed on it meditatively, like someone tasting an unfamiliar dish. Then he smiled. Big. "Know what, Cap'n? That's good. Real good."

"Now, if I can just close the deal tomorrow. But enough about world peace," she said. "Tell me about you. Any plans beyond Orient Air?"

Chevy grinned. "Honestly? To do some serious sailing. I've begun saving my shekels for an early retirement—not real soon, but when I do quit, my sights are set on one of those round-the-world things. Take my time about it. Grow a beard. Read all the books I've been meaning to. Put into port here and there."

"Don't you need a boat?"

"Got one. A fifty-footer anchored in Lahaina. If you finish off that champagne, I'll tell you her name. I've sailed her down to the Marquesas and Tahiti, and I'm thinking about maybe entering the Transpac this July, but first I've got to get her to L.A." Chevy shrugged. "Disappointed I'm not planning to start my own airline or something equally ambitious?"

"Nope. I think your plan sounds fantastic. Are you talking about single-handed sailing?"

"I could do it, rig self-steering gear. But I'm no loner. I need a crew mate. I'm kind of on the lookout for one who can't grow a beard. Interested?"

"Around the world? I'd have to check my Day-Timer."

Chevy leaned forward. "Seriously. I've got some time off here. After you close your deal tomorrow, we could celebrate by sailing around some of the islands. There's still time to catch the humpbacks off Maui, Jane. Maybe do some scuba diving, snorkeling, kayaking—you name it."

Suddenly Jane felt a shiver of desire awfully close to the one she'd felt that crazy night in Dhahran—and hadn't felt for quite a while. Hell, recently she scarcely had time to *think* about a romantic life—let alone have one. Just when she's closing in on her big breakthrough, along comes Chevy with his cowboy grin. *Awful good timing, Major.*

Jane tried to visualize them cruising the islands together. But when her mind had him bare chested and sweat slick, he was suddenly behind the wheel of that school bus, driving hell-bent-for-leather to the hospital.

"Please don't frown," he said. "Sorry if I pushed. I guess I was thinking of all the time we've wasted."

Jane sipped the last of her champagne. "I'd love to," she said.

"You mean you will?"

"Let's put it this way. If everybody signs on the dotted line tomorrow, and if I can find an appropriate swimsuit at the Hilton . . . why not? I'll give you a call."

"Her name is *Happy Days*—my boat! I must have been looking into a crystal ball when I named her."

II

All the parties had shown up at the Sea Pearl hospitality suite of the Hilton's Mid-Pacific Conference Center promptly at eight-thirty, but now they weren't exactly mingling. The dark-suited, diminutive men from Macao were sipping tea in one corner of the suite, occasionally exchanging a quiet word or two. The head-higher delegates from Dallas were clustered around the TV in another corner, gulping coffee, devouring sweet rolls, and trading quips as they watched a scrolling stock ticker.

Jane took the situation in hand. "We'd better do something to mix things up," she told Mark.

"Pass," Mark said. "I don't speak Mandarin *or* Texan."

"You'll do fine, Mark. Go!"

They began to circulate, weaving their way through the two groups, smiling and bowing and shaking hands. Cross-pollination was the next step. TranSonic's PR director, Regina Chellis, took the cue and began to escort several of her colleagues on similar social forays. It wasn't ten minutes before Jane discerned a definite conversational buzz. She found herself rather relishing the clash of dialects—Chinese-inflected British versus Lone Star twang.

Suddenly the focus of her attention shifted from sound to sight with the rather tardy arrival of an older Chinese who was not on the list of attendees. His wispy silver hair and pouched eyes in no way diminished the considerable dignity the newcomer radiated. But what was even more noticeable was the deference shown by the members of the Eastern delegation as they lined up to greet him.

"Tony Chan," Mark whispered in Jane's ear. "He left the late evening cocktail reception before you returned from your little outing with your old boyfriend."

Jane ignored the gibe. She'd made sure to get back in time to drop in on the reception. Turned out, few heavyweights from either side had bothered to attend.

She knew who Tony Chan was, of course, which made the extravagant courtesy being extended to him somewhat puzzling. For Chan, who owned several gaming casinos in Macao, was only a minority shareholder in South China Air, the Macao-based airline Jane had selected as TranSonic's Asian partner.

She would have expected *Chan* to be deferential to the four nondescript representatives of China's National Aviation Corporation, which actually controlled 51 percent of the airline. Special casino privileges weighing in?

Whatever the reason, with his arrival, the South China Air delegates began wending their way toward the adjoining conference room. The TSA team promptly joined the flow.

The countdown to signing was on!

Dave Ellsmere, the gangly, gravel-voiced chairman of TranSonic, welcomed everyone, then introduced his team, which included several high-ranking staff members as well as Jane and Mark. James Shen, South China Air's chief executive, did the same for his side of the table. Two of the Chinese men relied on interpreters; all the rest spoke English. The process was deliberately unhurried, partly because some of these people were getting their first look at one another, but also because, as Jane reminded Dave, Asians liked to observe the amenities.

Having had contact with all the parties already, Jane stole this time to scan mentally, one more time, the basics that had been agreed upon in numerous earlier meetings: TranSonic Airex would finance and build a state-of-the-art package sortation hub at the Macao International Airport—their engineers had already inspected the site. In addition to the location, South China Air would be providing landing rights and gates at other major Asian airports, plus ocean connections to the ports of Hong Kong, Yantian, Shanghai, and Taiwan's Kao-hsiung. TSA would be supplying planes—mostly DC-8s, but also 727s, 757s, and 747s.

The deal would constitute a considerable leap forward for Macao, the former Portuguese colony across the Pearl River Estuary from Hong Kong. Macao's airport was still only a few years old; before that the island's air service had been restricted to seaplanes and helicopters. But South China Air had already carved out an impressive regional niche, opening the first direct air link between Taiwan and the People's Republic, ending Hong Kong's long dominance of Taiwanese transit travel to China.

That last fact had convinced Jane that Macao was the right choice to be TSA's partner.

The hardest sell at TSA's headquarters in Dallas had been Macao's reversion to Chinese control. Doing business with what was essentially still an unknown quantity did not at first sit well with some TSA board members concerned about the safety of their investment. To overcome this resistance, Jane had made a full-scale presentation to the company's hierarchy, including a state-of-the-art slide show into which she'd subtly integrated the results of her conscientious research into the political realities of the area. In the end, TSA's management had been won over by the sheer number of multinationals already expanding into China, the exact customer base the air freight company was so hotly pursuing. Shenzhen, China's free-market boomtown across the border from Hong Kong, was now home to giant chains from Europe and Asia, not to mention the world's largest retailer from Bentonville, Arkansas. Hong Kong companies had been doing a thriving business in Beijing for years. Even Taiwan was rapidly shifting its own manufacturing base to the mainland.

But the real prize had been South China Air's routes into the mainland. Jane pointed out that under the current bilateral aviation pact between the U.S. and China, FedEx was limited to four flights a week into the mainland. By partnering with the Macanese airline, TranSonic Airex could make daily flights, thereby trumping their giant overnight rival.

Since Jane's last-minute search for hitches had revealed none, she turned her full attention to the conference table, where the introductions were finally being concluded. The attorneys from both sides had already hammered out the contract language, so today's high-level face-off was mostly to give all these executives an opportunity to familiarize themselves with the terms of the operating agreement—but even more to become better acquainted with one another's working style. The actual signing would take place this evening, after the attorneys reviewed any revisions.

The meeting had moved smoothly down the first page of the program, when, just before the scheduled eleven o'clock break, Mr. Chan asked if this would be a good time to raise an issue not on the agenda.

The faces along one side of the table, especially the representatives from South China Air, made plain that his request was not unexpected, and the Texans went along. So Chan began to describe what he called the "synergistic option" of converting TranSonic Airex cargo planes into passenger carriers on weekends, when some jets were idle. The method, once considered by FedEx and implemented by UPS, was to slide a platform fitted with carpeting and seats through the extrawide cargo doors.

"Our opportunity," Mr. Chan said ingratiatingly, "would be to institute charter flights to bring in people from the U.S. and elsewhere to our tourist properties in Macao."

You mean to your casinos. Jane winced inwardly. TSA's management was not going to be pleased; several of them, according to their profiles, were members of conservative religious denominations. They were willing enough to bet big time on a business venture, but they weren't about to agree to use company planes to feed a Chinese gaming resort.

She jotted down a single word—"Postpone"—and passed the folded paper to Dave Ellsmere. Dave glanced at the note and smiled the tiniest smile as he said, "That's an interesting proposal, Mr. Chan. Certainly one that merits looking into, after we mosey a little farther down the partnership road. I'll put you together with some of our best people. And right now, I think we've all earned a short break before we wrap things up. There's fresh coffee and tea brewing right outside. Suppose we reconvene in twenty minutes?"

"Why do all the folks from South China Air seem to curry favor with Mr. Chan?" Jane asked Mark as they grabbed a moment alone just before the break was over.

"Well, he *is* venerable. Plus I understand he has a lovely daughter, who happens to sit on the airline's board."

Fortunately, after the break Mr. Chan did not pursue his little proposal. Things went so well, in fact, that only a few items remained when they broke for lunch.

Jane had arranged with the Golden Dragon restaurant for the delegations to be shuffled among tables. She sat between a TSA international operations manager and a Chinese translator, who used the time between courses to study his notebook filled with baffling Texas idioms, among which Jane spotted "mosey."

Now, Mark caught her eye across the room and tossed her a fleeting smile of pending victory. He might not be so happy, she thought, if he knew that over delicious green tea ice cream, images of watching a tropic sunset on Chevy's *Happy Days* were floating through her mind.

Why not? God knew she'd earned it. After tonight's signing and a festive dinner, everyone would be flying out in the morning. Mark could handle the office for a couple of days, even three. He'd squawk, but it would do him good.

They had only a final page of timetables to go when Dave Ellsmere stood and said, "I don't know about the rest of you, but I ate too much of that delicious lunch. I need to walk around a little. How about a fifteen-minute break?"

Jane, who never ate much when she was excited, moved through a coffee queue, then along to the cream and cookies. As she was stirring, she glanced at the corner TV. The screen bottom was still side-scrolling stock quotes. The top two-thirds showed a bald, ruddy-faced man. "*ROYAL AKERS*," read the graphic. Below this were birth and death dates.

She sagged against the sideboard, quickly braced herself, managed to set down her coffee cup, then lurched toward the TV, her heart pounding. She found the volume control, turned it up, got a solemn voice-over: "—was killed instantly as his single-engine plane crashed onto his company airstrip in Ocean Plains, Illinois, center of his superstore empire. The death of the sixty-year-old founder and chairman of Akers Corporation, which occurred yesterday but was announced only today, comes at a time when company stock was already trading lower due to unexplained operational problems and a recent expansion into Asian markets. A company spokesperson said no further statement would be forthcoming until after the funeral, which is scheduled for Saturday. Meanwhile, in overseas financial news, Tokyo's Nikkei stock average rose point two percent on—"

Jane backed slowly away from the set, feeling a numbness overtake her that was almost paralyzing. She commanded her feet to move, to take her to the relative privacy of the ladies' room.

"I couldn't find Jane, Mark, and we need to have some input from you people right away. Wiley, tell him."

The tension in the conference anteroom was thick as Wiley O'Neill told about the phone call he'd just been paged for. "That was Silk Road on the line—Jhou's first deputy, no less," said TSA's chief operations manager. "Those guys must have very good hearing, because they've just upped the offer we passed on."

"And how they've upped it!" Dave snorted. "Listen to this, Mark. Go on, Wiley."

"Not only are they ready to offer us a more favorable financial arrangement than Macao. But"—Wiley ticked off the new elements on his fingers—"they're prepared to build the transsortment facility to our specs. Because Guangzhou is inland, real estate costs are much lower. Labor is cheaper there than in Macao. Plus Silk Road offers to contribute a larger air fleet to our joint operation."

It was clear that, despite the fact that the sign-off with South China Air was only a few hours away, at least some of the TSA people were wavering.

"Can you nail down those figures, Wiley?" Ted Hare, the company's CFO, was asking.

"I suppose we could make up some pretext to postpone signing for twenty-four hours until we get something in writing from Silk Road," Dave said. "But, frankly, I like to keep my negotiating lies to a bare minimum. I don't see any problem in taking a little more time before we reconvene, if we can reach a decision here and now."

He turned to Mark. "What about the logistical issues? What do you guys think is critical?"

Mark knew Jane would have his head if he hedged on Dave's direct question. "I think you'll want to hear Jane's view," he said, feeling his way. "But it seems to me she'll say that you've got an excellent deal right here with South China Air. She made *three* visits to Silk Road. It's all in our reports to you."

Ellsmere's eyes told Mark he wanted a more definitive answer.

"I've been looking high and low for you," Mark said. "Hey, what's wrong? You look like you've seen a ghost."

Jane's eyes widened momentarily before she said, "I was just coming to find you."

"Tonight's signing may be off."

"What! What happened?"

He filled her in quickly. "Dave is extending the break for fifteen minutes. He and his crowd are waiting to hear your opinion. I told them I'd get you."

"I'd say what you said."

"Jane, you're head of Malcolm and Associates. They want to hear from *you* that Silk Road's new proposal is not good enough for a switch."

"Where is Ellsmere? I do need to tell him something, but as far as Silk Road goes, you're going to have to finish handling that on your own, Mark. I have to leave. . . . Don't look like that. I'll make sure Dave understands you can speak for us both on this."

"The hell I can. I wasn't on those trips to Silk Road with you. Maybe Ellsmere can postpone reconvening for as long as an hour. Can you be back by then?"

"Mark, I'm not coming back. Listen up. The new stuff they've put on the table is all nice, but none of it eliminates the drawbacks I found, which remain decisive. Remind them that geopolitical realities always have to be part of logistical equations. For instance, here, Guangzhou doesn't have the ocean connections of Macao. They've no history of foreign partnerships—which Macao does have. They haven't got twenty-four-hour customs clearance. You can pick it up from there. Now, I've got to find Dave."

"I can't believe you'd leave now."

"I *have* to leave."

He looked so shocked. She *had* to tell him. "Did you happen to notice on the TV, Royal Akers died?"

"So? We weren't pitching them or anything. You didn't know him personally, did you?"

"Not recently." She took a breath. It hurt. "Oh, God, Mark. He's my father."

III

"So, you see, I have a funeral to go to."

At that very instant Tony Chan appeared, smiling. "Miss Malcolm, I wish to personally invite you and your associate, Mr. Gibian, to a special celebration at my suite this evening."

"I'm so sorry, Mr. Chan, but I won't be able to come. I'm sure Mr. Gibian will be delighted to attend."

Chan's smile remained, but his eyes blinked in puzzlement. "A family matter requires me to return . . . home immediately," she said. "Please forgive me."

The Chinese bowed and moved off, puzzlement intact.

"That went over big," Mark said. "Hey, look, I'm sorry about your father." He touched her shoulder. "You know I am. And I can see on your face that you have to go be with your family. But this isn't just one more job for us, and while they think I'm okay as second banana, it's you they rely on. What if they simply don't understand your leaving?"

"I wouldn't blame them. I'm not sure I understand myself."

The 747 roared steeply up from Honolulu International with Jane the lone passenger. Across the empty cabin, the hotels of Waikiki slid past the windows like pastel dominoes. Then, as the jumbo jet leveled and turned its tail on the late afternoon sun, Jane caught a starboard glimpse of Molokai and Maui, lavender smudges against sparkling sapphire.

She'd managed a thirty-second call to Chevy, who had been sympathetic and disappointed, but mostly mystified. But by then she was all out of explanations—and time.

Almost as soon as she'd heard her father was dead, Jane knew she had to make it home for the funeral. Or where home used to be. She hadn't been back to Ocean Plains since just after she'd returned from the Gulf. That time, her father had offered her a job monitoring logistics at Akers. When she'd turned him down, explaining that she needed to go out on her own and see what she could do with what she'd studied in school and learned in war, she could see he was hurt even though he said he understood. But when, realizing there was only one sure way not to trade on being Royal Akers's daughter, she'd taken her mother's maiden name for her company—and herself—he hadn't even pretended to understand. He was furious. They hadn't spoken since.

Not that she spoke all that often to her mother. But, suddenly, she wanted to know if Joycelyn Malcolm would be coming to her former husband's funeral. On any plane but this one, Jane could call her mother in New York and find out. But there were no air-phones on the cargo jet.

Besides the four seats bolted to the deck plates, the only nods to human comfort were a bathroom and a cardboard box filled with sandwiches and soft drinks—ample for herself and the three-man flight crew on the 747's little top deck. The bulk of the plane, the cavernous cargo hold below, was stuffed with pineapples, a hundred containerized tons.

A strange way to travel. But how kind—and surprising—of Dave Ellsmere to offer it.

Mark had been so afraid that Ellsmere would feel betrayed because she was leaving when TSA had a problem on its plate this close to the signing, but there was barely a beat between Jane's telling Dave Ellsmere what had happened and his offering condolences—and a flight home. He'd called TSA's Honolulu office and gotten her on their Pineapple Express, which was about to leave.

"I'm a schedule buff." He smiled. "You can't make Northwest's five-ten. The next flight, on United, isn't for several hours," he'd said. "I'm sure you want to be with your family as soon as possible. This way, you'll arrive at our L.A. terminal just after midnight and transfer to another cargo flight that'll get you to O'Hare around seven-thirty tomorrow morning. I'm only sorry your first TranSonic perk has to be associated with such a sad occasion."

As she was about to leave, he offered her a last gift. "Your Mark Gibian is a real smart fellow. I wish he had a bit more confidence, but I know he'll be able to see us past this stumbling block. And I'm sure our Asian friends will understand once I tell them what happened. I know they take respect within the family very seriously in their part of the world. Now, get going, Jane. The plane's just waiting for you to arrive so it can take off."

"How could this happen!"

The ponytailed young mother spoke softly, so as not to wake the infant she carried in a front harness sling. But there was no mistaking how upset she was. She pointed again at the baby formula shelves, where fourteen-ounce cans were stocked four deep, their labels depicting happy babies, teddy bears, bunny rabbits—except where periodic gaps indicated the absence of any lactose-free soy formula.

"I prefer mixing my own, but there isn't any liquid, either— not in any brand. Now you come back empty-handed. What am I supposed to feed her?"

"I wish I could tell you how it happened, ma'am," Phil Leavitt said. The manager of Akers Store #28 in Lake Reliance, Wisconsin, was breathing hard from a fruitless search of the back room. "What I *can* do is promise you we'll get it for you as soon as possible."

"I don't want promises. I want Amy's milk. She'll be crying for it as soon she wakes up, which could be any second. And if she doesn't get it—"

"I understand, really I do." Phil ran a hand through his thinning fair hair. "Unfortunately I'm not in a position to provide your little one with what she needs right this minute. But if you'll come back a little later—"

"You the manager?" Phil felt a forceful nudge in his right biceps. He half turned toward a scowling, large-jawed man in a dark green workshirt. Phil recognized the logo of the firm that had repiped his house.

"Right with you, sir."

"How soon?" the mother was asking. Her baby was growing restive in her pouch, wrinkling her tiny face, making mewling sounds.

"Any time after, uh, let's say eleven-thirty." During his hectic search of the stockroom, Phil had paused at a computer terminal and pulled up an inventory screen that showed stock on hand, in transit, or on order—by vendor, item, and department. An emergency shipment of half a dozen critical out-of-stock items, including the soy formula, was en route from their RDC—the regional distribution center for their area in Milwaukee—eighty-five miles away.

"We'll have several cans up front in Customer Service with your name on them." He took out his pen and a notebook. "You're Mrs . . . ?"

"Collier. You guarantee that?"

"Absolutely." Phil nodded confidently, then turned to the large man, who had assumed a cross-armed, belligerent stance. "Appreciate your patience."

"You guys are out of my antihistamine. Can't work without it. Dust kills me."

"Are you sure there isn't *any*?" Phil had caught this shortage himself just before opening and arranged a hasty purchase from a local pharmacy.

"The only kind you got knocks me out. I got a furnace to install today, mister."

"I'm sure Mr. Jones—our pharmacist"—Phil pointed to the "Corner Drugstore" department up front—"can provide you with a satisfactory substitution."

"Some expensive prescription? Besides, my doctor's not in till nine o'clock." The man glanced at his thick wrist. "Damn, I'm running twenty minutes late already."

"Tell Mr. Jones that Phil Leavitt said I'd take responsibility. He should call your doctor at nine and explain the situation—and tell him I said to charge the same as for an over-the-counter antihistamine."

With a dubious grunt the man lumbered off, but Phil realized the mother was still standing there, demanding his attention. "Because if you can't guarantee it, Mr . . . Leavitt, I'll have to drive clear over to Millville. There's nowhere else around here to shop anymore."

Phil flinched inwardly. Shoppers were usually thrilled to be able to find everything on their list under one roof—except when Akers happened to run short of an item. Which hardly ever happened—until a couple of weeks ago. Now even the young ones talked about ShopBest, which had relocated its supermarket from Lake Reliance to Millville a month after Akers opened, as though they never ran out of *anything*. The town's small grocery store, Twombley's, had held out a year before closing. A couple of gas station convenience stores by the interstate carried only snacks and sundries.

"You won't have to go anywhere else"—Phil glanced at his note pad—"Mrs. Collins."

"Collier. And I'll be back at eleven-thirty sharp."

Finally she was gone, and Leavitt ran back to the stockroom. "Steve!" he barked at his receiving foreman. "Call me the instant the DC truck pulls in."

"I was just coming to tell you, boss. Our emergency delivery got canceled."

Phil had spoken to Allen at the distribution center himself. "They can't cancel."

"Well, 'diverted' is the word they used."

"Where the hell to!"

"Graydon Forge's out of truck tires and Lac Nouveau's out of crop pesticide—there's still some planting going on. The DC had to pull our pallets to make room. We have to wait for our regular truck tonight."

"Dammit, Steve! We can't wait that long!"

Jane had just finished her fourth cup of coffee, two newspapers, and her ninth try to reach Mark's room back in Honolulu. And if he wasn't there, why wasn't he answering his cell-phone? She'd managed a few fitful hours of sleep on the cargo flight from L.A. Now she glanced around the O'Hare coffee shop, wall-to-wall with air commuters at 7:35 A.M. That Mark was out after three could mean everything had worked out well and he was at Mr. Chan's party. Or things had gone completely the other way, and he wasn't answering because for the first time in his life, he was sitting alone in the hotel bar well on his way to oblivion.

With only a few minutes before boarding her American Eagle flight to Champaign, Jane decided to try her mother.

"Hi, Mom."

"You heard."

"I just flew in from Honolulu. I was afraid I might miss you."

"At this hour? Darling, you know I never set foot out of my bedroom before ten."

"Sorry I called so early."

"That's all right, dear. We're all upset. God knows, I wished him out of existence a few times, but that was long ago. Now that it's true . . . frankly, it's hard to imagine. Royal *dead*? Whatever else he was, Royal Akers had more life in him than any other two men I've known."

"I gather you're not going to the funeral?"

"Is that where *you're* going?"

"Yes."

Jane wasn't going to *apologize*. She waited.

"You're a grown woman, Jane, and you'll do what you want. But after the way he treated you when all you did was take my name for business reasons—"

"Not now, Mom, please."

"I admit I've had a little cry. There were some good times, a long, long time ago. But I have a show opening next Wednesday at the Mid-Madison Gallery and I have to be here to oversee the hanging. I'm really counting on this show, Jane. You're a successful businesswoman now, while your mother is still trying to make a name for herself in the dog-eat-dog art world here."

"So you're staying in New York?"

"Speaking of which, someone called here wanting to reach you."

"What was the message?"

"They didn't leave any. I gave them your number in San Francisco, darling. How was I to know you were off gallivanting in Hawaii?"

"That's fine, Mother. If it was important, I'll find out soon enough. Who was it, do you know?"

"I think it may have been *her*. She didn't say, just asked for your phone number, and I didn't ask, of course. You do understand, don't you, darling?"

"Understand what, Mother?"

"That I can't come, of course. I talked to the boys and they both understand," she said. "I know you will, too, once you've had a chance to think about what it would be like for me to have to stand beside that woman at Royal's grave."

So that was it. "Yes, Mother. Got to run to catch my plane. Oh, have a great opening!"

"That's my girl."

"All the more reason to take care of yourself," Sunny Leavitt said as she spread low-fat mayo on Phil's turkey sandwich. "When things are crazy at the store."

"Got to admit, I'm glad to be out of there, even for a few minutes. You were right."

At the height of the morning madness, he had phoned home

to beg off his pledge to come home every day for lunch. But Sunny had insisted. "A slice of greasy pizza is just what you don't need when your stomach's already churning—yes, I *can* hear it from here. The way you walk, it'll only take you ten minutes each way. The exercise'll do you good."

"Who needs more exercise? I've been running around here all morning like Chicken Little."

But the walk home had restored his sanity, at least temporarily. Now Sunny was trying to put the crises he'd been describing in perspective.

"Soy milk and antihistamines, that's not exactly the end of the world."

"It is, I guess, if your baby's screaming because she can't digest cow's milk, or you're a plumber whose nose and eyes are running like leaky faucets."

"But you solved their problems, right?"

"Well, yeah, but just to get a few cans of that soy powder, I had to rush a stock clerk I needed for something else to the ShopBest in Millville. Told me he did ten miles over the speed limit in both directions to make it."

He took a bite of his sandwich, a gulp of milk. "But every day now it's like this. Can you believe it? We haven't got any lettuce?"

"You mean, you've got another one of those, whatchamacallits, goblins?"

"Gremlins, Sunny. Yeah, another one of those."

Akers headquarters in Ocean Plains had warned all its outlets to expect "gremlins," the unholy spawn of any brand-new computer system. The thing was, the nasty pranksters were supposed to go away after the first few weeks. Only they didn't. As soon as one got patched—or dispatched—a new one cropped up, like today's nasty formula and antihistamine surprises.

"If the new system isn't working, why don't they just go back to the old one?"

"Honest to God, Sunny, I almost said those very words during the last video management conference. But I chickened out."

Phil thought the old inventory control software had worked just fine. Tens of thousands of SKUs, or stock-keeping units, in all the Akers stores were continuously monitored and replenished. If any item fell below the minimum shelf level and started eating into safety stocks, alerts were generated.

Then, a couple of years ago, Royal Akers had started opening outlets in Asia and bought out Goodstuff, a rival discount chain in the northeast U.S. and eastern Canada. That would have been okay, Phil thought, if they'd kept the divisional structures. But somebody had talked Royal into combining all his operating systems into a single integrated system—and getting it done in two years. Scuttlebutt traced the idea to Flynn Emerson, a whizbang vice president who had since jumped to 4-Mor Stores. Phil remembered that at the start company memos had trumpeted this "common" system as a "Great Leap Forward."

Two months ago, and only six months behind schedule, brand-new terminals were installed in Lake Reliance—and at fifteen hundred Akers stores worldwide. Phil Leavitt and his staff were cycled through intensive software training programs. The two systems—old and new—ran in faultless parallel for a month. Then, over one anxiety-filled weekend, the old terminals were yanked out.

About a week later the first gremlins appeared—never messing with the same item and with no discernible pattern.

Even scarier, for Phil, no system alerts appeared. Shortages were discovered by irate customers, surpluses by bewildered stock clerks. Meanwhile inventory screens continued to show stocks at optimum levels.

Phil thought that since the whole damn system was obviously snakebitten, it ought to be scrapped—provided they still had the old system in a crate somewhere. Of course, nobody was asking him. He was just the fellow who had to face the public.

"Something wrong with the rice pudding?"

"It's fine."

"You took one spoon and the rest is just sitting there."

"I got a lot on my mind, Sunny."

She took away the rice pudding. "It'll keep. You will be home for dinner?"

"I don't want to promise. God knows what might go wrong between now and then."

"It's not fair! Doesn't headquarters realize you managers need help?"

"They got a few other things to deal with first. Like who's gonna call the shots now Mr. Akers is dead."

"Want a lift, mister?"

"No thanks," Phil said flatly without looking at the driver of the car that had slowed alongside the shoulder. He was only half a block from the store.

A minute later, as he turned into the parking lot, he heard, "You sure look to me like you could use a lift."

Phil turned to glare at the driver. "Wade! Man, am I glad to see you!"

"I told you, you could use a lift. Get in while I park this rental."

"Yes, boss." Phil clambered into the Taurus. Wade pulled into the far end of the Akers lot and turned off the engine.

Phil looked at his boss. "Nice tan. You look like you been at the beach, not knocking yourself out climbing Mt. Whitney." He looked out the car window. "You heard about Royal, I guess."

"Halfway up the mountain I met up with a really high-tech hiker. Had one of those briefcase-size, satellite-cellular PC terminals. We got to talking. When he found out where I worked, he got this funny look and pulled up an AP story about Royal. All of a sudden I felt . . . almost lost up there. I turned right around and headed down the mountain. That was yesterday morning."

"Going to the funeral tomorrow?"

"I figured to fly down today, but as soon as my plane landed in Milwaukee, I had a message that our inventory problems are getting hairier. The information technology people at Ocean Plains swore they had a good fix on things, or I'd have never started on that two-hundred-mile hike."

The IT guys didn't use to bollix things up. "Honest to God, Wade, I don't know what the hell Ocean Plains is doing, but we're getting real nervous out here in the regions."

"Give me a minute to use the john, and we'll talk in your office."

Wade Crain was used to meeting problems head-on. It was how he worked—and the method usually worked *for* him. Right now he was almost glad to have to focus on the inventory problems—*his* problem as regional manager for the dozen Akers stores half circling Milwaukee. Having to give it his best attention yanked him out of the emotional free fall he'd been in since learning of Royal's death.

Royal Akers had been a father figure to him, a role shared with the baseball coach at Ocean Plains High. Both men were close to Grady Crain, Wade's father, who died in a grain elevator explosion when his son was eleven. It was Royal who'd backed Wade's return to college after a torn knee ligament ended his baseball scholarship. Royal's generosity made all the difference, and after he graduated, Wade never seriously considered working anywhere but at Akers.

Two years after getting his marketing degree from Northern Illinois University in De Kalb, he was the youngest store manager in the Akers chain. Now, ten years later, with a masters in logistics from Ohio State, he was a regional manager—and a candidate for a top headquarters job.

Or he had been, before Royal's sudden death.

He gave a hollow chuckle as he flushed the urinal and headed for Phil Leavitt's back-corner cubicle. If someone didn't get a handle on these inventory-replenishment problems soon, it wouldn't matter how high Wade Crain had climbed in the hierarchy: it would all come tumbling down.

P A R T   T W O

I

"We'll be starting our descent into Champaign Airport in a few minutes. . . ."

Jane tightened her seat belt. Out the window, cloud shadows drifted across the great oceanic plain. When she was a little girl, her dad used to walk her out into the cornfields at night, tell her stories about the swarming stars. Standing there, feeling as though she were afloat in the middle of a vast sea, Jane first understood how Ocean Plains got its name.

The childhood memory scratched at walled-off emotion. Grief was what she ought to be feeling, she knew. Yet what she mostly felt was a gnawing emptiness.

Emptiness and something else.

A blood-deep anger at her father for dying now. She had worked so hard and for so long to land TranSonic Airex as a client, all without benefit of family connections. Hell, only a few hours ago no one even knew she was an Akers. She hoped the Silk Road offer had been disposed of . . . confidently. Dave Ellsmere had been understanding to the point of generosity— but they both knew he needed to keep the confidence of his whole team for the deal to go forward successfully. At the very least, her father had upstaged her on the threshold of her first major triumph. He better not have pulled the plug on her victory.

Jane slowed the car as she approached the house. The four-story brick Victorian with the gabled, slate gray roof sat foursquare on a big corner lot on Maple Street, whose name was come by honestly. Number 45 Maple was a little larger but otherwise not very different from the other houses on the street. It looked more like the residence of a small-town banker than of a Forbes 400 family. Filling the long driveway and along the curb were dozens of vehicles. No media van in sight, thank God, though doubtless the Champaign station would be turning up soon. It wasn't every day they got to report on the funeral of a man of Royal Akers's standing.

Thanks to Dave Ellsmere and his cargo fleet, Jane had gotten here with time to spare. Besides, she wasn't in any hurry to face the family. As she eased past the house in the car she'd rented at Champaign Airport, she mumbled, "Coward"—and set out to reacquaint herself with the town she hadn't been back to for what seemed a lifetime.

She'd already had one surprise. From her window seat on the short hop south from Chicago to Champaign, she'd noted that since her last visit home, suburbia seemed to have taken several more giant strides into prime farmland.

The neighborhood the house was in, on the other hand, had changed remarkably little since she'd last seen it. She cruised familiar precincts, noting old houses with new accessories— satellite dishes in the yards, Fisher-Price clutter on the porches—but still no boundary fences, just green grass between them.

The biggest change to hit town, Jane knew, had been the closing of Siler Air Force Base seven years earlier. Since the early 1940s, Ocean Plains had been a military boomtown in the middle of bean-and-corn country. The base closure hit the townsfolk like a death sentence.

Then Royal Akers had stepped in, relocating his corporate headquarters from Kankakee back to Ocean Plains, where he'd started. Despite last-minute inducements by Kankakee's chamber of commerce, the sudden availability of land and even an infrastructure that could accommodate Akers's headquarters while helping out his town made unarguable sense to him.

"*Royal Rescue for Ocean Plains,*" boasted the banner on the local paper, which had been forwarded to Jane by her brother Alex, to whom Roy Jr. had sent it. Old air force hangar space had been converted to offices, a data-processing unit, and a huge distribution center. But Jane saw that the Korean War-vintage Sabre jet remained perched on its pedestal by the base entrance, and the MP sentry box still stood. But the big tractor-trailer just nosing past the gatehouse had "Akers Discount" emblazoned on its aluminum sides, as did the water tower above the old barracks.

She drove to the outskirts of town and in the convenience section of a gas station drank a cup of terrible coffee. She stared at the phone, her brow furrowed in frustration. Where the hell was Mark? She'd left her cellular in the car, so she picked up the pay phone.

"How *many* pallets got mislabeled, Manny?"

"Don't know yet. We're working on it fast as we can, Jack."

The two men, one barely in his twenties, the other almost thirty years his senior, looked much the same even to their sandy hair. Both were in shirtsleeves and wore safety glasses, more or less the executive uniform at the Akers high-rise grocery warehouse in Plainfield, Illinois. Only their frowns didn't match, because Jack, as assistant distribution manager, had greater responsibility than Manny, a section manager, for anything that went wrong at the warehouse—and both of them knew it. But Manny was conscientious, Jack reminded himself. Give him a little space and he'd come up with the answer.

"Just get back to me when you know, okay?"

"Sure thing. But, Jack, there's more—"

But Jack had stalked off, headed for his office, where he could swear in private. How could this happen in his warehouse?

Jack always felt a touch of pride when he entered the warehouse. It wasn't exactly like seeing the flag go by in a parade, but it came closer than he'd admit even to Evelyn. The building was ten stories high and occupied two hundred thousand square feet of industrial parkland thirty-five miles southwest of Chicago. He worked diligently every minute of his time and then some, confident that every ounce of effort was worth it. He was helping Royal Akers make this the best merchandising chain on the planet.

Jack Dunn had started at Akers when he was eighteen, the week he'd graduated from high school. The day he'd celebrated thirty years with the company, a few months back, Mr. Akers himself had taken Jack for a drink. But now Mr. Akers was dead and a whole bunch of pallets were labeled to stop them from going where they were needed.

Damn. He'd always thought if you were watchful enough, nothing like this would happen. Now, suddenly, the place looked different to him. Too big, somehow.

Most of its cavernous interior was taken up by twenty aisles of floor-to-ceiling storage racks, fifteen pallets high and two deep. No humans entered these shadowy aisles with their soaring racks. They were traversed, horizontally and vertically, only by rolling and elevating robotic cranes that could quickly access any of sixty thousand pallet slots, fetching or putting away as needed. Computer-driven conveyers brought the loaded pallets to the robotic cranes from adjacent receiving docks and carried them off to replenish a smaller "selection warehouse" next door.

It was to look at one such pallet, just off-loaded from the conveyer, that Manny had beeped him five minutes ago. He'd asked Jack to read the flexographic printing on the square boxes under the transparent stretch wrap that held the load together. The pallet clearly contained twenty cases of laundry detergent. But the Akers warehouse label, slapped atop the stretch wrap, indicated twenty cases of a popular brand of cornflakes.

Jack glanced at the piece of paper on which he'd jotted down the ID number of the worker who, obviously, had failed to check the boxes against the label before placing the pallet on the inbound conveyer.

Halfway to his office, Jack turned on his heel and retraced his steps. He couldn't wait until the count was complete to take action.

"Here's the deal," he told Manny. "You go on checking the pallets. Meantime, I'm going up to the computer room and have them pull out four or five pallets keyed to this ID. If this is the only screwed-up label, we issue a warning. But if I find even one more variance, we cut the guy loose, I don't care what the union says."

"Jack, you don't understand. You're assuming the outbound inspection station caught this, right?"

"What else?"

Manny shook his head. "I tried to tell you a minute ago, but you walked off, and to be honest, you were already so upset I wasn't dying to tell you this part."

"*What*, Manny?"

The younger man sighed as if he'd aged ten years in the last half hour. "I just happened to catch it, as it was rolling by. I'm glancing at the label, see 'cornflakes' printed out, then I notice all these cases of detergent. *I* pulled the pallet. It *passed* inspection. You know why? Check the label again. That product code on the purchase order there, that's for cornflakes. But that SKU number, that's the same as on the cases. Detergent."

"You're saying our computer scrambled the SKU number?"

"Had to, Jack."

"That's impossible!"

"I don't know what else it could be. And we better check how many pallet loads of cornflakes the computer thinks it's got in the racks, then make sure they're not fucking laundry soap."

"Holy shit!" If Manny was right, and the computer had

scrambled one item number, what about the other thirty-five thousand items they stocked?

And how many wrong boxes had already been shipped to stores? Jack envisioned cases of laundry soap piling up on supermarket docks, with big holes in the cereal shelves inside the store—because the huge facility supplied dry grocery items to *hundreds* of stores in the upper Midwest, while two other Chicago-area warehouses handled the supermarket perishables—meat, poultry, dairy, deli, frozen foods, produce, and baked goods.

Fighting an impulse to panic, Jack looked around. Another pallet load went trundling slowly by, en route to one of two spur lines feeding the adjoining warehouse. The conveyers, like the huge SRVs—storage retrieval vehicles—worked from two in the morning till nine at night to keep the stores replenished on a daily basis. Those pallets were the lifeblood of the Akers grocery business and, like blood, had to keep moving constantly.

"Stop the goddamned conveyer!" Jack ordered.

II

"So, is the deal on or off?"

"Jane?" Mark asked groggily.

"You're still asleep? It's already nine-fifteen in San Francisco. And only about the fifteenth time I've tried to call you. The first fourteen were to Honolulu. First, you were out nearly all night. Then, around half-past six Honolulu time, the desk told me you'd checked out. I made a wild guess that you'd returned home. I've been en route most of the time since. Now, are you awake enough to tell me what happened?"

"Yeah. But I want you to know I earned the right to a little sleep."

"Mark!"

"Also the right to tell this my way. But never mind. The short of it is, you'd be proud of me, boss. I didn't get back to my room until half-past four because, after Mr. Chan's party, I spent a couple happy hours down in the Paradise Lounge with Dave and the other TSA honchos, hoisting a few. Kind of a guy thing."

"Yeah, right." Mark's drink of choice, Jane knew, was imported water. No point in hassling him about why he didn't have his cell-phone with him during the celebrations. Maybe having handled one big situation, he'd felt entitled to be incommunicado for a few hours. She'd felt that once or twice herself during the early months of Malcolm and Associates—although she'd never allowed herself to give in to the impulse.

"By then it seemed silly to go to sleep," Mark was saying, "so I headed for the airport and some well-deserved rest in my own bed."

"Fine, fine. So none of the other TSA people were angry that I left?"

"I didn't say that. Matter of fact, Linc Penworthy called your sudden departure, quote, emotional and unprofessional, unquote."

"That chauvinist!"

"Not to worry. Dave came charging to your defense. Of course, he didn't have as hard a time as he might've, because by then I'd done my thing. Pretty nicely, too, if I say so myself."

"You didn't have any trouble shooting down the last-minute offer from Silk Road Air? The TSA people bought my logic?"

"With a few embellishments from yours truly."

"You're preaching to the converted, Mark. But tell me what you said anyway."

"I started with geography. How the supplier base is much more convenient to Macao. Guangzhou, after all, is a hundred fifty kilometers inland. That's the figure I used, anyway. Which is large potatoes when you're talking about air freight service. And I played up the customs angle, like you suggested. Macao having this history of being trading oriented, like Singapore— used to fast in and out. Want more?"

Jane laughed. "Not just now. But if you ever need a job, come to me first."

"Very funny."

"You haven't told me how the Chinese reacted to my departure. Were they offended?"

"Oh, didn't I mention that? If you think Dave was understanding, the Chinese were *impressed*. I recall hearing the phrase 'dutiful daughter' more than once."

"They should only know."

"Now *you're* mumbling. If you've been traveling all night, you must be dead tired, too."

"I am." Jane remembered staggering into the first mall she'd found in Champaign, needing an appropriate funeral outfit, then staggering out fifteen minutes later with a simple black two-piece dress. "But get your . . . yes, well-earned sleep. Sleep the whole weekend if you feel like it. I'll call you from wherever I set my bag down in Ocean Plains. Bye."

Jack Dunn studied the intricate array on the oversize monitor. It was a moving graphical display of the Plainfield warehouse operation—or would have been, except that all operations were currently stopped dead in their tracks.

Seated beside him, in front of the screen, was Anita Matthews, who managed the ASRS, or automated storage retrieval system. She was mouse-clicking on a series of red squares, each square representing a pallet now motionless on the conveyer. Had the pallets been moving, he knew, they would have been green—and their locations continuously updated on-screen by thousands of photo cells installed alongside the rolling conveyers.

"Definitely another system glitch, Jack."

"I need to know for damn sure."

"I've double-checked our database, and we've got the correct item numbers for both products, which means so does the

Ocean Plains mainframe." He watched her click on a new pallet square, then zoom in to display database information fields linked to that pallet label.

"That pallet of detergent has been in the racks six days," she said, glancing over at Jack. "What probably happened is that, the day it came in, there was one of those errors in the download from Ocean Plains—you know, the kind we've been getting since the new system cutover? Two digits got transposed. Zap, we label soap 'cornflakes.' Next pallet of detergent ten minutes later, everything's back to normal. It drives you nuts, but it seems isolated, rather than systemic—know what I mean?"

"You're not telling me we gotta live with it?"

"No, we gotta fix it, and fast. Or *they* do. What *we* need to do is keep bitching. I'm firing off a red-flag report to Ocean Plains right now. In the meantime, if you want, I can pull more cornflakes pallets down to the inspection station."

"Do it." Jack got to his feet. "And go ahead, start 'er up."

As he reached the stairwell, the start-up bells rang. With a great, vibrating growl that rattled the steel treads under Jack's shoes, servomotors began to churn throughout the building, activating rollers along a mile of conveyer track. By the time Jack reached the floor, the ten-story SRVs were hustling down their great aisles, resuming their Herculean tasks exactly where they'd left off.

The reverberating cacophony should have been reassuring. But something in Jack's gut told him that before they were cleared up, the problems could get worse. A *lot* worse.

III

Jane decided to drive right into the center of town, thinking it had probably changed more than the old neighborhood.

More? Akers Store #1, where she'd worked after school and weekends, was now merely an annex to a new Super-Akers, the parking lot already filling up.

The rest of Main Street had changed as dramatically. The grocery, hardware, variety, and drugstores of her youth were gone, shuttered or converted to antique shops, crafts centers, thrift stores. The box office at the Lyceum was boarded up, the marquee announcing Wednesday night bingo.

My father, the "Royal Rescuer" himself, did this.

Be fair, Jane!

It was, she had to admit, simplistic to indict Royal Akers for the decline of the town center. There was more than a modicum of truth in the corporate defense: rural shoppers couldn't wait for the discount bandwagon to reach their town. They were weary of scanty local merchandise with hefty markups, of long drives to the nearest city, or settling for ordering from the Sears catalog. The truth was, the demise of Main Street may have been inevitable.

But passing so many shuttered storefronts flooded Jane with a sadness powerful enough to evaporate her entire stock of rational arguments. What had happened to the center of Ocean Plains was not merely some socioeconomic trend; this was personal. *Her* town had died. If that was inevitable, her father had certainly nudged fate along.

A block beyond Main Street, Jane rumbled over the tracks of the old Illinois Central and reentered residential terrain. Tour over, Jane thought as Maple Street loomed ahead, every house, every tree, conjuring memories. That ruptured sidewalk had upended her first tricycle, later become a daredevil roller-skating ramp. There was the tree where Mugs, her neurotic beagle, had caught and bitten the youngest Tucker boy, before being "sent away" despite Jane's grief-stricken protests. Under that same tree Chuckie Sizemore liked to park his beat-up Mustang after movie dates and talk about the future, while getting up his courage to make a move.

She stopped the car in the same spot now, fingered the ignition keys, kept the motor idling. When Chuckie had finally kissed her, not knowing any more than she did about where their noses were supposed to go, awkwardness easily triumphed over desire. He'd wanted to try again, but she wanted only to get away and had run into the house.

She could still get away now. No one knew she was here. Jane understood better than she had yesterday her mother's reluctance to go back to Ocean Plains. Given its size—and the place Royal's family occupied in town—there wasn't a person who didn't know why Joycelyn Akers had left, taking her youngest with her.

One of Joycelyn's closest friends had spotted Royal in a Kansas City dinner club with another woman. "Not the first woman he's had, probably not the tenth," Joycelyn, in a whirling fury, had tossed at fifteen-year-old Jane as carelessly as she tossed clothes in their suitcases. "But the last goddamned time for *this* marriage."

As it turned out, the woman, on the occasion in question, happened to be Royal's secretary, but Joycelyn's friend's perception was nonetheless accurate: Royal's relationship with Anne Glynn was by then no longer merely professional. That Anne Glynn was working class, two years older than Joycelyn, and no beauty only made this transgression worse than its predecessors.

Back then, too, Joycelyn had been certain that her sons would understand her position. But whatever they thought about what was happening, they clammed up. Royal let them be, said they should do what made them least uncomfortable. So Alex, who was seventeen, had stayed on to finish school in Ocean Plains rather than move to Chicago with his mother. And Roy Jr. had gone back to UI-Champaign. Today, again, Joycelyn assumed her sons understood her position, but Jane doubted that her brothers had accepted Joycelyn's decision not to attend Royal's funeral as readily as she thought. Jane's oldest brother, Roy Jr., would doubtless maintain a stolid front, while Alex, with a middle child's lifetime of practice, would pretend that this, like anything anyone did, was okay with him.

Jane, at fifteen, had had little choice but to move with her mother to Chicago. But the truth was, with her father refusing to say anything in his defense, she had, in her heart, sided with her mother. By the time, two years later, that her mother, following her artist's star, moved on to New York, Jane headed for the U of I campus. In that short time, Akers had grown from a local phenomenon to a discount chain with stores all over Illinois. And that had been only the beginning of Royal's empire building, which she'd watched from outside, feeling increasingly distant from both phenomena—Akers and its creator.

Jane glanced at the dashboard clock. Eleven-forty. Several cars that hadn't been there earlier were now parked curbside. It was time either to make her getaway before her presence was observed or go inside.

She'd risked walking out on the most important meeting of her life to get here. It didn't make any sense to back off now. Besides, it was only a brief detour from her life. For one day. Tomorrow she could be back in San Francisco, doing any patchwork that might be needed on the TranSonic deal.

Jane switched off the engine and got out of the car. As she grabbed the hanging dress bag off the hook in the back, she noticed a TV truck—from WCCU in Urbana—drive up and park across the street. A pretty young woman bounded out, headed in her direction. Jane moved quickly up the brick path to the front porch.

The door opened.

PART THREE

I

Jane had often wondered what her homecoming would be like. As it was, with Royal's death center stage, her appearance caused only a peripheral stir in the family. She was hugged and scolded by aunts, uncles, and cousins as if she had simply come home late (or been sent home early) from summer camp. Familiar rooms, little changed and only a little the worse for wear, enclosed her as though she'd never been away. Jane found herself in the kitchen, without knowing how it happened, helping her sister-in-law, Grace, arrange cold cuts on platters for after the funeral.

If Grace had slimmed down nicely, she obviously hadn't similarly influenced her husband. Roy Jr. was jowlier than Jane remembered, no longer cherubic looking. After all, she reminded herself, he was thirty-seven now, and only plaster cherubs failed to age. What was more surprising, Roy Jr. acted even older than he looked. He moved from room to room in his father's house, thanking everyone for coming with a proprietary graciousness. It wasn't so much that he seemed to have become his father's son as that he had shed the "Junior" label overnight and become . . . Royal Akers.

Oh Roy, it won't be that easy, even if that is what you really—

Roy interrupted her thought with a warm embrace, a gesture she couldn't remember him ever making before. "Got to keep the family together now, Janie," he said, letting her go— but not quite, holding on to her gaze and squeezing her hands.

"Yes, we do," she heard herself echo. It was only when Roy had moved on that she wondered: How? But by then she had spotted Alex across the room. Her spirits lifted. Younger than Roy Jr. by two years, the same number by which he was older than Jane, Alex was the one member of the family—besides Joycelyn, of course—with whom she had stayed in touch over the years.

He looked great as always, with his Hilton Head tan, generous smile, and dimpled jaw. "How you doing?" she asked when they broke their hug.

"Oh, you know me, Janie. Ever the same. Happy, lazy, and happy to stay laz—"

"Hey, this is me, Jane."

"Is it really you? Oh, well then. I was just wishing we could have gotten back together—all of us, even our mother the artiste—just once, before . . . " He shrugged.

Not knowing what to say to him, Jane was straightening his tie instead when he was tackled at the knees by Tyler, Roy's three-year-old. "Come upstairs, Uncle Arix!"

"Ty, I'm talking to Aunt Jane."

"Right now! You gotta see my Galactic Gorillas!"

"Bet they're terrific," Jane said to Tyler. "Go on, Alex. We'll talk later."

"Want to come?" Alex asked.

"Two's company," Jane said.

Alex bussed her on the cheek. "Later, then." He let himself be dragged toward the stairs, as content as she'd seen him since they were very young. She continued to watch as Alex pretended to be pulled upstairs, step by step, by Tyler. Suddenly regret knifed through her: she hardly knew her only nephew. Alex, but not she, had earned the temporary reprieve from the role of official mourner.

Then she turned, saw her stepmother approaching, and wished she'd gone up with them anyway.

Anne Akers was short, with a round face and a figure to match. On this unaffectionate inventory of her stepmother,

Jane tried to superimpose the woman eighteen years younger, out on the town with her father in Kansas City. Jane remembered her as Mrs. Glynn, her father's secretary, and the woman walking toward her now didn't look all that different from that Anne Glynn. What could have possessed her father to betray the elegant and accomplished mother of his children for—

"Jane, I'm so glad you're here! We tried very hard to reach you. It would mean so much to Royal, having you—"

"Please," Jane said, taking her stepmother's outstretched hand. "It's all right." But nothing about holding the hand of her father's widow felt right. Anne had made several overtures during the years of Jane's estrangement from her father, but God only knew what her motives were—then or now. Fortunately Grace Akers intervened, recruiting Jane to help set up a big coffee urn.

Jane had just finished that assignment when she was cornered by a crew-cut, bull-necked man who gave his name as Robert Grooms. As he offered condolences, Jane caught herself staring at his eyes—one blue, the other gray, an anomaly she'd seen before only in cats and dogs. She adjusted her focus downward. His nubby tan tweed suit looked like an English squire's. His accent didn't match.

"Roy speaks highly about you, Ms. Malcolm. Your background in logistics is particularly impressive."

"Roy said that?"

"No." Grooms smiled.

"I'm afraid you've lost me, Mr. Grooms." She took a step away. "Perhaps, another time, you'll explain—"

"It's my job to know, Ms. Malcolm." He handed her his business card. It identified Robert Grooms as director of security for Akers Corporation.

"Actually," he was saying, "I'm a devotee of military history. And I found some interesting parallels between Desert Storm and Montgomery's North African campaign, especially your movable supply depots."

"You're right. But then, we also borrowed ideas from Rommel, Napoleon, and Alexander the Great."

"Touché. I was in Vietnam, Ms. Malcolm, and I can tell you, Alexander would have done a better job with our logistics operation, especially our immobile firebases—"

"Excuse me, Mr. Grooms." Though he'd opened the door on one of her favorite topics, this plainly wasn't an appropriate time to pursue it. "I think I need to check upstairs on my brother Alex. He may be being held hostage by our nephew."

At the top of the stairs she paused to listen for Alex and Tyler and found herself facing the closed door of her old room. The murmur below faded, but instead of Alex's or Tyler's, the voices she heard were from the past. She'd been standing just there, on the landing beside her old blue suitcase, when she'd heard them shouting below, the outside door opening and slamming, her father's car screeching off, her mother continuing to shout. Trembling, Jane had shut the door to her room—and her childhood—and gone downstairs to comfort her mother while they'd waited for the taxi that would take them away from Maple Street forever.

She'd seen her father only a few times after that, on rather formal occasions. Joycelyn said, every so often, that Jane shouldn't be surprised; hadn't he always been too busy for his family? He had come to both her high school and college graduations, and both occasions had merited a handsome check. Then they'd had that meeting during her brief visit to Ocean Plains, soon after she'd come back from Saudi Arabia, when he'd offered her a job with the company—admittedly a good job, at a generous starting salary. When she'd said she couldn't accept, she'd seen the surprise he couldn't quite hide and hurriedly tried to explain why she felt impelled to go out on her own. He'd said he understood her decision to go into business for herself, and maybe he had. What he definitely did not understand was her decision soon afterward to stop using the Akers name—*his* name.

Now, Tyler's muted yelping told her they were in Roy's old room down the hall. But instead of going there, like a somnambulist, Jane slowly took three steps forward, twisted the cut-glass knob of the door to her old bedroom—and stepped through time.

Into *her* room—not as she'd left it, but the way it had been in her childhood. All the teenage trappings were gone, replaced by . . . her doll collection, her maple kiddie table and chairs, the table set for tea with her miniature flowered cups and saucers and the teapot with the reglued handle. She half expected to see her toddler self come running in, full of giggles or tears.

Jane shivered.

"He loved you very much."

Anne Akers stood in the doorway. "Forgive me for following you up here. But there's something else you should see."

II

When the red-flag report on the SKU snafu in the Plainfield warehouse arrived at Akers headquarters, it had a ripple effect. The lengthy e-mail distribution list began with vice presidents and ended at the night-shift operators in the data-processing center.

Because it was Saturday in Ocean Plains, and the day of Royal Akers's funeral service, many along executive row would not see the report till Monday. But others, especially in the data processing unit, were at work, Saturday or no, funeral or no, because the computer problems were obviously increasing. They would find Anita's trouble report all too familiar. Its technical jargon did not conceal the fear that was beginning to spread like a secondary virus throughout Akers: Would they be able to stop the problems before the problems buried the company? There was little if any comfort in the fact that full operation had resumed at the dry grocery distribution facility.

One person who got hold of Anita's red-flag report read it with a barely suppressed smile.

Thanks for the feedback. But we're just getting started here.

Out of politeness, and an eerie kind of curiosity, Jane followed her stepmother down the narrow back stairway to Royal's private home office. Inside, drapes were drawn. Anne switched on a floor lamp.

The walnut paneling, as she remembered, was covered with framed photographs commemorating the Akers family and businesses. Alex's gridiron glories took up several rows, while his golf and tennis trophies occupied an entire case. Roy's grouping was smaller, mostly schoolboy certificates and Scout merit badges.

But Anne Akers was pointing to the alcove to the left of her father's desk. There was Jane, not more than a week old, in her father's arms, his face beaming; Jane in a photographer's portrait when she was three; Jane dressed for her first day in kindergarten. But what about the one in the dress she wore to her high school prom? How Joycelyn had sighed when seventeen-year-old Jane had insisted the lilac with the ruffles was the dress of her dreams. Moving closer, she realized there were pictures from *all* the years since she'd left Ocean Plains with Joycelyn. There she was getting her second lieutenant's commission. And one in which she was wearing desert gear that was somehow more flattering than that hideous prom dress; she was flanked by Generals Powell, Schwarzkopf, and Pagonis—and that was definitely flattering. And another singular moment in her military career—an off-the-TV-screen shot, being interviewed by Sam Donaldson in Dhahran.

Her father's pride in her Gulf War service was no surprise. One of Alex's letters contained the news that Royal had tried to see her before she shipped out—missed her by less than an hour. How like him not to have mentioned that when they'd had that one meeting after she'd come back. God, how she wished now he'd made it in time. If only they'd had a few moments together then, just father and daughter, they might have somehow found a way to reach each other, bridge . . .

"I'll just leave you alone," Anne said from the door.

Jane resisted a fleeting impulse to ask Anne not to leave. She very much didn't want to be alone right this minute. But she said, "Thanks," and managed to hold on until the latch clicked.

Daddy, why didn't you ever tell me?

III

Wade Crain plunked the Tiger down in a pretty three-point landing on the Akers company strip. A sudden crosswind had shoved the AmGen four-seater to the left, but there was plenty of room on a runway designed for air force heavy bombers and cargo jets. There had been media speculation that crosswinds could have contributed to Royal's fatal crash three days ago at this very spot. Wade didn't buy it. The chief not only had thousands of hours of stick time, but quick reflexes—as Wade had often witnessed from the right-hand seat.

Maybe they'd never know what happened. The old Siler AFB control tower had been unmanned on that day, as it was on this.

He taxied to the hangar and gave the Tiger back to the company mechanics and fuel handlers. There were ten planes in the Akers fleet now, but only five pilots. Quite a few executives, like Wade, opted to do their own flying.

Make that nine planes. Royal's Beech Bonanza was now a ton-and-a-half heap of scrap metal, though rumor had it that Eliezer Grey, Royal's first backer and a company executive from way back, still on the board in his retirement, and Royal's closest friend, planned to rebuild the wreckage as a monument to Royal. That seemed kind of grotesque to Wade. Wasn't *Akers* Royal's mon-

ument? Then again maybe the symbol wasn't too farfetched—
the company itself might be headed for the scrap heap.

The funeral was at three. He still had a couple of hours.
Wade transferred his mountain gear to his Blazer, then entered
the general offices. Like the other twenty-three regional managers, Wade was based in Ocean Plains but spent two or three
days a week flying around his territory.

"Smiley, aren't you going to the funeral?" he asked the
elderly security guard as he signed in.

"Sure am, Mr. Crain. Going to close down here soon and let
the rent-a-cops hold the fort."

The secretary Wade shared with three other managers
wasn't in, of course. It *was* Saturday. But with the company having big problems, only Royal's funeral could account for the
whole place being deserted. Wade went into his office, flicked
on his PC, downloaded e-mail, stared at a screenful of high-pri-
ority messages. The senders matched up with voice mail he'd
checked earlier. Mostly supply-chain snafus: out-of-stocks and
overstocks, misrouted trucks, mislabeled pallets at the
Plainfield food warehouse, everything urgent. They'd have to
wait a few hours. And so would the prioritized stack of messages Sue Ann had left for him.

Wade's instinct was to clear out of this office for the duration
of the crisis—or whatever the hell it was—and work out of the
regional distribution center in Milwaukee. At the RDC, he'd be
closer to his stores, able to react quicker, make emergency supply arrangements.

He left the building and headed across the compound.

The old air force fighter-wing hangar now had a nine-
meter antenna dish on its roof and a colossal array of high-tech
hardware inside its concrete walls. State-of-the-art computers
continuously crunched one of the world's largest business
databases, while interactive satellite communications

linked every Akers facility by voice, video, and data. In contrast with the general offices, the computer center's parking slots were full.

Wade's coded-access card got him only into the lobby, where an armed guard signed him in, checked his ID, then released a second electronic security door. Beyond, in the hive of systems offices, people hustled from room to room. There was more traffic, at higher speed, than Wade had ever seen here before. If he didn't know better, he'd think the folks in Information Technology were unaware that it was Saturday—or the day of Royal's funeral.

A long corridor window offered a view of the master computer room, an area that, Wade often thought, bore some resemblance to Houston's Mission Control. There were rows of computer consoles, all occupied by folks wearing headsets, talking to stores, facing an electronic tote board blinking out daily transactions (by region, district, and store), credit card approvals, and dozens of other running totals. Had they set up new categories, Wade wondered, for systemic screwups and their hemorrhaging costs?

Next along the corridor was the War Room, two-year home of the project-management team in charge of designing, installing, and debugging Royal's new inventory system. The setup was supposed to have been dismantled a month ago, right after the project wrap party. But, peeking in, Wade saw it was still in high gear. All four walls still held white-boards crowded with Magic Marker scribbles—formulas and schematics, arrows and acronyms. The long conference table was littered with doughnuts and Styrofoam cups. Wade recognized most of the anxious faces—reps from the consulting firm, software and hardware vendors, a panicky-looking phalanx of Akers IT managers. Only a few techie types were unfamiliar.

Getting glares and glassy looks, about half and half, he ducked out. Not a happy group. If they had a clue, it didn't show on their faces.

Outside the lunchroom he bumped into Leonardo Cristofaro, their systems administrator. Lenny—even in calmer days no one had taken the time to get out all four syllables of Leonardo—was probably pushing forty by now, Wade figured, yet his chipmunk smile and goggly glasses kept him looking like one of the young wire heads.

"And how did it look from up there?" Lenny asked.

"You lost me already, friend."

"The top of Mt. Whitney. Highest point in the continental United States. Don't tell me Mrs. Fullerton taught us wrong?"

"I didn't quite make the top, Lenny. I heard about Royal and came down fast."

"How'd you hear? Pager, cellular modem, smoke signals?"

"Believe it or not, another hiker was Web surfing with one of those dual-mode satellite-cellular modems at ten thousand feet." Wade gestured back toward the War Room. "Why aren't you in there?"

"Somebody has to do actual work." Lenny sipped his coffee, rocking back and forth slowly.

"Who are the bearded guys?"

"Unix wizards. They're flying 'em in."

"Nobody gets time off for the funeral?"

"Doesn't look like it."

"Any hope?"

"That's the mantra, but it's like killing cockroaches. The little buggers keep popping up in different programs—and on different platforms. Just today, one fear became a fact."

"What's that?"

"Seems our roaches have migrated into some of our vendors' systems."

"How?"

"That part's pretty obvious. Through VMI via the EDI."

"Son of a bitch!" The implications were mind-boggling—

and extremely scary.

VMI was shorthand for vendor managed inventory, a system whereby inventory information was exchanged electronically between retailers and suppliers; the method was EDI—electronic data interchange. Now the highest-volume vendors were using VMI to replenish their own items on retailers' shelves, whenever stock dropped below agreed-upon levels. The process was triggered at the checkout counter, when the barcode label on a single tube of toothpaste, say, was scanned. The blip of sales data fed back through the computer pipeline, back through regional distribution centers, all the way to the manufacturer, which then knew to replenish that store's inventory by one unit and generate a manufacturing cycle to replace that unit. By cutting costly inventory-and-reorder links out of the product supply chain, manufacturer, retailer, and supplier all profited. The key was mutual trust—not easy to generate or maintain, even between trading partners. If corrupting data was now being transmitted through EDI links, Wade thought, Akers's biggest suppliers might very well pull the plug on the whole process.

"You're describing a potential disaster here, Lenny."

"More like an actual disaster."

"I gather you don't think the folks knocking themselves out in there have got a handle on it."

Lenny shook his head. "You really want to know what I think? At the rate we're going, we'll be having a funeral for the whole damn company next. Sorry. Wish I could be more cheery."

"The hell with cheery. If we're sinking, it's time to spread a little panic. For God's sake, tell whoever's in charge what you think."

"I did, I have. But they're trying to keep a lid on it. Afraid Wall Street will get wind and start a sell-off."

"Seen our stock lately? I think Wall Street has some pretty good suspicions."

"I mean a real nosedive."

"You're just chock full of good news. Think I'll go to a funeral."

Forty minutes later Wade was standing in a crowd of Akers employees, squinting into the sun beating down on Burleigh Memorial Park. After showering at his condo, he'd changed into a dark suit. But his mood was darker, not only because of how he felt about Royal Akers, but because Lenny might be right. The company Royal had built from nothing might soon end up worth just that much. Dust to dust wasn't supposed to apply to companies.

At the microphone, behind the flower-blanketed casket, stood Eliezer Grey, the man who bankrolled Royal's original discount store in Ocean Plains. He was telling a story about the early days. Apparently one of the first Akers downstate got built on unstable landfill. The store foundation itself settled only a few inches, Elie recalled. "But, on the morning of the grand opening, this big sinkhole opens up in the parking lot, and I mean right in front—where the balloon man was going to stand. But that doesn't stop Royal. No, sir, he hustles into hardware, grabs a five-pound sledge, and starts busting apart some steel shelves. 'Help me, Elie,' he says. 'Whaddaya doin'?' I ask. 'Gonna build a bridge to the front door,' he tells me.

"Well, the fire department's truck rolls by about then—and stops. They don't think much of Royal's bridge-building idea. But by then Royal's got another idea. We all grab hand trucks and start wheeling merchandise into the back room. Now I really don't know how many thousand dollars we sold off the loading dock, but I recall we had a pretty good day . . . "

Elie's own chuckle was echoed in the crowd. Right about now, the good old days seemed very good indeed to a lot of the mourners.

Every now and then Wade would pick up a faint, insectile

sound, the whirring of camera motors from a small media contingent, roped off under sycamores thirty yards away. It brought home to him that they weren't there just to record the funeral of an American business legend.

Wade was thinking that it might be useful to get Elie Grey's take on the current disaster, when his attention was caught by someone unfamiliar among the Akers family grouped behind the podium—immediate family on folding chairs, cousins and such behind. Royal's older sister, Louise, was being comforted by Royal's wife, Anne. Roy Jr., the next to speak, was glancing at his notes. Tyler, Roy's son, was seated between Grace, whose hands were working a handkerchief, and his uncle Alex, who had his hand around the boy's shoulder. Wade had seen Roy's brother only a few times since high school, mostly at alumni banquets. But Alex didn't look much different from when he'd been the varsity quarterback, just with a deeper tan and trendier clothes.

But it was to the woman on the other side of Alex that Wade's eyes kept returning.

"I'll be goddamned!" he said softly, which got him a quick elbow from Stanley Gladden, regional manager for southern Indiana. "Sorry, Stan. I just recognized Royal's daughter. I went to high school with her."

"Yeah, somebody mentioned Jane had turned up."

Nobody had mentioned that bit of news to Wade. He hadn't even recognized her at first glance, not in that slim black dress. *His* Jane Akers was a skinny tomboy with a big laugh, the girl with more items and activities under her name than anyone else in the yearbook. A girl who stocked shelves faster on weekends than most of the guys at the old Akers store. Then, in the middle of her sophomore year, she was gone. The gossip was all over school, all over town. Joycelyn Akers had moved out, taking her daughter up to Chicago. Whatever the reason—Royal's tomcatting got the most votes—Wade had never seen Jane again until today.

But he'd never quite been able to forget her, either. Over the years he'd wangled a few updates from Royal. The man's distress at his estrangement from his daughter was as obvious as his pride in her, so Wade was reluctant to ask about her except very occasionally. He had managed to find out that when, thanks to Royal, he'd been in college in De Kalb, Jane was getting her degree not far away in Champaign. Word was, for some reason she'd gone through on a full ROTC scholarship. Later on, while riveted to Gulf War TV coverage, he'd known she was over there, involved in logistics. Like her dad, Wade had hoped she'd come home and go to work for the company, which was by then cloning itself all over the map. But Jane had continued to follow her star elsewhere. And after this ceremony she'd disappear into her own constellation again.

But what business was it of his? It wasn't as though that was *his* family up there. Most of them, including her, probably had no idea he'd looked to Royal as a kind of surrogate dad.

Wade made himself refocus on the podium, where Roy Jr. was now speaking.

"On behalf of the family, I want to thank each and every one of you for coming. I know what my father meant to many of you. 'Least I thought I knew before this tragic event. But these last few days we've been hearing from people, right here in Ocean Plains and from all over the world, wanting to share stories about how Royal Akers touched their lives. I'd like to share just a few of their stories with you today.

"The first one has to do with Cal Mallory, whose wife—widow—Marie called me. Most of you knew Cal as one of the best regional managers we ever had down in Alabama, but only a few of you know he was a man who had hit rock-bottom before he happened to sit down beside Dad in the waiting room of Union Station during a snowstorm twenty years ago . . . "

Wade was familiar with the anecdote of how Royal had listened to this unemployed salesman's drunken ranting for an hour before, on impulse, he'd told the man that if he'd go into AA, Royal would give him a job. The man had yelled at Royal for taunting him with vain promises, but by the time the weather let up enough for the trains to run, the two men had shaken hands on their deal. Cal Mallory kept up his end, and Royal, of course, kept his.

There were plenty of similar testimonials, Wade knew, probably most of them true, like his own. Just as there was no shortage of tales about Royal's failings. If his loyalty to his workers was well-known, his lack of loyalty to his first wife was almost as well-known. As for his occasional mercantile ruthlessness, people were about evenly divided whether that was a sin or a virtue. Wade's guess was that the people Royal had helped up far outnumbered those he'd plowed under. Of course, Wade thought, chuckling, Royal had probably figured out a way to profit from both.

Which could almost serve as the punch line of the story Roy Jr. was telling now, how his dad had kept a Danville manufacturer out of bankruptcy with timely contracts. "This isn't charity," Royal had told the man. "I expect quality products and on-time delivery." And he'd gotten it.

Just as he'd gotten Wade's best efforts in college and, later, when it was time to repay Royal's generosity.

Roy Jr. wrapped up his remarks, and they all joined in what Anne Akers said had become Royal's favorite hymn, "Amazing Grace," but there seemed to be almost as much nose blowing as singing. Still, sadness wasn't the only emotion in the air. There was a sense of celebration, too. After all, Royal Akers had lived a large-scale life and written his name large across America. Hell, he'd written it all those places across the world that had an Akers. Most of those gathered in the park for his funeral felt he had affected them, and American business, for the better. It was quite a legacy.

Unfortunately it was not the only one Royal left. His final expansion moves may have been too reckless, incompletely thought out. Or perhaps the man had simply overreached himself. Whatever had caused Royal to take the actions he had in the year before his death, the fact was, his company was now in turmoil.

Wade scanned the faces around him and on the podium, wondering if anyone among them was equal to the task ahead.

I V

Walking from graveside to the family limo, Jane wobbled slightly, her heels sinking into the damp grass. A moment later she felt a steadying hand at her elbow and turned to find at her side Mark Twain incarnate, lavish droopy mustache framing his kind smile.

"Gotcha!"

"Oh, Uncle Elie," she said, reverting to her childhood name for Eliezer Grey. "I was so cruel to Dad. I'd give anything if I could tell him how much I loved him."

"I told him for you, child. Told him you'd be back, too, eventually."

"But I waited too long."

The man, blood-level honest, nodded. "Stubborn, just like him. Two from one cloth."

Just ahead, a chauffeur held open the rear door of a black stretch Lincoln. Roy Jr. and Alex stood there, waiting patiently for her to join them.

"Ride back with me, child," Elie said. "Got some things to say."

Jane signaled her brothers by pointing to Elie and said, "I'm ready to listen."

He steered her down the line of cars to a sparkling sky blue Buick with grandiose fins, opened the passenger door, handed her in.

"Advice usually comes too late," he said as he settled himself behind the wheel, "but here it is. Don't leave things unsaid. People you care about, go tell 'em. Nobody lives forever. Worst kind of grief on earth, child, is what you're feeling now. Something on your heart left unsaid."

As if his words held the key, the tears were loosed that she had managed to contain when so many eyes were on her. "Uncle Elie, tell me what to do!"

He didn't answer her right away. He was coaxing the Buick, gentling her, like the beloved old horse she was, to show signs of life. She didn't. Elie patted the wheel as if he understood and said, "Talk to your daddy, he'll hear. But don't expect your grief to go away. My Doris, she's been gone two years, but I wake up every day with a hole in my chest the size of her heart. Good Christ, but I miss her. I tell her. Sometimes, I holler it at her." He smiled weakly. "She doesn't answer me, 'specially not when I yell. Never did answer me when I yelled. But sometimes, I just . . . sense her with me, telling me to simmer down, the way she used to."

Elie turned the key again, and this time the Buick turned over immediately. Elie smiled, as though both he and the car had kept their sides of a bargain. He held the Buick idling, waiting for the limo to pull out first.

"You know," Jane said, "the few times I saw my dad, he never said one bad thing about my mother. Never defended himself, never tried to explain his side."

"No, Royal wouldn't."

"So I only heard Mom's version of what went wrong."

"Can't help you there," Elie said. "Seems to me the person to ask would be your aunt Lou—if you're sure you want to know. Royal was no saint, but you already know that.

"Anyway, that's not what I wanted to talk to you about. I wanted to talk business. You mind?"

"Now?" Jane asked.

"*Now's* when the company's in bad trouble. Worse than bad,

with Royal gone and Roy Jr. in charge. You know about that, do you?"

She nodded. Alex had told her. In an emergency meeting the day before, the Akers board had named Madeleine Archer, a long-time member, acting chairman of the board and their older brother CEO.

"Roy Jr. did a fine job of politicking the board, and I give him credit for that eulogy just now, and God knows he's been waiting all his adult life to sit in the big chair." Elie sighed a sigh of seventy winters. "But he can't run a company."

"He won't have to," Jane said, "if he names Don Landsman president."

For at least a year now she'd been reading speculation in the business media that Landsman, executive VP of operations, was the heir apparent at Akers. The speculation probably got as much attention as it did because Don Landsman was black. There were African-Americans at the top of some very big firms, but those were usually companies that the black man at the helm had started. What the media didn't know was that Royal Akers didn't give a damn what color Landsman or anyone else was: all Royal saw was what a man produced. Without ever having met the man, Jane knew that if her father favored him, Landsman had to have the goods.

"Don resigned yesterday," Elie said. "Right after the board meeting. It'll be announced tomorrow."

"Why?"

"Roy told him Wally Conover of Goodstuff is gonna be his president and join the board. Expect Don was pretty angry— had every right to be—but he kept both his mouth and his face shut tight. Heard his letter to Roy Jr. was one polite sentence. Why not? Man like that can write his own ticket just about anywhere. Just like Flynn Emerson did when he left."

"What about this Conover guy?" Jane asked. "He any good?"

Elie shrugged. "After the takeover, Royal brought over most

of Goodstuff's first team. Some pretty good talent there, Wally included. That don't mean he's got what it'd take to fix our problems. How much do you know about the problems?"

"Alex sends me stuff, and I read the papers."

"Then you don't know how bad things are. The new computer system is brain damaged, and it's getting worse, not better. What we got is the kind of logistical nightmares you and your army bosses were probably battling in Saudi Arabia. We not only got shortages and outages, we got stuff being shipped to the wrong places—you know, like guns one place, ammo another." He winked. "Snafus like that sound familiar?"

"All too!" Jane laughed. "For the first few months, we didn't have enough guns *or* ammo! Not to mention tanks or trucks. Thank God Saddam Hussein didn't figure that out."

The long limo started up, followed by other family cars, all at a cemetery crawl. Elie waited his turn, jaw outthrust, liver-spotted hands caressing the wheel. He spoke without turning. "But you figured a way *out*. That's the kind of talent this company needs."

"Elie, I didn't run Gulf War logistics."

He ignored her protest. "Someone who understands the big picture, everything from source to shelf, who can get all departments working together, save time and money, service customers better. That what a logistician does? Am I close?"

Jane couldn't contain a smile. "More like a bull's-eye. Since when did you become an expert on cross-functional management?"

"That what you call it, huh? Sure don't know the jargon, but some of us old-timers do manage to keep up with what's happening in the world."

"I came back from the Gulf years ago, Uncle Elie, but I've been back here only a day. You mentioned Flynn Emerson. The systems guy, right? Could Roy get him back?"

"What if Emerson sabotaged the system on his way out? What if he tampered with the inventory-and-replenishment code? That might explain the kind of problems we're getting."

Elie shook his head. "No, Jane, you're the one we need."

They had stopped, waiting for another funeral procession to cross theirs. When they were rolling again, Jane said, "I like what I'm doing."

"About what I figured you'd say." He was nodding slowly. "Like when your dad offered you that job after you come out of the army. If Royal had his *real* druthers, it wouldn't even be Don running the company now, but you. Or maybe you and Don together."

"Uncle Elie, c'mon!"

"No, you listen. I remember one time Royal and I went fishing up to British Columbia. You would have been at U of I then. We were around the fire, frying up some sockeye we'd caught, looking up at a zillion stars. 'Jane's the one, Elie,' he says to me. 'Trouble is,' he goes on, 'we're too much alike.' I think he was right on both counts."

"That means a lot, Elie, knowing Dad thought that. Thank you for telling me."

"But?"

"But what you're suggesting, it's just kind of . . . well, irrelevant. Roy's running the company. I doubt he'd want me around any more than he wanted Landsman. And why should he? I haven't worked for the company since high school. And I'm not a family partner. I don't even have a stake."

"Hell, you don't!"

"You've lost me."

"It's not my place, but you'll hear soon enough. Royal made you a full family partner. He figured you'd want no part of it, so he set it all up in trust, making himself the trustee. I witnessed. Full family shares, dividends reinvested. It's all yours now. You'll have the same voting rights as Anne, Louise, Alex, your mother, even Roy Jr."

Jane didn't say a word, but Elie's words kept detonating inside her. There was some kind of giant irony at work here. She had tried mightily to escape her father's orbit. The family's great wealth, coming later, had only increased her resolve to be independent. Now it turned out he'd thrown a big fiduciary net around her anyway.

But no one could make her take the money. All she had to do was say no.

"Guess I dropped a bomb on you," Elie said. They had reached the main cemetery gate. "You do understand why I told you? You needed to know you *have* a stake in the company."

Jane shook her head. "There's too much to take in all at once. Dad's death, Roy taking over, major problems in the company, now this news. What am I supposed to do, Elie?"

"Want some financial advice?"

"No, that's not what I'm asking."

"Well, I'm going to give you some anyway. Hang on to your stock."

"Why? I mean, just for argument's sake, if what you say is true, it's likely to keep going down."

"Because I think Roy Jr. is looking to sell out fast. In which case, the family partners could all do very well." He paused to let that sink in before he said, "'Course, everything your dad worked to build will cease to exist."

"Is that why Tom Soverel was at the funeral?"

"You noticed him, eh?"

"He came up and introduced himself—as if I wouldn't have recognized him." The homely-handsome, big deal maker was a frequent business magazine cover boy. Soverel Corporation seemed to be into everything these days, including—with its takeover of 4-Mor Shopping, a big East Coast discounter—mass merchandising.

"What's his angle, Uncle Elie?"

"Well, I guess it was late last year we first heard he was picking up little pieces of Akers common here and there, so as to keep a low profile and not drive the price up. Then about three months back—around the time we somehow hired one of his employees as our new security chief—he came out of hiding, began picking up some sizable chunks, and launched his campaign to get proxies."

"How much do you think he controls?"

"Oh, between him and his group, I'd guess maybe thirty percent of voting shares. Enough that I'm afraid Tom just needs to swing some family shares to take control of our board. However, with you in the picture, there's moves we could make here and there to maybe block his play."

Jane laughed. "We're back to that, are we, Uncle Elie? After all these years away, you'd really expect me to jump into a snake pit like you're describing? Have a little mercy. I have my own business to tend to, not to mention my own life—though I have to admit, that's pretty skimpy these days. Got a few dreams of my own, too."

"Musta heard you wrong a while back. Thought you hinted you wanted to do something for your dad."

They drove on, tailing the big Lincoln and the other family cars. Jane turned from the old man to watch the furrowed fields roll by, all the crops in now. People were praying for rain—but not too much of it.

Silence prevailed till they were back on Main Street. Then Elie said quietly, "Your daddy saved this town once. But something happens to his company, won't be anyone left to save the town again."

"Uncle Elie, that's not fair!"

"No, it isn't, is it?"

It was two A.M. when the overnight delivery from the regional Akers distribution center arrived at Akers Store #974. At three-fifteen, more than three hours earlier than usual, Faye Stevens, the store manager, pulled in, parked all the way in back, and climbed onto the loading dock. She was in running clothes, her face naked. The receiving area was full of pallets, only partly broken down. Faye took a good look. "Shit," she said. "Shit, shit, shit."

What she had were cases and cases of upscale merchandise: $70 bottles of Chateau Lafite-Rothschild; smoked-salmon platters; imported cheeses. Her store, a converted Goodstuff bargain-basement outlet on the outskirts of Bridgeport, Connecticut, a city that had filed for bankruptcy not too far back, didn't even pretend to have a gourmet section. Oh, they sold wine all right, but mostly screw-top. As for cheese, Cracker Barrel Extra Sharp was as sophisticated as they got.

She found her receiving manager in the corner office, on the phone, nodding dejectedly.

"Who you talking to, Pete?"

"Brookline."

It figured. The brand-new Brookline Akers, in an upscale Boston suburb 150 miles up the interstate, would be able to move this tony stuff.

"You said it!" Pete hung up and turned his tortured face to his boss.

"Good thinking," she said.

Pete shook his head. "I thought so, too, till they told me they got *our* delivery."

"They got our truck, we got theirs?"

"Yep. But only by sheer luck."

"What's lucky about drivers switching routes?"

"Didn't happen that way, that's the scary part. All those pallets out there have our store number. Nine seven four's on all the paperwork. Only it's not our order. The switch had to happen in the main computer."

"I really don't want to hear this." Computer-generated problems were turning Faye's job, which she'd absolutely loved for nearly seven years, into something she was coming to hate. Basic items she needed—stuff her customers desperately

needed—were suddenly and mysteriously not being replenished. And every day a nasty, new shock.

"Are both drivers on their way back to make the swap?"

"The Brookline driver just turned around. We're still trying
to reach the guy who dropped off here."

"Damn!"

"Boss? I got another question."

"I got a whole list of 'em!"

"What about all the pallets on this shipment we *can* use?
We're real low on janitor supplies, car batteries, paper goods—
do we gotta send all them back till they get it unscrambled?"

"Hell, no! Start stocking the shelves. Let Ocean Plains sort
out the mess. They created it! Look, I'm going home to change,
be back in forty minutes. Don't beep me unless war breaks out."

She backed out of the office, knocked against something
hard, felt it give way behind her. She whirled as half a dozen
shiny bicycles built for two clattered to the floor.

She swung back toward Pete. "What in hell are those doing
here?"

"Part of the shipment. French tandem bikes. They go for fifteen hundred apiece."

"Not in *my* store they don't. Get those bikes the hell out
of here!"

V

The meeting belonged in her father's den, Jane thought. That's
where family business had always been talked. The swaybacked
leather sofas and brass banker's lamps set the proper, serious
tone. But Roy had led them back to the family room. The vinyl-
floored expanse contained what seemed like half the inventory
of an Akers toy department. Clearly Roy's son, Tyler, usually
had the run of the place. Jane was wondering if her father
had gotten to spend more time with his grandson than he had
with his own children when Roy asked the boy to take his

games outside.

"Can Uncle Arix come, too?"

Alex stooped and hoisted the boy high to delighted screams. "We'll play later!" he said.

"Promise?" Tyler asked.

Alex put his hand over his heart. "Promise."

Satisfied, Tyler grabbed a box of Legos and let his father lead him from the room.

"I can't believe how big he's gotten," Alex said. "I should be horse-whipped for not seeing him more often—no offense, Janie."

"None taken," Jane said.

But the truth had found its mark. Tyler's birth announcement had come when she was in Malaysia, scouting high-tech manufacturing sites. She'd sent birthday and Christmas gifts, but she and the boy were strangers, their relationship one more casualty of her break with her father.

She didn't even know where to sit. So she just stood and watched as Anne Akers scooted over to sit beside Roy's wife, Grace, on the davenport. Royal's sister, Louise, pulled a straight chair over to the sliding glass doors, giving her good light for her needlework. Alex took the ottoman beside the coffee table. Finally Jane sat—in the old love seat, by herself. They had all avoided the big green La-Z-Boy—Dad's old TV chair. Were they leaving it for Roy?

He was back now. Managing a bare bones smile, he rested his palms on the back of the big recliner. "Sorry for the delay. Speaking about timing, I apologize for the timing of this meeting. I wish like hell—excuse me, Aunt Lou—I wish like heck we didn't have to go through all this stuff now. But from a scheduling standpoint, there was just no way around it. I hope everybody understands."

"We all understand, Roy," Anne said.

"Thanks, Anne. All of you. I'm sorry my mother couldn't be here. In any case, Jane has volunteered to take notes for her and let her know whatever we decide. I've also FedExed her a packet with the materials all the family partners will be getting."

Roy moved around the La-Z-Boy, grabbed a chair away from the card table, and straddled it backward. Jane remembered her father doing the same thing at employee gatherings, his tie loosened and shirtsleeves rolled just like his older son's were now.

"I'll try and keep this as brief as possible. We've all had a terrible few days. Aside from our private tragedy, as most of you know"—he glanced briefly at Jane—"we've been having an ongoing crisis in our company distribution systems. Which was one of the reasons Dad was out flying again so much the past weeks, visiting stores and warehouses, trying to get a handle on things.

"But it's gotten worse, especially the last few days. We're starting to get some pretty negative coverage on local news shows—kids in Akers towns not getting their asthma medication, gasping for breath on camera. Three of our West Texas stores out of electric fans during a big heat wave that sent some old folks to emergency rooms—horrendous stuff like that.

"We're doing everything we can to stop the tailspin." Roy shook his head in what looked to Jane more like resignation than determination, then added somberly, "But I have to tell you today that our corporate options are closing down fast."

Bingo, Elie.

"Akers has to make a move pretty darned quick," Roy was going on. "We can't afford to wait while we grieve Dad's passing. I guess that's the way things work sometimes." His voice perked up a little. "At least we have the comfort of being together."

In our grief—or in your hopes? Jane was tracking her brother's words carefully, but what fascinated her were his eyes. A startling turquoise, like their mother's, they seemed almost garish in Roy's round, pink face. When they were teenagers, Jane had often thought how much more use of them she or Alex—both of whom had inherited their father's level gray—

would make. But now, she observed with growing interest, Roy was using his eyes effectively, scanning the faces of his relatives with sensitivity, while his voice meted out soothing phrases.

Still, with such a momentous matter at hand, he was being too damn vague to suit her.

"Help us understand the crisis, Roy," she said quietly. "You say the problems are getting worse. Why would that be happening?"

She spotted a new glint in Roy's eyes; his glance hardened. But he kept the impatience in them out of his voice as he said, "Wish I knew."

Elie thought *he* might know. Jane said, "Are you talking about sabotage?" And if Elie's suspicion about Flynn Emerson was off base . . . "Or just some colossal chain of blunders?"

"Maybe both. We're trying to find out, we've *been* trying to find out, Jane. Believe me, everyone here wants to know the answer just as much as you do." He looked around the room as if embracing his family. "We *mustn't* blame Dad, but I think we can say that we made a serious miscalculation when we took over Goodstuff, in the middle of a monster computer upgrade, thinking we could keep all those additional stores supplied with our own trucks out of the distribution centers we already had. It seems clear in hindsight that the extra strain on our supply routes was too much for us to handle, even without all the haywire computers—"

"I don't see why we have to talk about all this," Louise said, not glancing up from her stitching.

"Because it's important, Lou." Grace leapt—if only figuratively—to her husband's side.

"We've had problems before this, Grace, and we came out of them stronger than ever," the older woman said firmly. "Remember, Junior, when we opened that first mechanized warehouse over to Danville and everything just went crazy, and Royal and Don and some of the other men had to go down there and drive forklifts and move cartons all night long? Royal may not have made the right decision initially in every situation, but whenever Akers was in a tight spot, he always figured a way to get us out safe and sound. And I trust Junior to do the same."

"That's very sweet of you, Aunt Lou," Roy said, trying not to wince at the boyhood name she alone continued to use. "And I hope to justify your faith. But Jane asked a good question, and we do need to discuss the business—as a family, just like we're doing. Unfortunately it's a little easier to recover from a bunch of conveyer belts out of control than an entire business system crashing and burning."

"Is that what happened?" Jane asked. If so, things were even worse than Elie thought.

"I'm exaggerating," Roy said. "But only a little. Right after we made the decision to go live, we ran parallel four weeks, at one hell of a cost, before we had the confidence to pull the plug on the old system. But something—a virus, corrupt code, bad design—derailed us. Flynn Emerson and his team might have made a difference, but maybe not. In any case, they're gone, and we've been in damage control for weeks now. But like I said before, the situation's only getting worse. It's a pretty scary list." He started ticking items off on his fingers. "We've got all the inventory and replenishment problems, we've got trucks mis-routed, goods lost, demand forecasting that suddenly doesn't match historical numbers—"

Aunt Louise stood, stuffing her needlework in her bag. "You vote my proxy like Royal always did, Junior. But I can't listen to any more of this."

After the door latched behind her, Roy's tone became even bleaker. "God, I hate this. But facts have to be faced. Akers common closed at forty-seven and a half yesterday. I'm afraid it's going to go lower, a whole lot lower, as we get more bad news coverage—how shortages are hurting Akers towns—and especially after we report substantially lower quarterly earnings with a bleak forecast for the rest of the year. Our PR people are shell-shocked from overwork already, our institutional investors are deserting in droves—no way they're going to want Akers to appear on the quarterly performance reports to their clients. And I don't need to tell you how much we've all lost on paper.

"If we had more time, could we fix these problems? Eventually, yes. But it's my opinion, and Wally Conover's, and most of top management's, that we don't have that luxury now. However, we do have another option."

The only thing Elie Grey hadn't foreseen was how fast Roy would make his move.

"Some of you know Tom Soverel, who was here yesterday to pay his respects," Roy was saying, his tone lifting like movie music as the hero and heroine run toward each other through a field of uncrushable flowers. "Dad was a fan of Tom's 4-Mor Shopping. Incidentally, Tom recently picked up Holley's in the Southeast and Key-Buys in California and Arizona—"

"Excuse me," Jane interrupted. "But how is selling out going to help us now—today? Today is when our stores have to be restocked. That can't wait for a sale to go through."

"Perhaps you'll let me finish." Roy smiled through a twitch of anger. "I'm calling another special meeting of the board for tomorrow night, at the bank. Jane, I assume you're aware that Tom has been taking a sizable position in Akers for some months? Through his own holdings and proxies, he and his group already control around three hundred thousand shares of Akers voting stock—enough to force representation on the board. All things considered, his proposal seems generous. I've already discussed some of the financial details with the rest of the family partners. More important, I think if enough of us sign our proxies over to give Tom voting control, all his distribution centers and trucks will pitch in immediately to help supply our stores. By the end of the week, we should be out of the woods."

"Just like that, huh?" The challenge in Jane's voice was unmistakable. "Tom Soverel will wave a magic wand and make our problems vanish. Roy—everybody—please listen to me. If we haven't got enough trucks already, we can hire more. I assume we've got plenty of goods in the pipeline. The problem, as I understand it, is too many goods are in the wrong place. No one's magic wand can fix that. If it were that easy, you, Roy, and the people you've hired, would have done it already. So until we *can* figure out why that keeps on happening, today's task—and tomorrow's—is to get all the misdirected things to where they're needed, *fast*. And that's not Soverel's job, it's ours. It's an Akers problem, and Akers will solve it."

In the silence that followed, Jane could hear her heart thumping angrily in her chest. Why had she stuck her neck out like that? Far enough, for sure, to tempt Roy to lop it off. She didn't know whether to curse herself or Elie, so she concentrated on trying to reclaim a normal heartbeat.

Roy finally broke the silence with a sigh, heavy as a concrete block intended to keep a body submerged in water. "Jane, I appreciate your passion on behalf of the company . . . belated though it may be. But I'm going to have to correct you." A telltale vein throbbed below his ear.

"Fact is, Soverel *can* wave a magic wand. To understand how, I'm afraid you'll have to step outside your logistical mindset for a moment and into the big bad world of corporate finance. The magic wand is called investor confidence, Jane, and we've lost it. Big time. But the instant Tom Soverel comes on board, we'll get it back. Overnight, whether our shelves are fully stocked or not. Among industry analysts, brokerage strategists, up and down the Street, Akers common will go from 'sell' or 'avoid like the plague' to 'buy.' I call that magic. And it's what we need right now, folks—and I mean desperately—to buy precious time."

"And how long before all Akers become 4-Mors?" Jane asked.

Roy shrugged. "I imagine they'd be phased in."

"But, eventually, all of them?"

"I think that's fairly obvious." Roy's smile this time was mostly grimace. "With all due respect, Jane, don't you think it's pretty unfair of you to come back here, after all these years, and accuse me of letting down the family?"

"I haven't *accused* you of anything."

"That's how I take it. Maybe it's true that none of us—not me, not Alex—has the drive and the passion to carry on where Dad left off. If we did, maybe we wouldn't be so damn close to losing everything he built. But we have one last chance to guarantee that the family members will be well taken care of. And that's one way of preserving Dad's legacy."

"Family aside, Roy," Jane shot back, "how are *you* going to be taken care of by Soverel?"

Roy flushed, whether from guilt or righteous anger it was impossible to tell. "I'm going to ignore that insulting question. Anyone have anything *constructive* to offer?"

Silence. Jane made a quick visual reconnaissance of the room. Glances seemed evasive, body language rigid—except for Alex, who looked as nonchalant as usual. Then Jane noted the unlit cigarette in his hand. He was obviously just waiting till he could decently slip into the backyard and light up. Did he merely want to inhale lavishly as he contemplated how much money he'd get from Soverel's takeover? Or could it be that, under his detached mien, he actually cared what choices were made?

She couldn't afford to be distracted by the sudden temptation to think there might be a more complex Alex hidden behind all that charm and warmth.

She returned her focus to Roy. "What does Don Landsman say?" she asked with an inward nod to Elie.

"Don's left the company." Roy shrugged. "So his opinion

would be pretty damn irrelevant. Look, I'm as sorry as I can be about this. If it makes you feel better, blame me for what went wrong. But I won't sit by and see us wiped out financially. Goddammit, we have to move on this."

Roy heaved to his feet, went to a side table, grabbed a stack of manila envelopes, began handing them out. "You'll find full financial details of Tom's offer in there, along with a proxy statement and form for you to sign . . . assuming you agree with me."

Talk about magic! As his eyes went the round of attentive faces, the cool turquoise was warm enough to constitute a hug—even when his glance reached Jane. "Let's all think about what's best for the family," he said softly. "Okay? I'd like them back by, let's say, noon tomorrow."

Jane pedaled double time out of the spell. *Thank you, God, for 20-20 hearing.* The sudden desire to believe that Roy's primary motive was love of family had nearly shut down her other senses—and good sense.

"It's too fast," she said.

"I'm sorry, but I need them by noon."

"I don't mean that. I mean the whole deal. It's too fast."

"How dare you!" Roy stood over her, his pink face darkening.

"Calm down, Roy," Alex said. "As for you, Janie, what's the point?"

"Here's the point. Okay, we have terrible problems in the stores. But as Aunt Lou pointed out, Akers has had terrible problems before. But what Dad would say—remember?—he'd say, 'Let's fix 'em!' He spent twenty-five years of his life making his name—and ours—mean something. Let's not auction it off the day after he's buried."

"Seems to me you tossed his name in the wastebasket years ago, didn't you?"

Jane flinched, then straightened her shoulders. "I know Dad believed I did just that. But even if I'm not an Akers by name anymore, I am *here*." She put her hand on her heart. "And *here*."

She tapped her head. "I'm still Royal Akers's daughter and damn proud to be."

"Too bad your timing's so lousy, Jane. Dad might have liked to hear that little speech."

"Right twice in a row, Roy. I waited too long to tell him, I know that. But I'm here to tell you . . . all of you . . . that I'm not going to let my father's name—your name—be taken down from store after store like so much burnt-out neon."

"You've really got a helluva lot of nerve, you know that, coming in here after so many years and preaching to us—the people Royal *could* count on as family."

"That's quite enough, Roy," Anne Akers said in a composed voice that brooked no dissent. "Jane, I for one think you're courageous for saying what you did. Maybe, just maybe, your father is listening. But whether *he* heard you or not, *I've* been listening. To you, Roy, and to you, Jane. And I must say my love for Royal Akers and regard for what he achieved make me want to believe we can save Akers. But, Jane, while you've moved me, you haven't convinced me it's possible. The question, to my mind, is: What can we do? Do you have an answer, Jane?"

"Of course she doesn't," Roy said confidently.

"Jane?"

Jane took a deep breath. "Look, Akers has always used logistics—it's how Dad supplied his stores, and it's how we keep them supplied now. Logistics can solve the problems we're having with that—it's all *about* getting the right things to the right place at the right time. Okay, so we can't use some of the technology we've come to rely on—but logistics has always been an art as well as a science. And I think using both together is what can save us now."

"Of all the goddamned arrogance!" Roy burst out. "Jane, you think you're such a goddamn hotshot out there in California, you can just walk in and spot what all of us hayseeds have missed?"

"The truth is, I don't think a problem of this magnitude can just be 'spotted' by anyone. But, as you said, you've been trying to solve it for weeks now, and the problem not only hasn't been solved, it seems to be growing. What's more, judging from your enthusiasm to sell as fast as possible, I'd say it's a fair inference that whoever you've hired, Roy, hasn't been able to make any noticeable headway." She paused to make sure she had everyone's attention.

"I'd like to try," she said.

As she heard herself say the words, TranSonic Airex and several of her other clients barged into her brain. *I must be out of my mind.*

"This isn't the army, Janie," Roy said. "We don't have the entire Pentagon at our disposal."

"Sometimes in Saudi Arabia we didn't, either. So we'd hire a bunch of old beat-up trucks driven by Bangladeshis and Pakistanis and Afghanis and load 'em up. And you know what that reminded me of?" Jane found herself on her feet. "It was like when we were kids, helping out on weekends at the first store down by the old Ocean Plains depot. Dad used to bellow, 'Get the stuff out of the back room and onto the shelves fast!' Well, that's all I'm saying. There's something that needs doing— let's do it!"

Jane thought she might be having an effect, then realized the others were all looking toward Roy. He shook his head again.

"Jane, I can be as moved as anybody by a rah-rah speech when it's genuine—as yours seems to be. But you're too late. Why don't you get it? We've run out of time."

Time! How fast could she hope to accomplish the apparently gargantuan task she'd rashly volunteered for? Given the added pressure on the family of Soverel's offer, she couldn't possibly ask for the four weeks the job could reasonably take. Three weeks would still be realistic. Two weeks would be nearly impossible.

For my daughter?

Dad!?!

Could it be? Or had she without realizing it become so like her father that she thought she heard his voice inside her head when she was really only talking to herself?

"Jane, we're waiting," Roy said.

He can wait another minute. Remember, whatever you tell them now, that's all the time you'll have, Janie. But it seems to me that two weeks—

Even that will seem too long to them.

Then tell them you've changed your mind about trying.

No!

Then pare the time a little more.

That really is impossible.

Not if you succeed.

Now I know why they called you the king of negotiators.

"We're *still* waiting, Jane."

"Ten days, Roy. Give me full authority for ten days to bring in my own people. Give me access to Akers staff and facilities. Give me the resources I need. Ten days is all I'm asking."

She took a shallow breath. "Surely Tom Soverel can wait that long. There'll still be plenty of value in the stores. If there's a chance I can turn Akers around . . . "

Her eyes canvassed the group. No one was looking her way. She focused on Alex, but he had opened his folder, was staring at columns of figures. He didn't look up. Of course, the last time she'd turned to him for help, she was twelve. Roy had challenged her to a race across the lake. Less than halfway she'd gotten a cramp, but Roy was too busy beating her to hear her cries for help. Alex, sunning himself on the dock, had jumped in, reached her, and swum back with her on his back. So long ago. Since then, how many times had she given him—all of them—plenty of reason to think she didn't want help from anyone?

More to the point, why on earth would Alex think *she* could help them all now?

Jane's eyes moved on to Grace, but Grace was watching her husband. Practice makes perfect.

"I say yes."

Jane whirled at Anne Akers's voice. "I say let her try," she repeated.

Roy threw up his hands.

Jane made a lightning calculation. Anne voted Royal's shares. If Jane could talk her mother into voting against Roy—far from a given . . . No, she'd still need one more block.

Roy must have been doing the same addition. He was staring expectantly at his younger brother. Slowly Alex slid the financial documents back into their envelope. "I was trying to think back," he said. "Janie, have you ever actually failed at anything you've set your mind to?"

A nanosecond, then: "No." Clearly.

And clearly a less than wholly honest answer. But in order to save her father's company, she needed Alex's vote.

Alex laid his envelope carefully on the coffee table, as though it contained something combustible. He looked at Roy, then at Jane, then down.

Bad sign.

"I guess I could wait ten days," he said, his voice so low that Jane wasn't sure she'd heard him right.

"What was that, Alex?" Roy said sharply. Was he hinting Alex could still—had better—change his mind?

Maybe that shard of threat in Roy's tone was what firmed Alex's voice as he said, "I vote to give Jane her ten days, Roy." He shot Jane a swift grin. "She did make two bars."

Roy Jr. sat down heavily in the La-Z-Boy, then slammed his fist on the armrest.

"Easy, bro," Alex said. "It's just ten days."

"Goddammit, Alex, why the hell don't you just stick to your

golf game? I'm trying to save your fucking dividend check!"

"I heard you, Roy. And I still hear you. But it seems to me, what Jane's going to try to do is save more than that, and I think she deserves the chance."

Roy shrugged. "I just pray to God you don't all live to regret this decision, I really do, folks." He stood. "Grace," he said, and led his wife from the room.

Did you hear Alex, Dad? He came through!

Your turn now, girl.

PART FOUR

I

Jane was propped against the quilted pink headboard, shoes off, CNN muted, checking Bridey's home number. How far away San Francisco seemed, and not just in miles.

She had just closed the pale green cotton drapes on the view from her room at the Welcome Inn. The warm-hued prairie sunset, the rural serenity, were beautiful to look at—but disorienting, so out of sync were they with her current state of mind. It was less than an hour since she'd made her audacious offer, and Roy had grudgingly agreed to give her ten days to turn Akers around. The ten days were to begin officially at six o'clock tomorrow morning, but already Jane could hear the clock ticking toward her deadline. She'd need to hit the ground running, so everything else needed to be taken care of before then.

Jane punched in Bridey's number and asked if she would look after things in her Telegraph Hill apartment for the next ten days. Bridey had done it before, when Jane was gone on a business trip, but never for as long as a week and a half.

Bridey said, "Sure, no problem."

"Thanks."

"Mark's called me twice this afternoon, asking if I've heard from you. I don't know what to say about your father. Will 'I'm sorry' do?"

"Perfectly. As for Mark, he's next on my list to call. Listen, do me an extra favor. If you notice him starting to go over the edge in the next few days, give him some of your chamomile tea, okay? Speaking of the next few days, I know the *last* few have been full of surprises, and even now I can't explain why I have to stay on here. Wish me luck anyway?"

"Jane, only people less smart than you are need luck."

"Remind me to think of a more appropriate title for you than office manager when I get back," Jane said.

"I'll jot down a list of grandiose possibilities and have it waiting on your desk. Bye."

Bridey was efficient, meticulous, upbeat, a one-person support group. But occasionally wrong. Jane would need luck, and she knew it. Starting right now, as she dialed Mark's number. *Please, God, don't let him see this as just one more opportunity to flunk the real world.*

He picked up on the first ring. Jane broke the news to him in bite-size pieces, taking a full two minutes to work up to her decision to stay on in Ocean Plains and tackle the Akers logistical problems. Only then did she begin dictating a list of TranSonic Airex contacts and issues he'd have to keep after on a daily basis.

"Time out!" he cut in. "You can't just tell me you're not coming back for a whole week and a half, then go on like everything's fine. Give me a second to find the smelling salts."

"I can't, Mark. I've got at least a dozen phone calls to make. Until this is over, I'm going to be working twenty hours a day, squandering the remaining four on sleep. You'll just have to bear with me—and my schedule. The plain fact is, I'm going to need you to keep all the balls in the air out there."

"Jane, I know this isn't what you want to hear, but what makes you think I can do all that? You know me . . . "

"That's right, I do. I'm only going to say this once. If I didn't have full confidence in your ability, I wouldn't have taken on this challenge. I've put too much into Malcolm and Associates to be willing to forfeit it now—not for all the Akerses in the land. Got it?"

"Why can't you face the fact that, good as I am at what I do, that's not as many things as you're good—"

"I know, I know. You don't do windows."

"—at. Like client hand-holding."

While she was getting her master's in business logistics from Penn State, Jane had worked part-time for the Philadelphia branch of Rohmer International, a logistics consulting firm. Then, six years ago, when Rohmer transferred her to its San Francisco office full-time, she'd taken additional courses at Stanford's Graduate School of Business. Associate Professor Mark Gibian had lectured on international logistics strategies. Jane had found his nontraditional thinking a valuable counter to her own down-in-the-trenches approach to problem solving. They'd kept in touch by e-mail. Then, last year, when she had finally gathered enough courage to strike out on her own, she'd rented her own office and dialed his number.

"The gold paint is still wet on the door," she'd announced proudly. "It's called Malcolm and Associates. Except I don't actually have one yet—an associate, that is. I'd very much like him to be you, Mark."

She'd offered him half again what the university was paying him. Mark had hesitated—for all of three hours. Then he'd called her back and said he could start part-time right off, full-time once the school year was over. But it was as though that decision to quit academe had used up all his nerve. Ever since, whenever she'd needed him to try his wings, he had balked. But Mark *could* fly. He'd proved that again in Honolulu.

"Look," she said. "You're a lot better with clients than you give yourself credit for. Dave Ellsmere praised you even before I left Honolulu. Besides, it's only ten days," she added, echoing Alex's line to Roy. "Mark, I'll help you all I can, and so will Bridey. Okay?"

"Did Ellsmere happen to say *how* good he thought I was?"

Was there a grade-hungry student stuck inside every teacher?

"At least an A minus," she said.

"I can live with that," Mark said. She could almost see him smiling. "For now. Shoot. What do you want me to handle?"

Jane began to prioritize client issues for him. "I know that's a long list," she said finally. "I don't want anyone to feel neglected, but if you need to postpone or reschedule some of that till I get back, I don't think it'll be a problem."

"Except TranSonic Airex, right?" he asked.

"Right. TSA doesn't get moved off the front burner for even half a day."

"They love to hold meetings."

"You can survive a few meetings. Take careful notes, e-mail 'em to me, I'll fire right back. Or as soon as I can."

It was time to move on. "Now, there are some things I'm going to need your help on here. I've got to assemble a first-class logistics team—"

"Not so fast. What about our other clients? Remember Healthy Hills?"

"What about them?"

"Remember that big order of canned soups we helped them with?"

"To?"

"Japan. Eight thousand cases?"

"That one. Something happen?"

"Their factory in Puyallup screwed up on where they printed the expiration date. Japan has a different requirement for ink-jet coding. Instead of putting the expiration date on the top of the can, the factory printed the production date there, the way they do for domestic product. Then they loaded the eight thousand cases in ten ocean containers and delivered it to the pier."

"And?"

"And they all got merrily shipped off to Yokohama. Honmoku Pier, jetty D, wharf number three."

"I was afraid you were going to say that," Jane said. "You're going to tell me that Japanese Customs spotted the dating discrepancy."

"Yep. The date told them all that soup had already expired, which of course it hadn't, but ever win an argument with a customs officer? They refused to clear the shipment. So Healthy Hills has got eight thousand cases of their gourmet soup sitting in a Yokohama free trade zone, while their new distributor over there is getting extremely unhappy, unable to fulfill deliveries to his big retail customers. Meanwhile Joe Dawes, Healthy Hills's sales manager, has been on the phone the last two days trying in vain to get his product released. Finally he gave up and called you. Eight times. I called him back soon as I got the messages. Jane, he's about reached *his* expiration date. They're not experienced with this stuff."

Mark finally took a breath. "Frankly, Jane, neither am I."

Jane thought quickly. "Call Harada Terminals in Yokohama. It's, what, around seven tomorrow morning there now? Guy named Jerry Ueda runs their stevedoring operation, oughta be in by now. Mention my name. He can give you the name of a local co-packer that can remove the ink-jet coding and put on the correct expiration date. He can also tell you what it should cost, then get somebody to haul the containers to the co-packer, then back to Customs. Tell Healthy Hills they'll have to pay some extra freight, plus gate charges and whatever it costs to fix the plant error in date coding. Now here's *my* problem—"

"Wait! I'm still scribbling." Sigh. "Ready."

"I need an expert on computer viruses."

"What about that young guy you told me about who worked with you in the Gulf? The Saudi prince's son? Wasn't he kind of a whizbang on computers?"

"Not bad, Mark. Khalif's at Michigan State now, doing his Ph.D. If he can get away for a few days, I'm sure he'd be a useful member of the team. But we're talking world-class cybersleuth now. E-mail around to your academic pals and get back to me as soon as you can."

"Will do."

"Anybody else you can think of who might help?"

"How about me?"

"Can't spare you. But prepare to have your brains picked on a regular basis."

Wade Crain had no intention of waiting for Monday morning to get hit with bad news. So he'd called ahead, and Big John Brinkema, the Milwaukee distribution center's night shift manager, had come in early and was standing by when Wade got there at quarter of eight Sunday night to brief him on the latest computer inventory problems. The short answer, Big John had told Wade on the phone, was that they'd been hit hard in Wade's absence. Now, as Wade pulled into the RDC parking lot, he was ready to hear the worst.

He thought.

But not this disaster. All the receiving dock doors were open. Several truck-trailers were backed up to the dock, but Wade couldn't see any unloading going on. He parked and was walking toward the dock when he spotted two people leap from it to the asphalt and sprint away into the night.

"What's happening?" he yelled after them.

"Gas!" came the answering yell.

An instant later and two steps closer to the loading dock, Wade's nose caught a noxious trace in the air. Pure instinct kicked in, telling him to turn and run after the others and ask questions later. But this was *his* facility—he had to know what had gone wrong. And do something about it ASAP. He vaulted onto the loading dock. Immediately a burning sensation assaulted his eyes, as if he'd just dived into a pool with way too much chlorine.

Chlorine gas! Oh, Jesus!

John Brinkema came rushing up. "Wade, bad news, real bad!

We got a chemical spill."

"Pool chemicals?"

"Don't know for sure. Because of all the fucked-up orders—returns from stores, wrong product from vendors—we're shifting pallets and boxes all over the floor just to make room. What I'm guessing is, a brand-new fork driver hit a rack corner where we store pallets of household ammonia next to bleach, and the whole section just collapsed, the pallets crashing down, the containers smashing open, mixing up the chemicals right on the floor—"

"When? When did this happen?"

"Maybe ten minutes ago. I called the fire department. Their hazardous material team's on the way. Meanwhile I got fans running, and we're trying to wash down the area with hoses."

"You got people still in there?"

"Yeah, me and three other guys."

"Get them out of there right now, John! Come on! Where the hell are they?"

John turned and charged down an aisle. Wade hurried after him. The fumes were worse now. Not only was the smell more intense as they ran through the storage racks, but Wade's lungs were starting to feel the burning as well as his eyes. Factoids from college chemistry flashed across his mind—bleach and ammonia react violently, produce toxic gases, including chlorine. Which, Wade knew, happened to be one of the gases used to kill soldiers on the western front in World War I. He also recalled a TV news story about an entire neighborhood being evacuated when a chlorine tanker truck overturned and ruptured. The two household chemicals shouldn't even be stocked in the same warehouse, Wade realized, let alone be part of the same pallet stack.

"There!" John yelled. Just ahead on the outbound dock, a trio of workers, with towels over their faces, were manning a fire hose and a couple of big squeegee mops, pushing water across the concrete slab toward the wide-open dock doors. Nearby, two big propeller fans were pointed the same way. The stench was intense, and every breath seared Wade's lungs.

John, two feet ahead of Wade, called out: "Deke, turn off that hose! Jorge, you and Sabo pull the plugs on the fans! Then, all three of you, out of here!"

Not more than thirty seconds later, all five men were scrambling off the outbound docks and running into the parking lot. They kept going until they reached the far end, truly out of breath. There, one by one, they risked inhaling deeply. Again. And again.

"What about building security?" John asked finally.

Wade shook his head. He was worried about a lot of things right now, but pilferage wasn't even midway up the list. The health of any workers exposed to the chlorine gas was of uppermost concern. Still, he thought as he heard the sweet sound of oncoming sirens, as bad as the fumes had been at close range, a chemical reaction between the contents of several household-size plastic containers wasn't likely to pose the threat of, say, a ruptured industrial tank. With a little luck—which he felt he was due about now—the men would come out of this with no lasting aftereffects.

Chalk up one more to whoever was attacking their electronic supply lines. Though the actual spill might have been accidental, happening more or less the way Big John guessed, the computer-driven chaos on the warehouse floor had obviously been a big contributing factor.

The fire department truck was pulling up on the inbound side of the huge structure. As Wade and John hustled around to meet it, they gave the building a wide berth. When they got to the truck, the space-suited hazmat team was already on the ground, forming up to go in.

The team leader, his hood off, had already smelled the chlorine and checked wind direction. Once Brinkema answered his specific questions, including the probable source of the gas, the head man decided to evacuate the two-story Fiesta Inn that bordered the RDC's outbound truck lot. He also instructed a team member to radio the local police and have them go house to house in an adjacent neighborhood downwind, telling people to stay indoors and shut their windows.

"Here." The hazmat leader handed John a walkie-talkie. "Once we're inside, you're gonna tell us where to go."

Wade shifted his attention to the fire department paramedics, checking the three warehouse workers who'd been in closest contact with the fumes, then others who'd been working a distance away from the spill. Several ambulances were on the way, Wade was told.

For a moment, Wade wondered about himself. He'd been close to "ground zero" for maybe half a minute. He could still feel the effect in his lungs—"breathing complications," they'd probably say if he elected to take a siren ride to the hospital—but he doubted he'd incurred actual damage. Maybe they'd all be real lucky.

That, of course, didn't rule out hefty disability claims down the road for respiratory problems, especially on behalf of the three workers John had ordered to clean up the spill. Big John should have damned well known better!

Next order of business: How soon would the hazmat team give them an all-clear, so they could get the DC moving again? Wade hoped for only a few lost hours. Whatever, he was in for a very long night.

As he stood there, bathed in the rotating glare of emergency lights, he suddenly thought of the pristine ridge-crest trail along the California Sierras, which he'd abandoned at news of Royal's death. Could it have been only three days ago?

II

Jane was waiting for a return call from Dan Sammons, her

favorite logistics professor at Penn State. Anyone he'd recommend as a possible team member would, she knew, be first-
rate. She hoped he might have some ideas for dealing with the
Akers supply-chain crisis himself.

Jane checked her list again. Another campus logistician,
Georgia Helm, according to her secretary at the University of
Tennessee, was en route to a lecture in Boston but would get
Jane's message the minute she called in from Logan Airport.

She decided to put off making other prospecting calls for a
while. She'd been on the phone constantly for nearly an hour.
Time to free it up so that if Sammons or Helm called back, they
could get through to her room. Besides, she had one more
thing to do before morning erupted.

She went to her briefcase and retrieved an eleven-by-four-
teen-inch manila envelope her stepmother had handed her,
saying, "Some things your father kept for you."

Jane undid the clasp of the envelope, feeling an
apprehension as powerful in its way as the one she'd felt about
the morning and the official start of her challenge. The Akers
challenge, she realized, was two-sided: a challenge *from* her
and a greater one *to* her. Would the contents of the envelope also
challenge her?

Jane put it down.

*Hey, Dad, if you're really there listening, let me tell you, this
timing's not great. I'm not up for any more challenges at the
moment.*

Well, if he *was* listening, he wasn't letting her off the hook.

Jane picked up the envelope again and tilted it. Out slid a
rubber-banded packet of envelopes. Letters she herself had
written, she saw, thumbing through—all to her father. She
looked at the return address on the bottom one. Girl Scout
camp. On top was one from late in 1991, just after she'd left the
army—and turned down her father's job offer. She'd written,
she remembered, to soften the blow. She didn't open the letter,

but thumbed to the next one. The truth was, she had no desire to read any of her letters to her father. Pain was one thing, pointless pain another. She tipped open the manila flap again, to put them back—and another sheaf slid out.

This packet was considerably thinner. The first envelope had an embossed Akers return address—a letter *from* her father. Why hadn't he ever mailed it? Jane pulled the letter from the pack. The postmark was unreadable, but the stamp had definitely been canceled. The envelope was addressed, in her father's own jagged handwriting, to Jane at her first apartment in State College, Pennsylvania. So it must have been written shortly after their argument, when she'd gone to work part-time for Rohmer and enrolled in Penn State's MBA program. He could have gotten the address from Joycelyn, but more likely from Alex.

And, defying her name change, he'd written to Jane *Akers.*

The envelope, which was still sealed, bore a smudged red stamp: NOT AT THIS ADDRESS. Dear God! Only a couple of weeks after moving in, Jane had added AKERS below the MAL-COLM on her mailbox, realizing that all her military correspondence would be returned otherwise. She hadn't thought of her dad, never imagined he'd refuse to honor the name change even in correspondence.

All of which only added somehow to her current misery.

But even now, Jane didn't second-guess her decision to become a Malcolm. The professional and personal reasons remained compelling all these years later. But the timing was awful, and the mail carrier's laxness in not checking had doubtless aggravated the pain she'd given her father.

She put that letter aside to come back to. The next three, though properly folded, were without envelopes. She unfolded the first one. A carbon copy of a letter, typed on her dad's old Selectric. She opened the other two. The same. All three were addressed to Jane, in care of her mother's Chicago apartment, all dated shortly after Joycelyn had moved out of 45 Maple, taking fifteen-year-old Jane with her.

Jane had never seen the letters before.

Several explanations were possible, but one carried immediate conviction: Her mother had kept them from her.

Joycelyn Malcolm Akers always knew what was right—for everyone. The betrayal, if Jane's intuition was right, had happened nearly two decades ago. No matter, the hurt downed her as forcefully as a huge wave. In its wake came fury.

Only moments later, other emotions were sweeping over her as her eyes grabbed on to two words atop the first page: "*Dear Janie*."

She put down the letter and looked for the minibar. No minibar. Welcome Inn suddenly felt less welcoming. Wait! Jane opened her carry-on case, pawed about, and came up with a fancy miniature of some Hawaiian liqueur. She'd pick up some other gem for Bridey's collection before she got home. Her hand was steady as she splashed the dark liquid into a water tumbler. Amazing. She took the glass and the letter and sat herself down in the room's only armchair. When the sweet but fiery liquid ignited in her throat, she began to read what her father had written to her eighteen years earlier:

> *Dear Janie,*
>
> *I'm heartsick at what has happened. I still can't believe you and your mother are gone. How empty this house is! I* know *your mother blames me for everything, and maybe I am to blame for it all. I'm certainly sick over what I* know *was my part. But I've never been less than 100% honest with you, Janie, and I'm not willing to start now. The fact is, I'm also sick over your mother's extreme reaction, which takes you away from me and your brothers, and us from you. I'm hoping your mother will change her mind about this before very long and give me another*

chance. But right now she's too angry at me, I know, to even consider that.

I won't defend what happened. What I did was very wrong. Someday we may be able to have a talk that will help you understand more about what happened, but I realize that can't be now. For now I can only admit that I haven't been the kind of husband your mother wanted me to be, even when I tried to be. And your mother wouldn't, or couldn't, forgive my failings. Probably I haven't always been the kind of father I should have been, either. Spent too much time minding the store, as they say. If so, I hope and pray that you will forgive me—if not now, then one day soon, Janie—and come home and give me another chance to be—

Your Daddy, who loves you so . . .

The last line blurred, and Jane clutched herself, waiting for the spasms of grief to pass. They didn't.

There was so much time, calendars full of days and years, when she could have reached out, could have called her father on the phone, written to him, gone to see him. And she'd done none of those things.

All those opportunities were as gone now as he was.

The shrilling telephone cut through her sobs. Jane swiped at her damp face as though the phone had eyes. She had to piece herself together and answer; it might be Dan Sammons or Professor Helm. She blew her nose, then lifted the receiver.

"Hello?"

"Is this Jane?" The voice was female but lacked the Maine twang Professor Helm had never lost.

"Yes. Who is this, please?"

"This is your aunt Louise."

"Aunt Lou! Are you all right?"

"I'm fine. I heard about what you did after I left the meeting.

Grace told me."

Jane braced herself for a lecture on her upstart behavior.

"You showed a lot of spunk there," her aunt said. "I just called Junior and I told him to change my vote, too."

"You did? Thank you, Aunt Louise. Thank you so much. You don't know how much that means to me. I just—"

"I haven't finished, Jane. There's one more thing. I know you're going to be very busy, but I think we need to talk."

DAY 1

AKERS STORE #1, OCEAN PLAINS
MONDAY, MAY 10
8:34 A.M.

The story was in the *Chicago Tribune* business section: "AKERS EMPIRE HIT BY LOGISTICS WOES. Small Towns Complain of Shortages." Jane read it quickly, standing by the news racks near the store's front entrance. The gist was that in the wake of Royal Akers's death, nobody was really minding the store. As a result, the empire was said to be in swift decline.

But a glance back at the parking lot showed a respectable number of cars even at this early hour. Jane jettisoned the *Trib* and entered the huge store, which was easily ten times the square footage of the original Akers she remembered vividly from her youth. She walked up and down the aisles, pacing herself to gauge the state of the shelves. There seemed to be plenty of product, customers were filling their carts—everything looked quite normal. She looked into the cold cases up front: packed with milk and juice and sodas. Five checkstands were lighted, shopping carts already backed up three and four deep, items being scanned.

But Jane knew that the appearance of normality at the checkout points was deceptive. It was right there, at the checkout—at the threshold of an invisible, electronic realm—that their problems were centered. The point-of-sale information was being vacuumed into a polluted datastream.

Only moments later, as Jane was walking by the *Corner Drugstore* department, she saw firsthand how the pollution was impacting operations. The pharmacist was explaining to an elderly gentleman with a hearing aid why his blood pressure prescription could not be filled. Apparently his particular beta-blocker medication was temporarily out of stock. Jane recognized the brand as one of the most popular antihypertensive drugs.

"Then give me whatever you do have," the old man was pleading. "You want me to drop dead right here in the store?"

"We're trying to reach your physician right now," the pharmacist said—his voice taut, as though he were being forced to repeat himself. Apparently the customer didn't understand that his doctor's okay was needed for any prescription change.

The assistant store manager, a sharp-featured, redheaded woman Jane didn't know, was offering the man a glass of water and trying to steer him to the nearby couch provided for customers waiting for a prescription.

The fear in the man's face was too much for Jane. "What about the Ocean Plains Clinic?" she said in a low tone to the redhead. "I bet they have samples of propranolol."

The woman turned with a startled smile. "Thanks, miss. I've already sent someone."

Jane backed away. What she had just witnessed brought into sharp focus what was happening to Akers—and in Akers stores all over the world. Not only were the implications for operations ominous, she'd just seen how inventory tallies gone berserk could be life threatening.

The night before, she'd reviewed the stocking process, looking for glitches. On paper there weren't any. Akers prided itself on maintaining smaller inventories than its competitors, even for fast-moving items. Their distribution centers now carried only two weeks' supply of high-velocity products. But by using "just-in-time" deliveries, they were able to replenish their stores with any particular item the day before they were due to run out.

It was a textbook case of how logistics could add value by getting the right product to the right place at the right time for customers at a low price. By accelerating inventory turnover, capital and operating costs were reduced all along the supply chain—purchasing, production cycles, transportation, warehousing, retailing. When goods and information were flowing smoothly through the pipeline, the benefits could be dramatic. But with all safety stocks cut to the bone or nonexistent . . .

Jane rounded a corner and came upon a small crowd circling an employee in the sporting goods department and thrusting Akers newspaper insert ads in his face, demanding rain checks for a missing exercise rider. Those preprinted specials probably went out statewide. If all the stores were out of product—a reasonable assumption, under the circumstances—she was witnessing in "micro" yet another "macro" debacle. As she watched the harried employee try to assure the crowd that of course they could have rain checks, Jane recalled a comment on "just-in-time" by one of her Penn State logistics professors. "The 'just-in-time' system," he'd said, "works wonderfully in theory. In the real world, however, it's wise to hedge your bets. The cost of inventory is not as great as the cost of losing a sale or, worse yet, a customer as a result of an out-of-stock situation."

Amen, Professor Leeds.

In fifteen more minutes of aisle prowling, she encountered a dozen more outages, perhaps twice that many SKUs at or below minimum inventory levels, and a similar number of unacceptable shelf-presentation levels.

Most stock-outs were merely vexatious, Jane thought, but at least one more she encountered was potentially dangerous—definitely a nightmare scenario for company attorneys. A woman had just picked out a training bicycle for her son's birthday, only to discover there were no helmets available in boys' sizes.

"But his birthday is *today*," the distraught customer was saying to the clerk. "What am I supposed to tell him—'You can ride it next week when they get the helmets in'?"

Dear Lord, what have I taken on? If all this was happening at one of the company's showcase sites, less than a mile from headquarters and the main distribution center, what scenes were being played out at other Akers stores? She'd better get to work.

Jane headed out of the store, stopping only long enough to introduce herself to the manager and the redheaded assistant manager. Glenn Hawkes, the manager, confirmed how bad the situation on site was, quickly ran through half a dozen examples of particularly bad outages they'd managed to survive.

"You're the gal who tried to help out in pharmacy," the red-head said. "Want you to know we did get Mr. Taylor his medicine from the clinic. But these one-by-one solutions can't work for much longer, Ms. Malcolm."

"Once people start losing faith that they can find whatever they need at an Akers, we're cooked," Glenn Hawkes said. "I'm not much good in the kitchen, but even I know, once something is cooked, you can't uncook it."

"I wish I could tell you we'd have the problems solved tomorrow," Jane said.

"It helps to know you took the trouble to see it from our point of view as well as headquarters'," his assistant put in.

"What's your name?"

"Pattie Daran."

"You have my word, Pattie Daran. We're going to lick this nasty animal." She glanced at her watch. "Better get on it. You want to talk to me, either of you, you can reach me over at headquarters 'most any time."

Jane walked out of Akers #1 filled with fighting resolve.

No one who knew Jane Malcolm for more than five minutes would ever accuse her of wearing rose-colored glasses. In fact, seeing a problem without *any* distorting coloration was among her major strengths. The current situation wasn't just really bad, it was critical. It was seeing that for herself with stark clarity that triggered Jane's determination not merely to make Akers's computers work right again, but, as soon as possible, to get its logistics systems back on track and to solve the replenishment chaos—any which way that worked.

That promise she'd made in Store #1 just now wasn't only to Pattie Daran, it was to herself as well. And to her father.

After all, she'd been to war, hadn't she? Could this challenge be any tougher? During Desert Shield, Pentagon wizards had moved four hundred thousand troops halfway around the world on short notice, into harm's way and an infrastructure of mostly superheated sand. The daily job of supplying those troops with food and water, shelter, sanitation, clothing, weapons, ammunition, and transport fell to logisticians like Jane and her 22nd SUPCOM teammates. And they'd done it.

Before her time, she knew, people in her own family had figured out a way to do what others regarded as unfeasible. Was the task ahead more Herculean than that accomplished by her grandfather Colonel James Malcolm and his fellow traffic officers during the Berlin airlift of 1948 and '49? Each day thousands of tons had been loaded into a cobbled-together armada of aircraft and flown into the blockaded city—fuel, food, medicine. That meant up to a thousand landings every twenty-four hours, Jane knew, day after day for eleven months—until the Soviets gave up and reopened the roads and rails.

Lack of computer technology was no excuse. Much of the logistical wizardry in Saudi Arabia had been worked with three-by-five cards and ring binders, charts, and graphs. In those first weeks in-country, as Jane well knew, supply officers often operated out of Bedouin tents or the backseat of vehicles. Deliveries for the Berlin airlift had been tracked on blackboards, though flight-and-cargo information was transmitted ahead via wireless, employing a simple electronic code—a pioneering use of EDI.

Some people, Jane knew, regarded these accomplishments as miracles. *But God doesn't do the legwork. That's man's job—and right now it's this woman's.*

There had to be plenty of ways, old-fangled as well as new-, to lick these logistical problems, even if they did involve a global supply line, with tens of thousands of products flowing to a thousand destinations.

As she pulled out of the lot, the Akers water tower loomed dead ahead, like a giant golf ball on a tee. But Jane pointed her rental car the other way—toward Eliezer Grey's farm. Jane had to have one person she trusted implicitly, and Elie was it.

On the phone this morning, Elie had agreed to join her

team—on a limited basis. "Don't expect me to work around the clock," he'd told her firmly. "The days I could do that are long gone. Besides sitting on my hands at Akers board meetings, I got real work—responsibilities around here. But I'll do what I can."

Jane was willing to bet that his "what I can" was more than most people's "I'll give it my all." With his help, Jane believed, and the input of a first-class "log team," they could slam together some fast supply-chain fixes—and do it well ahead of her ten-day limit. *That* was one Akers story she couldn't wait to see in the *Trib's* business section.

AT A COMPUTER, SOMEWHERE IN THE MIDWEST
9:08 A.M.

Timed how long it took me this time, just for kicks. Three careful minutes plus twenty very careful seconds, to type it in.

Done. Ready for launch.

No time like the present, right?

Type the name at the system prompt, hit Enter. Another little bug on its merry way. Joining the happy horde. Lilliputians, bringing down a cumbersome giant.

One—my favorite so far—has already been halfway around the world. It's on its way back right now.

That one even I couldn't stop from going off. If I wanted to.

What a dumb expression. People should know for sure what they want before they start something. Then they won't have to even think about changing their mind.

What I want is to hear that bang.

Very. Big. Bang.

Music to my ears.

THE GREY FARM
9:08 A.M.

Eliezer Grey lived just south of town on four hundred acres inherited from his father. But Elie had never been much of a farmer, Jane knew; his acreage was worked by a local family on a contract percentage basis. She made the turnoff into a lane alongside an apple orchard. Brown corduroy fields stretched to the horizon. All planted now, some with field corn, probably, the rest soybeans, like most of the land hereabouts.

But there was something different about the Grey farm. Adjoining the main parcel was a small demonstration farm—forty acres of cropland, vegetable gardens, cow pasture, and duck pond. It was strictly a nonprofit venture, worked by young people with developmental problems. Elie had about a dozen kids right now, he had told her, from all over the state. They attended special-ed classes, bunked in a dorm behind the grain silo, raised chickens, slopped hogs, and milked cows, all under the tutelage of Elie, two paid teachers, and numerous community volunteers.

Parking behind the Grey farmhouse, Jane saw Elie was already working with his kids. One stood behind him as he drove a vintage tractor in an adjoining field. A moment later Elie spotted her and waved. When he switched off the engine, his companion jumped down first, then helped the old man. She watched them approach, one shuffling with age, the other on tiptoe, in a lurching gait Jane associated with cerebral palsy.

"Trisha, this is Jane Akers," Elie said. "Or is it Malcolm? I keep forgetting."

A gap-toothed teen pump-handled Jane's arm. "Malcolm's a boy's name," she said. "You don't look at all like my cousin Malcolm. *You're* pretty." She grinned.

Jane grinned back at the girl and said, "Why, thanks, Trisha. I think your pigtails are *very* pretty. And you're right, my name's Jane. Looks like you all have been busy this morning."

"We're planting corn," Trisha said. "We're not done yet."

"On that old John Deere," Elie said, walking them toward some big aluminum sheds, "we can do two rows at a time. My big planters seed a dozen rows using computers to dole out just the right mix. Got a little TV in there, too, so you can watch the Cubs. Then I think about my daddy. Started out farming walking behind a horse plow. Now that was *work*." Elie mopped his forehead, glanced around. "Zach? Do I hear you, boy?"

Ten feet away, a freckle-faced youth popped up from behind a pallet stack of herbicide. His soft body shape, high forehead, and slack smile seemed to fit the Down's syndrome profile. "Zachary there is my grandson, Jane. Zach, I want you to meet Jane. She grew up in town, but she spent quite a bit of time out here with us. She loved riding ponies, just like you do."

"Didja have to do all your chores first?" Zach asked her, approaching slowly.

"You bet," she said. "My brothers and I used to try to do a very good job so we could get permission to come out here. What a great place to grow up!"

"You can come out here anytime—when you've got all your chores done, can't she, Grandpa?"

"Sure can."

"Thank you. You too, Elie. I've got a lot of chores right now, Zach, but I'm going to do my best to take you up on that."

"Speaking of chores, Zach," Elie said, "how about if you and Trisha feed the chickens?"

"Sure, Grandpa! Bye, Jane! C'mon, Trisha."

As the two kids hurried off together, Jane turned to Elie. "Heck of a retirement you got here."

"I won't deny it. Like old Ronald Colman said in that movie, 'Tis a far, far better thing I do . . .' Couldn't get me up on a tractor 'cept for those kids. Truth is, I've never been paid so well in my life for any kind of work."

"Well, I'll try to steal as little of your time as I can."

"I want to help. Talk to your mother yet?"

Jane was caught off-guard. "Yes. This morning."

"What'd she say?"

"She, uh, agreed."

"To give you voting control?"

"Not exactly. She agreed to wait the ten days before signing her proxies over to Roy Jr. He got to her before I did, apparently, asked her not to go along with my plan. So I guess that's a small victory for our side. I just hope ten days is enough to turn the corner."

A yellow Labrador came out of nowhere and bounded along beside them. Elie patted its head.

"If it's a victory," he said, "why do you suddenly sound discouraged?"

"There was something else I wanted to ask her. Only I chickened out."

"What was that?"

Jane stopped and bent to pet the Lab. Continuing to stroke the dog, she told Elie about her father's letters, letters she'd never seen before.

"I just know in my gut she decided I'd be better off not reading them. She had no right—"

"If Joycelyn did that, she shouldn't have," Elie said.

Jane stood. "I have to confront her, get her to admit it."

"Think so?"

"I'll never forgive her."

"Whoa there! Let's not start one of *those* cycles again." Elie shook his head sadly. "I've got to admit, your theory has the ring of truth. Joycelyn could've made a decision like that, and like I said, if she did, it was wrong. Period. You can confront her or not. But do Joycelyn—and even more, yourself—a big favor, and don't harden your heart against her."

Which was exactly the plea her father had made in one of his letters: *"I hope and pray that you will forgive me . . . "*

I'm not a forgiver.

Could she learn to be one?

As they passed an open shed, Jane heard a hiss, and the dog bolted away. The sound came from inside the shed. She stepped to the doorway. A spearpoint of bright blue flame cut off, revealing a helmeted, gloved figure beside an oxyacetylene rig.

As Jane's eyes adjusted to the gloom, she saw a giant, brick red Massey Ferguson combine whose cab almost touched the shed's high ceiling. Her eyes traveled down, past one of the huge tractor tires, and she saw the twisted pile of metal to which the figure had been applying the torch.

With a gasp she realized what that scrap metal was.

AKERS IMPORT DISTRIBUTION CENTER, CHINO, CALIFORNIA 7:24 A.M.

Norma Sandoval was having a tough time. It was her first morning on the job, and Jae-Min, the young woman assigned to be Norma's trainer, explained everything so rapidly that Norma couldn't remember half.

"You haven't asked many questions," Jae-Min observed at break time.

"I don't know where to start."

"Too much coming at you, huh?"

"You could say that," Norma admitted cautiously.

First off, the DC itself was so huge—three buildings in one, Jae-Min had explained, with separate sections for Fashion, Small-Ticket Imports, and Cross-Dock Imports. Norma had glimpsed them all on her orientation tour but had been led through so many stairwells and catwalks and crash doors that she couldn't have drawn a map of the place if it meant being able to tear up her mortgage.

One thing she was sure of, though. She was glad she was going to be working in Fashion and not in Cross-Dock with its noisy conveyers—so noisy that she and Jae-Min had worn earplugs and had had to shout at each other.

Here in Fashion, the biggest part of the DC, it was actually quiet. Just the clickety-clack of garment trolleys sliding along overhead metal rails. All the clothing, like everything else in the distribution center, came in from Pacific Rim countries, Norma had been told, and got shipped out, not to individual stores, but to Akers regional distribution centers.

"See, that's why we're called a hub DC," Jae-Min explained. "The RDCs, they're like smaller hubs, with each one serving its own regional stores. Get it?"

Norma nodded vaguely, hoping she'd get it eventually. But she was having a hard time just remembering the names of all the countries where the clothes were made—like Pakistan and Bangladesh. Millions of pieces every year, but how many millions she couldn't remember, either. Or how many thousands of ocean containers they came in. One statistic was easy, though. There were exactly six receiving doors in Fashion. You could see all six from just about anywhere on the floor.

Four doors had been in use this morning, unloading containers trucked up from The Port of Los Angeles forty miles away. Norma had expected each container to be packed with cartons. But behind the first wall of cartons were racked clothes. GOH—short for "garments on hangers"—was the term Jae-Min used.

"You mean they cross the ocean like that?" Norma asked.

"Sure. One hanger, one garment, all the way from factory to store. Makes it easier."

Norma could see how. You wouldn't want to have the job of sticking hangers inside all those blouses and coats and shirts. Not with truckload after truckload rolling in and trolleys backing up all over the floor. "And this is nothing," Jae-Min had warned. "Wait till next month, when we start getting in our back-to-school lines."

Luckily, just about everything in Fashion was automated. Each garment came not only with its own factory hanger, but with a bar code that showed everything you needed to know about it—vendor, item number, size, price code. The only manual operation was transferring garments from overhead trolleys to individual hooks so they could travel through the sorter, which was where Norma would be working.

This automated system, with photoelectric sensors, "knew" all the Akers store orders and sorted each containerload of garments into lanes, one lane for each Akers RDC, and then into individual store orders. These lanes then got shunted off toward outbound dock doors, one for each regional distribution center. "It's like a giant railroad switching yard, only for garments," Jae-Min said. "Ten thousand pieces an hour, a hundred thousand a day."

"Wow!" Norma said, yet she could believe it, watching all the clothes go clicking by—little-girl party dresses, denim overalls, wool skirts, nightgowns. It was like a gigantic and never-ending dry-cleaning conveyer. Click-click-click, and off they went, to Reistertown, Maryland, or Sweetwater, Texas, Stockton or Raleigh-Durham, Utica or Milwaukee.

Jae-Min took her to see the outbound trailers being loaded, brown paper underneath the hanging garments, carton goods stacked behind. They watched a door slam and a truck pull out. A two-man sleeper team would have it ready to unload in Maywood, New Jersey, three days later, Jae-Min said.

By the time the lunch bell finally sounded, Norma's head ached from trying to stuff so much information inside. She was too mixed up even to care that one of her favorite TV soaps was on in the lunchroom.

"You know, Jae-Min, some of it just doesn't make sense to me," she admitted over her peanut butter on whole wheat.

"Like what?"

"Well, like that big order that came in from Bangla-wherever it was."

"Bangladesh."

"Right. It seemed like a lot of those culottes and sundresses and stuff were going to stores in Maine and Minnesota. Isn't it still cool there?"

"Gotta beat the season, Norma."

"Right, right." That did make sense, but . . . "But for those stores down south—you know, like Fort Worth and that one in Alabama—"

"Montgomery."

"Yeah, and even Phoenix, didn't it seem like *they* were getting mostly parka-type coats and bulky sweaters? I got a cousin in Tucson, called me on my birthday last week, said it was like ninety there day after day after day."

Jae-Min shook her head. "Norma, I told you. It's all figured out by these big computers. All these orders come in from Ocean Plains way in advance. Everything's automatic. You don't have to think. Just keep those lines flowing. Try to figure out the big picture, you'll go nuts, believe me."

Jae-Min was right, of course. There was plenty enough to worry about without trying to figure out how any particular store's orders made sense.

"Thanks, Jae-Min." She grinned. "I'll get it yet."

Norma turned her attention to the lunchroom TV. The new blond nurse was sneaking into the private room where the mystery man was still in a coma. Only maybe he wasn't, because when the nurse changed the IV bag for another and ominously whispered, "Nighty-night," into his ear, the mystery man's clutched the sheet.

THE GREY FARM
9:36 A.M.

"What's Dad's plane—or what's left of it—doing here, Elie?" Jane asked. "And why is that man working on it? I thought the

NTSB had finished its investigation."

She recounted what Anne Akers had told her about the National Transportation Safety Board's version of the fatal crash. The preliminary report stated that in the absence of witnesses, it appeared that the landing gear had collapsed, according to the investigator, causing the right wing to strike the ground, which in turn caused the single-engine Beechcraft to nose down hard, crumpling the cockpit framework. Verdict: pilot error. Royal either had inadvertently retracted the gear during the landing roll or hadn't slotted the handle into full down-landing position.

"What they told Anne is really only a preliminary finding," Elie said. "The official report, now that could take up to six months, what with how the government loves paperwork, one thing and another."

"But the investigation part is over, right?"

"Yes and no. See, I'm kinda carrying it on, paying Mr. Harlow there out of my own pocket."

Jane glanced back inside. "Mr. Harlow" flipped up the visor on his welding helmet and stared straight at her, but without expression. His was a bony face, hollow cheeked and shadow eyed; that sepulchral glance, and the fact that the man was obviously autopsying the metal remains in which her father had died, gave Jane an eerie feeling.

With a hands-free head jerk, the man dropped his visor and rekeyed the hissing blue flame.

"Is he an NTSB investigator?" she asked.

Elie shook his head. "Used to be. Worked for the FAA before that. Comes real highly recommended. He was able to pull some strings and get the wreckage released temporarily to me—which, I gather, wasn't exactly easy. Supposed to go to the insurance company. Probably will, if and when they figure out exactly where it is." Elie tossed her one of his sly grins.

"What's going on, Uncle Elie? You don't believe the pilot-error verdict?"

"No, child, I don't."

Jane looked at him, then at Mr. Harlow, who was again crouched over his work. A shiver of fear shot through her.

AKERS REGIONAL DISTRIBUTION CENTER, MILWAUKEE 10:12 A.M.

The toxic spill had been cleaned up, every trace of chlorine gas dispersed. The pallets of household ammonia had been relocated two aisles away from the bleach. A dozen cases of toilet paper, sitting on the slab floor in the forward pick area, had been water damaged, as had the break room carpet. Fifteen workers had been treated for minor breathing problems at local hospitals and released. None of the seventy-four guests evacuated from the adjoining Fiesta Inn had required medical treatment, but all were demanding compensation.

"Now we just gotta wait on the next piece of good news," John Brinkema had told Wade about midnight. The hazmat crews had finally left. "And I don't mean days or even weeks. Gotta see how many workers' comp claims we get before the statute of limitations runs out."

Big John had stayed on, wanting to finish the cleanup before handing off to the day shift manager, Ernie Escalante. Wade had gone home around one-thirty, only to find he was too keyed up to sleep. He'd been staring at the ceiling when Big John called. "Better get down here, Wade."

"You going to tell me what for?"

"You're not gonna believe this, but a situation we got this morning could be ten times worse than what happened last night."

"What situation?" Wade had yanked back the covers, checked the clock. Six-fifteen! Damn, it was too soon for another emergency. "And what the hell are you still doing down there, John?"

"We just got in two truckloads of Hyon computer monitors."

Wade had tried to think. Hyon was one of Korea's handful of giant industrial companies and Akers's main supplier of household electronics. "I don't remember them on our spring promotion schedule." He'd made a point of checking the upcoming special promotions before heading off on his hiking vacation. "When did they get added?"

"They didn't. And even if we had a promotion, we don't have any computers to match up with them. Wade, they're all over my fucking floor!"

"How in hell—"

"Wait. You're not my first phone call. I decided to find out if God was just playing gotcha with us for some reason. But guess what, they got the same damned thing in Maywood, in Baltimore, in Raleigh-Durham—"

"I'll be right down."

Brinkema hadn't exaggerated. The off-loaded pallets of PC monitors blocked two dock doors and part of the main receiving aisle. Each carton bore ideographs and flexographic printing identifying "HYON GROUP." Wade lifted the stretch-wrap cap of one pallet to get a better look. "John, look here. These monitors didn't come from Korea, they're from Mexicali."

"Yeah, right after I called you, we got a third load in from Hyon's Mexican plant. But those other pallets came from their factory in Suwŏn. That's south of Seoul somewhere."

Wade shook his head. Nothing about the order made sense. Like a lot of Asian firms, Hyon had been building factories just across the Mexican border to take advantage of NAFTA's tariff-free entry to the U.S. market. More and more Hyon product in Akers stores—"white goods" like washers and dryers, TVs and display monitors, VCRs and camcorders—were being trucked north from these *maquiladoras* rather than being shipped from Korea. But, sure enough, two-thirds of this monster load of monitors had crossed the Pacific.

Wade did some quick math in his head. Like most electronics manufacturers, Hyon didn't maintain warehouses full of finished goods; they built to order. Once they got a purchase order, they'd process, assemble, and ship. Figure several weeks for that. For the Korean product, add ten days on the water from Pusan to L.A., at least another week to reach the various Akers regional DCs. This order had to have been in the works at least a month, probably more.

And nobody had caught it till now?

"John, you got a legit purchase order on all this stuff?"

"Yeah, the PO checks out in Ocean Plains. But it still doesn't make any goddamned sense. They're going nuts down there, trying to figure out what happened. Meanwhile I'm going nuts trying to figure out what the hell to do with all of these. Our racks are full. The only place I can put 'em is in bulk overstock, which will jam the aisles—"

"Hey, don't park there!" John wheeled to shout at one of his part-timers, a college kid who'd left his forklift in the aisle.

"By the way, boss . . . " John turned back to Wade. "Something else I forgot to tell you last night."

"What is it?" Wade braced for the next catastrophe.

"Welcome home."

AMETHYST LAKES RESORT,
ORLANDO, FLORIDA
3:28 P.M.

This is crazy.

Here it was, halfway through Day One of her ten-day countdown, in crisis-management mode where every minute counted—and here *she* was, sipping chardonnay under a Cinzano umbrella to protect her from the central Florida sunshine, listening to Muzak-style calypso and watching four guys in Izod shirts and plaid pants lining up their putts.

What made it even crazier, she'd spent most of the day in transit—private planing to O'Hare, Delta jetting to Orlando, then driving a rental car thirty-five miles west to what looked awfully like a retired-executive pastureland. All to recruit one person for her team. Who, for all she knew, meant to live out his life on the green.

"Gotta have Don," Elie had said. "I can be useful to you here and there, sure, but Don Landsman, he'll be downright indispensable. *If* you can get him. He won't be easy. He's already down in Florida, hired a realtor to pack up his stuff and sell his house over on Porter Lake. You won't be able to talk him back over the phone, I'm sure of it. I don't think you should even try. After the way he was treated, Don deserves a face-to-face."

"What makes him so indispensable now?" she had asked Elie. "Would it be the same reason Dad chose him to be next in line?"

Elie's bushy gray eyebrows practically merged. "I never did like explaining other folks' reasons. Don't feel any more comfortable about it just 'cause they're dead. But I can tell you why *I* would've picked him. The man knows what he's doing and knows *that*, too. Doesn't make a fuss about it—nothing cocky about Don Landsman—but it's there all the time with him. That kind of confidence *inspires* confidence—something I'd say the company needs a whole lot of right about now, wouldn't you?"

Elie had called ahead to arrange their meeting. Landsman said it'd be rude to back out of his late morning foursome at Amethyst Lakes, but they ought to be getting to the 19th oasis around three-thirty. He wouldn't be hard for Jane to spot, he'd told Elie. So far, he was the club's only black member.

And there he was, holing out and handing in his putter, paying off his caddie, then striding slowly up to the terrace and acknowledging her wave by a tip of his White Sox golf cap. Landsman looked to be in his mid-fifties, with a lean, cocoa-colored face and a plump gray mustache half bracketing an easy smile. He offered condolences, and they settled on first names.

"I'm sorry, Jane, that I missed the funeral," Landsman said after ordering a Dos Equis. "Obviously you know my situation. Even so, I intended to come, but . . . well, let's just say I opted to say my good-byes to Royal privately."

"I understand, really. From everything I've heard, you were my father's clear choice for a successor. So I'm not just sorry it didn't work out that way for your sake, I'm sorry for the company. Akers needs you desperately, as I guess you know."

"Well, thank you for that. Royal used to talk about you, you know, hoping you'd come back someday. I'm sorry he couldn't have seen it. So . . . " Landsman leaned back in his director's chair. "Elie tells me you're trying to save the company, over Roy's objections. Somehow that didn't come out right, did it?"

"I'm afraid it did. Thanks for agreeing to listen to my sales pitch."

"I tried to get Elie to save you the trip. I've said my good-byes to the company as well as to your dad. I was hurt by how it ended, I don't deny it. But things do end, and what the heck, I gave it my best shot." He grimaced slightly. "Ten years of my best shot." He shrugged. "Must've been a pretty good shot, after all, because I've had quite a few phone calls the last couple days. Some pretty interesting. I seem to be rather in demand, and it feels good, you know? Makes me feel like a guy with a great jump shot or a home-run stroke." He drank some of his beer, the very picture of a contented man.

"You deserve to feel that way," Jane said.

"Thanks. Anyway, Darla, my wife, we've been talking. And . . . well, we're not in any desperate hurry to decide." He took another swallow of the Dos Equis. "What we're gonna do, I think, is take a nice little Caribbean cruise"—Landsman gestured toward the outdoor speaker, which was still dispersing steel-drum music—"kick back, play a little golf on some of those island courses, read some books we've been meaning to read. Make up our minds about our future when we come back."

"Wish I could go," Jane said.

"Why *are* you getting involved after all these years?"

"Because it looks like we're about to lose the company, and I suddenly discovered I don't want that to happen—after years of telling myself I didn't care what was happening at Akers, good news *or* bad. Turns out I've got enough regrets in regard to my father without adding another real big one."

"I've got regrets, too. Yeah, I still care about Akers, in case you were wondering, and I always will. But not as much as I used to, or as much as I would have, let's say, if I hadn't been kicked out on my butt, which is putting it politely. I sure can't see myself flying up there to rescue Roy's bacon. Frankly, from what I saw my last couple of weeks there, maybe I couldn't save it. All the distribution systems seemed to be breaking down. Elie tells me it's been getting worse day by day." He shook his head, but there was an obvious distancing there.

Great, I just wasted a day! Plainly she had no leverage to use on Landsman. She looked past him, at the meandering, blue-green fairway, until the scene was eclipsed by a passing cocktail waitress bearing a tray of tall iced glasses. Overhead, pine boughs rustled in the afternoon breeze. She looked back at Landsman. He was smiling.

And why not? The man was leading a damn nice new life here. Who could blame him for not being willing to exchange it—even temporarily—for what was waiting back in Ocean Plains?

She could leave now and catch an earlier flight to O'Hare.

Except for one thing. She hadn't made her pitch. She'd let Landsman run the meeting and turn her down without even being asked. She was quickly reviewing the strategy she'd decided on during the flight down when he preempted her strike.

"Tell me," he said, "you really think you can turn this mess around?"

"I do. I have to—not *think* I can—*do* it. But Elie tells me I shouldn't try to pull it off without you. He made a pretty good case." Jane leaned forward. "Don, when does your cruise leave?"

He laughed.

It was a lovely, deep, musical laugh. Damn! She was *not* going to like this man unless he said yes.

"Still looking at brochures. Darla wants one of those quality-shopping itineraries, and I'd like to take one that calls down at La Guaira, in Venezuela. From there you can fly a couple hours over the jungle and see Angel Falls."

"Give me ten days," Jane said with sudden intensity. "It'll be a full-blast, all-out situation, working with a strong management team. Don, with you at the helm right beside me, I know we can clear up all these supply-chain nightmares and get our stores stocked and full confidence restored—I'm talking about customer confidence, employee confidence, Wall Street confidence, the whole works."

Was she making headway? Jane couldn't gauge Landsman's reaction but plunged ahead anyway. "You won't be working for Roy. He's not on the team. Doesn't want anything to do with us. He'll give you—us—a wide berth, believe me. Tell me your consultant fee, and I'll get the board to authorize it today, plus a fifty percent bonus when we have the problems licked. Ten days from now, I guarantee you, you're going to enjoy that Caribbean cruise a whole lot more than you would now."

With her last words, Landsman threw back his head and gave a chuckle so resonant that heads turned all over the terrace. "*You're* the one who won the war, huh?"

Jane's comeback was dead serious. "The Gulf was an all-out team effort." She locked her eyes on his. "With fantastic leadership at the top."

His chuckle subsided, and he shook his head—sadly, Jane thought. *Here it comes: Thanks, but no thanks.* At least she'd given it her best try.

"I was just thinking," he said, "how we could have used you six months ago. Heck, six years ago."

"So it's 'no,' huh?" She stood.

"Did I say that?" Landsman's mustache twitched, but he didn't smile. "When you say ten days, you're talking the *next* ten days, am I right, not starting day after tomorrow?"

"No, the next *nine* days. Today is the team's first day."

Landsman stood, too. He took out a cell-phone and punched in a number. "Can you squeeze in half an hour to come home with me and explain all this to my wife?"

It took Jane a second to realize he'd said yes.

AKERS HIGH-RISE GROCERY WAREHOUSE, PLAINFIELD, ILLINOIS 6:27 P.M.

Two days ago the assistant distribution manager had thrown the switch to stop all the conveyers. Now he was forced into an even more drastic and gut-wrenching action: stopping the operation cold.

"Stop processing orders! When they've exhausted the selection warehouse next door, that's it. Nothing leaves this building till we reinventory every goddamned pallet!"

"Jack, that'll take days!" his section manager protested. "And we'll be dead in the water the whole time."

"I damn well realize that, Manny," Jack fired back. "But if we don't do it, we won't know what the hell we've got up in those racks. Anita, you tell him."

"He's not exaggerating, Manny." She swiveled from the console of the automated storage retrieval system. "We've lost the entire inventory database. What the computer *tells* us we've got out there"—she gestured at the long, windowed slice of the ten-story-high storage racks—"and what we've *actually* got, are two different things."

"There's no other way," Jack said dourly.

Twenty towering aisles of pigeonholed products suddenly inaccessible! The SRV robotic cranes could still fetch down any pallet in a matter of seconds, but who knew what they would bring back? Send them after chicken soup and you could just as well get cases of dog biscuits or rat poison.

This was exactly the nightmare Jack had dreaded on Saturday, when they'd discovered the first scrambled SKU— cornflakes instead of laundry soap. He had desperately wanted to believe Anita when she'd said it was probably only a glitch in the mainframe download—and not a harbinger of system doom.

But doomsday had arrived.

In the last hourly download, apparently damned near *all* the SKU numbers in Plainfield's ASRV minicomputer database had been corrupted. Anita had alerted Ocean Plains, which was now in panic mode, checking their own mainframe database and the integrity of other Akers distribution centers. In most conventional DCs, Jack knew, physical inventories could correct the problem without the facilities actually shutting down their operations. Warehouse workers could comb the racked storage with forklifts and pallet trucks and order pickers, verifying SKUs.

Not so here in Plainfield. There was no way for mere human beings to conduct an inventory. So gargantuan were these steel racks that the girders had been bolted together first, then the building walls and roof built around them. There was only one way to "reinventory every pallet," as Jack demanded.

Manny groaned. "You want me to pull every single damn pallet off the racks and run it past the scanner?"

"That's exactly what I want."

Sixty thousand pallets, give or take a few hundred, had to be "pulled" by twenty SRV robocranes—one crane per aisle—lowered to the conveyer, and rolled to the scanners.

"Someone has to eyeball each pallet as it comes along, matching the SKU on our label with the SKU number on the actual cartons. And get that look off your face, Manny. It doesn't help. This situation stinks—for all of us."

"Sorry. Boss, we can do it, but suppose the minute we're

done, the computer goes haywire again?"

"That's exactly what Anita and I have been asking our-selves—and the folks down in Ocean Plains. And we may have an answer. Suppose it takes three days for us to rescan. That's three days with hundreds of Akers and Super-Akers not getting product from us. A whole lot of empty shelves and a whole lot of angry customers. But that's also three days for Anita to work with Ocean Plains on what she calls a firewall or something to protect our minicomputer from another garbage download."

"I'll tell you this, Manny," she said. "Unless they can absolutely prove twelve different ways that this can never hap-pen again, I swear to God I'm not going to let them update our database. Not till I feed their data into a dummy database first and double-check every SKU. I'll rekey every order if need be. Jack says he doesn't care how many data-entry clerks it takes."

"I did say that, and I meant it." The procedure Anita had just sketched—printing out thousands of replenishment orders for hundreds of stores, then retyping them into the ASRV—was truly nightmarish. But not as nightmarish as another wholesale scrambling of inventory numbers would be.

This time, as he exited the computer room and went down to the warehouse floor, there was no ominous silence of con-veyers to underscore the pall that lay over his warehouse. Robocranes still elevated a hundred feet in a matter of seconds, pallets still moved along the rollers. But nothing moved on the outbound conveyer to the selection warehouse. Everything coming down from the racks was being routed toward the pal-let scanner, then rerouted back for put-away.

The task was monumental and felt never-ending, like bail-ing out the ocean with a teacup. Working round the clock, they'd be damned lucky to finish in four days, Jack thought. In the meantime they were dead in the water, as Manny had said—out of the grocery-distribution business.

He could shoot himself, but it wouldn't help.

OCEAN MEADOWS CONDOMINIUMS,
OCEAN PLAINS
9:21 P.M.

In order to catch the last Delta flight out of Orlando, Jane had had to cut her time with Darla Landsman a few minutes short of the half hour Don had asked for. In fact, the last five minutes Jane and Darla had spent alone while Don threw some things in a suitcase.

That was when Darla had told her. "There's always another cruise ship. But that look on my Don's face—the thrill he feels when he's about to head into the wind of a big challenge—well, that look was half the reason I fell in love with him twenty-seven . . . no, twenty-eight years ago."

"I think I know what you mean. My dad had a lust for adventure, too. The wilder the better, I gather."

"You must have inherited that. Trying to save Akers in a week and a half sounds pretty wild to me."

Jane shook her head. "Doesn't *feel* like an adventure."

"Take my advice. Think of it that way. It may help when the whitewater floods your raft. And it will."

Jane laughed. "I wouldn't bet against that happening."

"If you two don't stop chattering, we're going to miss that plane," Don said, coming back into the living room, carry-on over his shoulder.

Jane and Darla shook hands. "I'll ship him back in nine days, I promise."

"I'm not a package, Jane."

"Oh, my dear," his wife said. "You most certainly are."

An Akers pilot met them at the Delta gate at O'Hare and flew them back to Ocean Plains. They had talked there for less than an hour when Don said he needed to think and would see her at eight the next morning. He drove off in a company car to his Porter Lake house—which luckily hadn't sold yet or even been packed up. Jane headed for the company condo complex, which she'd managed to move into at seven-fifteen that morning.

Ocean Meadows had been built to house guests of Siler Air Force Base. With furnished suites—the furnishings included a desk and a good bed light—and ready-to-use kitchens, the condos offered a few more comforts than the Welcome Inn. Not that Jane figured to be spending much time away from the Akers war zone.

Right now she was ready for full collapse. She craved a hot bath, with just enough wakefulness afterward to prep for tomorrow morning's meeting. A wiser use of the rest of the evening would probably be to change into sweats and jog around the complex a couple of times, restoring circulation after too many airborne hours, then a quick shower and oblivion.

But neither option was open. Jane had one more appointment in her daily organizer, a meeting she'd actually tried, but failed, to postpone. She finger-brushed her hair, freshened her makeup, pocketed her key, and went out. On impulse she decided to walk there.

67 OCEAN PLAINS AVENUE
9:28 P.M.

Tired as she was, it felt good to walk. The evening was pleasantly cool, and Jane relaxed, knowing there was no place safer in the world to walk alone at night. Aunt Louise's directions were clear and very precise, Jane thought as she approached the entrance of the two-story apartment house. Up the single flight of stairs. Last door on the right.

Jane glanced at her watch. Waited forty-five seconds. Nine-thirty on the dot. She grabbed the lion's nose ring and rapped.

"Jane," said her aunt Louise, undoing the chain. "Come in!" Jane smiled, catching the scent of rose cologne she remembered from her childhood.

"It's lovely, Aunt Lou," she said, following her aunt into the rather small living room. The room *was* Aunt Louise. Against walls painted a warm but not flagrant yellow stood an array of furnishings in two quite different styles. There were several needlepoint-covered, Regency-style chairs, two with needle-point-covered footstools, a needlepoint-covered love seat, and a bevy of framed needlepoint botanicals. Then there was the Orientalia. Some really lovely antiques—or very good repro-ductions. Painted screens, lacquered chests, ivory carvings, a lot of rosewood and brass. The Asian motif even extended to Louise's silk brocade housecoat—and her slippers were needlepoint!

All right, Aunt Lou!

"Well, it's my own place, and that's what counts. Roy and Grace want me to move in with them at the lake. They try to tempt me by saying I'll get to spend more time with Tyler. I adore my grandnephew, but the truth of it is, after a few min-utes of trying to keep up with him, I feel ninety. Now poor Anne has invited me to move in with her, and I suppose that would be the right thing to do. But I'll hate to give up this place." Louise smiled, with more pride than ruefulness. "I'm kind of a loner, you know."

Of course! Jane had always admired Aunt Lou's upright bearing, her quiet dignity. But she had never felt close to her "maiden aunt"—the old-fashioned sexist title had been applied to Lou even in her thirties! There *was* a certain severity about her. Yet Louise Akers had been a handsome woman and remained so now, despite the Akers jaw (something Jane and her siblings had been spared). There had been rumors of long-ago suitors; none, apparently, had stayed the course. Louise's loyalties remained to the family.

"I've wanted to have this talk with you for years," she began over tea and butter cookies served on flowered Limoges.

"I broke a cup of this china the first time you ever invited me to tea."

Aunt Lou shrugged. "You were four."

"I was *desolate*. I begged my mother to help me find another one to buy for you. She said my piggy bank wasn't big enough. How I wept. Is it too late?"

Aunt Lou laughed. "Much. I've still got seven cups and hardly ever need more than *one*. Speaking of late, I'm sorry to ask you to come over after what I'm sure was a very long day for you. You said you were going to try to shanghai Don Landsman back from Florida. Did you succeed?"

"He's here." Jane smiled.

"Good man, Don. But the one I really like is Darla. Did you know she's published two books on Chinese art?"

"No, but it doesn't surprise me. I liked her, too. But what did you want to see me about, Aunt Louise?"

"I insisted on tonight because I was afraid to postpone it. Knowing you, and your busy life, we might never get the chance again. And this *is* important."

"I gathered that, and I'm here. What more can I say?"

"Exactly what you think, dear. That little rule's failed me less often than most others. I certainly didn't invite you over to make polite chitchat."

"What I think about *what*, Aunt Lou? I've been thinking so many things since my father died. I hardly know where to start—even *if* I should start. That's the terrible part. It's too late to go back and tell him all the things I should have told him. But I've also been thinking about the family. How it wasn't just Daddy I was cutting myself off from, but all of you—and the last few years, that includes Tyler. I not only deprived myself of a darling nephew, but with Grace having no sisters, I deprived Tyler of an aunt."

Louise shook her head slowly, perhaps sadly. Her tightly wound silver hair gleamed in the light of a ginger-jar lamp. "We all have our regrets, Jane, our cold ashes to rake through. Perhaps if I'd been a better aunt myself, you'd have had someone to turn—"

"Aunt Lou, I didn't mean—"

"Jane, let me finish. I've had a long time to think this through. I know you feel guilty about staying away so long. But, looking back, how could you have done otherwise?"

Jane thought about the arc of her life: the obvious milestones; smaller, surprisingly decisive events; major way stations; what had beckoned her to move on—or away. Jane could not really imagine an alternate course, despite the pain strewn in her wake. "The regrets are definitely there, but you're right, Aunt Lou. What I did I had to do."

"Royal wasn't an easy man to love, Jane. I'm not sure any of us truly succeeded at it . . . before Anne. She just wore Royal down with her devotion. All the rest of us, including me, got worn down by Royal, or were at least tempted to give up on him. Your mother was the first. She's a smart woman, Joycelyn, and doubtless she calculated how far down the list she was in Royal's life. Each new store he opened pushed her farther down. The children came in above her, too. There wasn't much of Royal left for Joycelyn."

"So it wasn't just the affair with Anne?"

"Nor all the affairs before her," her brow furrowed. "I'm not telling you anything new, am I?"

"Afraid not. Once we left here, Mother made it her business to tell me about those."

"Unfortunately, I doubt she could have exaggerated much. My brother would cross a room, or a street, or the county, to charm a woman, and he certainly had a way about him when he wanted to turn it on. He thought he was irresistible. And, God help any woman he took a fancy to, too often he was. Look at Joycelyn. Now who would have figured those two?"

"You make Daddy sound like Alex."

"Royal wasn't quite *that* handsome. But that didn't seem to slow him down."

"But Daddy didn't have to go far to find Anne, did he? She was right there in his office."

"When things came to a head, yes, by then she was his secretary. But your father had known Anne for years. She was the wife of George Glynn, the company's first bookkeeper. George dropped dead of a heart attack at forty-four, and just as shocking, there wasn't much money for Anne. Out of simple kindness—yes, he had that in him too, and this time the person to whom he was kind never forgot it—Royal gave her a job down at the office, just typing, filing. When Eunice Banks finally retired, Royal—"

"I remember Eunice. She taught me to type when I was ten or eleven."

"Well, Royal promoted Anne to take her place. Eunice had been a fine secretary, but Anne became indispensable to him. I think she was just this incredible buffer—protecting Royal from interruptions, and folks in the company from the first flaring of my brother's famous temper. So when their affair began, and I have no idea when that was—it's nobody's business but Anne's now—they already had a relationship that mattered to Royal. When your mother found out, I suspect she intuited that this affair was essentially different from the others. Royal needed Anne, and at some point that need turned to genuine love. Forgive me for saying this to you of all people, Jane, but you should have seen them together! I truly believe he was faithful to her till the day he died." Louise smiled just a little. "I can't swear to it, mind you, but I believe it to be true."

"Why didn't anyone ever tell me any of this, Aunt Lou?"

"Maybe others felt as I did, that when you got a little older, you'd come and visit and find out for yourself. And Royal had put Joycelyn through a lot. One wouldn't want to seem to be defending him by setting the record straight. I don't condone what Royal did—not to your mother, not to you children, by not being a better husband and father. But Anne was so good to him, he began to change. You mustn't blame her for what happened."

"I'd be a fool to do that now, wouldn't I, Aunt Lou? You know, Anne hasn't pushed herself on me, she's left me plenty of room to choose how I want to relate to her—even if I want to have any relationship with her at all. But she has tried to help me see my father more completely. She gave me some letters Dad wrote me long ago. Letters that were returned here unopened, that I never even knew came. There wasn't much bad about Dad that my mother held back. But she kept those letters from me."

Jane had thought she could tell that story cleanly now, without becoming very emotional, but she failed.

"I surely understand your anger," Aunt Lou said when Jane was done. "But you should understand that your mother wasn't herself back then. She was just furious with Royal, and had every right to be. She wouldn't have done what she did to hurt you, it was Royal she was wanting to hurt."

"*I* know that, Aunt Lou. But Daddy must have thought that I got the letters and didn't want to answer them. That's what I can't forgive. Even worse, it happened to him again." She told Louise about her father's letter returned as undeliverable from State College, Pennsylvania, because of Jane's delay in putting both names on her mailbox. "All I wanted was to make my own way in the world."

"Except you didn't take just any name, dear."

"It seemed so logical then. And Grandpa Malcolm *was* a kind of hero to me—the reason I went into military logistics. I guess that probably hurt Daddy, too."

"You were always so like him, Jane. Stubborn, proud, no giving up—and not much give—in either of you. You both *had* to succeed at whatever you tried. Royal wasn't much of a father, but he doted on you, you know that. Remember how he called you 'princess'?"

Funny, Jane had walled that off. But there it was. Princess.

Well, she wasn't going to get teary over it now.

"I realize endearments do not a father make, but he did try with you. Royal might have harbored high expectations, but he never came down hard on you, the way he did on Junior and Alex. That's one of the reasons I wanted to have this talk, Jane. I admire what you're doing, and your father would be cheering you on. But, you know, you were very hard on Junior the other day, and I thought maybe you didn't understand how hard things have been for him."

"Maybe not," Jane said, welcoming any change of topic. "Being the only one who stayed on, you mean, while Alex and I left home?"

"Not just that. Junior tries so hard, Jane. He always has. He was forever trying to get his dad to take notice. Alex usually used some kind of shenanigans to get attention, but Junior was always coming up against you and how well you did in school."

"But Roy was four grades ahead of me!"

"Didn't matter. He still had to compete with your report cards. And Alex could still beat him at games—any game."

Jane nodded, remembering. Alex won so many trophies that finding a place to put them all became a family joke. But maybe Roy hadn't laughed that much. His few plaques, buried in the family displays, were inscribed "Most Improved" or "Honorable Mention." If Roy's grades were slightly higher than Alex's, they never approached Jane's straight A's.

"That's why Roy got into trouble, wasn't it?" Jane said. During his senior year, she recalled, he had been expelled from school for cheating on a trig exam. After threatening to pack him off to military academy, her father had gotten Roy reinstated.

"He was trying to impress his dad. It was a pretty desperate act, and of course it boomeranged. Even years later, Royal would bring up the disgrace of it. I don't think he ever forgave Junior—at least, he let us all think that was how he felt. I don't think Junior ever got a real chance in the company. That's why, if you notice, I go out of my way to treat him with respect. I want you to succeed in saving the company, Jane, but I also want you to try to be more respectful of your big brother. Will you do that?"

"I'll try," Jane said. "I promise." She sighed. It wasn't going to be easy.

At the door to Aunt Lou's apartment, Jane took her aunt's veined hand in both her own. "Aunt Lou, I'm glad I came. But, remember, you set up the rules. You said I was to say what I think."

"Haven't you?"

"There is one more thing."

"Well, get it out, girl."

"Don't you think it might help him if you stopped calling Roy 'Junior'?"

The woman in front of her stared for a moment. Then the hard glint in her eyes turned to a sparkle. "I must admit it feels a little strange to be corrected—no one but your dad has ever had the guts to set me straight about anything—but I do believe you are one hundred percent right. Roy it'll be. Starting tomorrow. I promise, and I always keep my word."

Jane hoped her chances of keeping the promise *she'd* made about Roy were half as good.

DAY 2

THE BOARD ROOM,
AKERS HQ, OCEAN PLAINS
8:00 A.M.

Jane had notified the board of her intention to recruit Don Landsman as a consultant. Nonetheless there was a ripple when she walked into the boardroom with him. Plainly they hadn't expected her to succeed.

Elie Grey winked at her. Too many eyes were on her to wink back, but she risked a modest nod in his direction. There was another stir when Jane introduced the fourth member of her delegation, a plump, sleepy-eyed young man with tawny skin, totally unself-conscious in his Armani clothes. Prince Khalif al-Marzouki had turned out to be not only sharp but hardworking after she had been asked to take him under her wing in the Gulf. When she had reached him late the night before in his apartment off the Michigan State campus, he had expressed delight at hearing from her, said of course he could get away for a few days, and had promptly zipped down from East Lansing in his Maserati to make the meeting.

Jane scanned the surprised faces and had to make an effort not to chuckle. *Good start, Janie-girl.*

From the far end of the long conference table, Madeleine Archer greeted them warmly, indicated croissants and coffee on the credenza, and said, "I've asked Roy to chair this meeting."

"My compliments, Jane," said Roy, who was seated at the near end. "Looks like you've recruited an all-star team here, and awfully damned quick."

"We'll be adding a few more folks," she said. "But I'd rather think of all of us as *one* team. If that's okay with you?"

"Absolutely," Roy said without conviction.

He was right, of course. *His* crisis management team—a dozen or so faces around the conference table—had just been put out of business. At least, that was how they had to be feeling.

Roy introduced them one by one—several department

heads and assistants; two consultants from the main software vendor, Positek; two from a Texas firm, Bowerman, that had overseen the computer project from preliminary design to final installation.

The fact was, there was a heap of expertise among Roy's people. Jane, whose sense of the enormity of the task she'd taken on was growing, hoped to be able to call on that expertise; but who knew how they would respond, relegated to advisory status? The vibes in the room were not promising. She had expected to find folks defending their turf. And, God knew, she was familiar with the collective siege mentality of any group under prolonged enemy fire, as Akers had been these last weeks. But even allowing for all that, the group gathered in the boardroom seemed particularly grim.

Not for no reason.

"We've had more bad news," Roy was saying. "Really bad. *Geometrically* worse. Our dry grocery distribution warehouse for this whole region has been forced to shut down. Apparently our system fed wrong SKU numbers into their database. Until that's straightened out, they can't move groceries to our stores."

"Nothing?" Don Landsman asked.

"Nothing."

Landsman swore. "What about all the other DCs?"

"Problems there, too," Roy said. "But so far, thank God, what happened to Plainfield hasn't occurred anywhere else. Our perishable DCs—meat, dairy and deli, frozen foods, produce, bakery—they're all still up and running. The other main disaster to hit in the last couple of days concerns Asian imports—principally fashion and home electronics. And there are other critical issues we need to discuss. I suggest we go around the room and give you a status report. Chip?"

Chip Bragan, director of information technology, nodded and plunged in. "We can now say with certainty that our system problems do not originate from the software itself—as our

Positek people have maintained all along. We're dealing with software sabotage—deliberate, virulent, and extremely sophisticated. Not that we haven't had some genuine glitches."

He turned to one of the two Positek reps at the table, a woman with a neat cap of hair that was obviously prematurely gray. "Betty?"

"Thanks, Chip. When we ended our parallel run, we had our daily list of unresolved software bugs down to just"—Betty Bentley glanced at her notes—"thirty-three items." She paused. "Every single one of those was resolved."

"Wish that was purely good news," Bragan said. "Unfortunately, during our last debugging phase is when we may have gotten infected. We had troubleshooters and programmers dialing in from several locations. Those dial-in lines have all been secured, the log-ins removed—but, of course, the damage seems to have been done."

"Excuse me," Don Landsman said. "Shouldn't Lenny and Mimi be in this meeting?" He turned toward Jane. "They're our computer systems administrators."

"Their names weren't on the list," said Wallace Conover, the former Goodstuff executive Roy had named president.

"We keep 'em pretty busy with hands-on chores," Bragan said.

"Ask them if they could spare us a few minutes," Jane said, politely but firmly testing her mandate. Like Landsman, she thought the meeting heavy on "suits," light on "techies."

Roy Jr. hesitated, then nodded to Chip Bragan, who passed on the request to his assistant director, Norma Klein.

"From our standpoint"—Betty Bentley picked up where she'd left off—"the installation was routine—until the sabotage was diagnosed. No matter how thoroughly design specs are written, and how much time is allotted for remote testing and so forth, once installed on-site, *any* new or customized software is going to have problems, which have to be addressed one by one." She straightened in her chair. "It recently became clear, however, that we were being asked to 'debug' problems that were being generated by deliberate corruption of our code."

The door opened, and Norma Klein introduced Lenny Cristofaro and his assistant, Mimi Takuda. Their casual attire confirmed that they hadn't expected to be there.

"Please get yourselves some coffee and join us," Chip Bragan said. "We're about to discuss what kind of viruses or worms we're fighting. We're definitely talking in the plural—right, Lenny, a whole host of viruses?"

"Yeah," Cristofaro said. "It's an infestation all right, and there's all kinds, and they do all kinds of tricks. They replicate and migrate to other systems. They encrypt and disguise themselves, change their process numbers, attach themselves to innocuous files. Some even kill themselves off periodically. What happened in Plainfield, though—wow! Those are the meanest strains yet."

"We've been able to isolate several examples of these invaders," Bragan said, addressing Jane. "Thanks to Dick Lai of Positek here and Sammy Shah of Bowerman, who've decompiled them, line by line—"

"Aren't we getting a little technical here," Conover said.

It wasn't a question, but Dick Lai opted to answer him anyway. "Not really," he said politely enough. "Think of it as reverse software engineering. You start with machine language instructions and put them through a disassembler, try to re-create the source code. That's so you can see exactly what the program is trying to do, its various routines and subroutines."

Bragan, obviously more experienced at reading faces than Dick Lai, said firmly, "Just tell us what you found, Dick."

"One thing we *haven't* found," Lai said, "are 'logic bombs.' They're instructions designed to go off at some future time and destroy records or entire databases or crash a system. So that's the good news, relatively speaking. The bad news is, these invaders attach themselves to an existing code and alter it, ever so subtly.

For instance, they're wreaking havoc in your distribution systems by changing the algorithms for calculating replenishment."

"Give us a concrete example," Jane said.

The Bowerman consultant spoke up: "Sure. Let's say a particular product is scanned at a checkout scanner. A bottle of the best-selling aspirin, say. That's pretty critical to a lot of people, right? If they have migraines. If they're taking aspirin to prevent a second heart attack. But, instead of the computer recording that inventory has been reduced by one, it says inventory has been replenished by two. Multiply that error by your fifteen hundred stores. Within days, the shelves in all those stores are empty of the aspirin people rely on, while your distribution centers are glutted, and all reorders to the manufacturer are halted." He paused for effect. "You've got one hell of a headache."

"Hang on," Jane interrupted. "You're saying it doesn't affect all SKUs—except it just did, in one of your dry grocery DCs?"

"Right. Until yesterday, the virus was targeting individual SKUs—on a given day, say, five or ten out of, what, seventy-five thousand stock-keeping units?"

"And as fast as we would kill one off," Chip Bragan added, "another one would crop up somewhere else."

"Like cockroaches," Mimi Takuda volunteered.

"You know what this illustrates?" The speaker was Reid Smith, just named to the vice presidency vacated by Don Landsman.

"What's that?" Wally Conover, who was his boss, turned to him with an expression that said he was pleased Smith had something to contribute.

"The fragility of interlocking technology. Like the way an electrical outage in one area can take down an entire power grid across three or four states."

"What do you suggest we do about that, Reid?" Roy Jr. asked testily.

"I was just making an observation."

"Duly noted." Roy raised an eyebrow in Conover's direction. Conover discovered a hangnail.

"So how does the virus spread?" Landsman asked Lenny Cristofaro.

The Akers system administrator shrugged. "The 'how' is easy—through EDI. Generally, these programs piggyback on another program or attach themselves to a datastream. This is your classical virus. They don't seem to travel on their own power, like network 'worms' do."

"But EDI travels both ways," Jane said. "Is that how we keep getting reinfected? Or could it be that viruses are being continuously reintroduced right here in Ocean Plains?"

There was an awkward silence, during which people assiduously did not look at anyone else. Finally Roy Jr. said, "I'll let Chip address that one."

"Unfortunately, Ms. Malcolm," the IT director said, "what you're describing is a plausible scenario. So, yes, it could be happening right here. We're pursuing our own investigation."

"Who's 'we'?" Jane asked.

"Our new security chief, Robert Grooms, is handling it. Mr. Grooms has a good deal of law enforcement experience, as I understand it." He glanced at Roy Jr., who nodded. "I'll leave it right there for the time being, if you don't mind," Bragan said.

Jane didn't want it left there. She recalled Elie's suggestion that an ex-Akers executive, Flynn Emerson, who had been deeply involved in the computer project, could have had a hand in the sabotage.

"Mr. Bragan," she said, "if you're absolutely sure this is computer sabotage—a federal crime last time I looked—shouldn't the FBI be at this table? They have more experience with this kind of crime, I daresay, than your Mr. Grooms."

"I'll answer that one," Roy said. "Say we bring in the FBI Computer Crimes Division, like you suggest, Jane, and tell them exactly what we're fighting. How our computer security has been breached, our entire distribution system corrupted, how the damage may have already spread to our suppliers. How long do you think it will take the feds to leak that not merely privileged but *explosive* information? A day? An hour? Hell, you all know how that works. One leak like that, after what's already happened to our stock, will sink us financially."

Jane had expected something like this. Some very big corporations, she knew, had been victimized by ingenious hackers, who destroyed data, pirated software, crashed or gridlocked whole systems, causing the loss of millions of dollars in computing time. Yet the corporate reaction was invariably to hush up the incidents, not to call law enforcement. Even when culprits were caught, some companies refused to prosecute. The damage caused by negative publicity—to the corporate name and corporate stock—was considered far worse than any criminal damage. In this case, however, Jane wondered if that held true.

"It's hard to imagine headlines worse than the ones I saw yesterday," she said.

"Wally, tell her what's happening with Hyon," Roy said.

Conover nodded. "Yesterday, our regional distribution centers started getting in Hyon's newest PC display monitors. About a third were trucked in from their Mexican assembly plants, the rest came over from Korea in ocean containers—a tidal wave of product. All the paperwork is there, everything matches up with our purchase orders, but nobody here can figure out how those particular POs got generated. These are just monitors, you understand, no actual computers to go with them. Plus, they're apparently very high-tech . . . " Conover looked around for help.

"Superscan, high-res," Chip Bragan added.

"Hey, those are great monitors," Cristofaro said. "I could definitely use one of those."

"But our customers can't," Roy said sharply. "That's the point. We can't move 'em. And Hyon will be demanding their cash. This represents a very sizable investment for them."

"If most of this product came in from Korea in container ships," Don Landsman said, "the order has to have been in the works several weeks. I don't remember it—and I would."

"At least a month," Roy said. "And you're right, Don. Nobody remembers it. But it *was* in the pipeline. If you're asking why we didn't catch it, I can't tell you. That's another in-house investigation we're launching."

"I thought all computers were air-freighted these days," said Sammy Shah. "Because of the short shelf life."

"That's absolutely true for the assemblies," Chip Bragan said, "like CPUs, memory, CD-ROMS, disk drives, and so forth. But we still import some display monitors in containers."

"There's another phantom Hyon order headed our way in containers," Wally Conover said, reclaiming the floor. "Roy and I got alerted to this one right before the meeting. Twenty thousand karaoke machines, for God's sake, just about to hit our West Coast import hub."

"What in creation are those?" Elie Grey asked.

"Something else we don't sell," Roy said curtly.

"We need an emergency summit with Hyon ASAP," Jane said.

"A planeload of Hyon executives arrives day after tomorrow," Roy said dryly. "That's gonna be real fun.

"They're definitely the worst case so far," he went on. "But there are at least a dozen other EDI trading partners out there requiring major damage control. We've had suppliers fire the executives who agreed to link up with us. We've had others pull the plug on us—some electronically, some totally. They all say it's only till we get our act together, but who knows how long they'll wait before they hitch their wagon to another retailer?"

"There's a new wrinkle on that," said Norma Klein, the assistant IT director. "I mean besides Plainfield no longer taking our downloads direct. Yesterday we actually had one of our own regional managers make the same request, to sever all computer linkage between headquarters and the stores in his region. Says he wants to go back to using conventional methods with more manual control." She shook her head in disbelief. "We could have a mutiny on our hands."

"What's this regional manager's name?"

Norma glanced through her notes. "Let's see, that was, um . . . southeastern Wisconsin, Ms. Malcolm. The twelve stores ringing Milwaukee. The manager would be, um . . . "

"Wade Crain." The name was supplied by Don Landsman, seated beside her. "He started out here. Came from here, seems to me."

"I think I remember him," Jane said. "And that's the most sensible idea I've heard all day. Except that we should be the ones pulling the plug. On the whole system. And fast."

SUPER-AKERS, GRAVESEND, ENGLAND 2:38 P.M. GREENWICH TIME

"I don't flippin' wanna hear this!" was the anguished outcry of Nick King, manager of the mammoth Akers discount store in the old Thames River port ten miles east of London. He looked despairingly at the phone in his fist, then spoke into it again. "Georgie, how many more muck-ups can we survive?"

"Ocean Plains don't seem to know," replied George Parrish, "and that's a sorry fact." Parrish's office was less than a hundred yards away in the adjoining distribution center, which serviced both Gravesend and two more Akers stores in Birmingham and Sheffield.

The "muck-up" in question was a containerload of Taiwanese-made small appliances—electric fans and shavers, hair dryers and curling irons—which had just arrived from the port of Southampton fifty miles to the southwest. At a glance, all the items looked identical to those they were intended to replenish. But the SKU numbers were one off, and all the power plugs had the wrong configuration. Which meant they wouldn't fit English sockets. Good thing, too, since the electric motors were designed to work on 110-120-volt American current instead of the 220-240 volts standard in the U.K.—making the entire load of nifty, labor-saving devices utterly useless to Nick's customers, unless they wanted to buy expensive voltage converters and plug adapters along with them.

And this was the third consecutive day on which the Gravesend distribution center had taken delivery of utterly useless imports. Yesterday's very large and nasty surprise had come all the way from Singapore. That ocean container had been filled with kitchen appliances—toasters and coffee grinders, blenders and food processors—all with similarly wrong plugs and voltage requirements. Nick and George had each spent half a morning on the transatlantic phone to Ocean Plains, trying to sort out the mess. And both had come away with the maddening conclusion that further screwed-up orders could very well be en route somewhere along the Europe-Suez-Far East run.

The day before *that*, the DC's manager had gone volcanic when he'd slit open cases of brand-new VHS-videotaped movies, discovering all to be encoded with the American National Television System Committee standard rather than the British PAL standard. Which meant all were un-bleeding-playable on English videotape players.

Thank God all the wrong shipments had been caught on the receiving dock and not on the sales floor. But the resulting inventory holes had Nick ready to yank out by the roots what little hair he had left.

And judging from the phone survey he'd done yesterday, the replenishment virus had infected Akers stores all across Continental Europe—both stores in Spain, three of the four in France, and all seven Italian stores. Ocean Plains might not know who was behind the sabotage, but Nick had a strong hunch the culprit might be working for a certain French retailing conglomerate that was fighting Akers tooth and nail for the European market.

Now, thanks to these latest outages, he was truly fearful of losing market share here in England. God only knew it had been hard enough getting a toehold, battling lawsuits from native retailers along with competition from other American power retailers who'd crossed the pond now that the U.S. market was nearing saturation.

"I'm coming over, George," Nick said into the phone. "The two of us gotta figure out something and fast."

On his way out, he was nearly run down by a stout woman sprinting head down through the spring drizzle, looking rather like a charging bull except for her Akers shopping bag.

"Problem with one of our items, madam?" Nick said, retreating under the overhang as the automatic doors flew open. "You'll find exchanges and refunds just on the left."

"You can *bet* I want my money back," the woman said, looking tear-streaked from the rain. "I got this electric carving knife here yesterday. My son, Neville, had to practically pound the prongs into our kitchen socket. Then, when he finally did get it in, it knocked the power out all over the flat. If you don't make it right, this is my last trip ever to a blinkin' Akers."

AKERS REGIONAL DISTRIBUTION CENTER, MILWAUKEE
8:43 A.M.

Wade Crain had been up all night at the DC, working with John Brinkema, who had stayed on into the third shift. That was when Akers trucks were loaded up and moved out. The two men had been double-checking replenishment orders with store managers, trying to ensure that no critical inventory items were overlooked.

At the top of the critical list were all the dry grocery items normally supplied by the Akers Plainfield DC. With that key facility shut down temporarily, hundreds of Akers in several states—not only Wade's dozen stores—were scrambling to find alternate sources for thousands of essential food products. Wholesale food distributors, meanwhile, alerted to the crisis, were jumping in with offers—unfortunately at the hefty rates mom-and-pop outlets were forced to pay. Wade had fielded several such calls during the long night, referring them to the buyers in Ocean Plains.

"It looks like a whole new ball game," he had said on the phone to his Lake Reliance store manager, Phil Leavitt.

"Feels more like the old ball game, Wade. I'm back with my department managers, walking the aisles at four A.M., aisle by aisle, checking shelf levels. But, hey, whatever it takes, we gotta get it right."

And the laborious double-checking of auto-replenishment signals was paying off. Several computer-generated errors had been caught. Seven of Wade's stores, including Lake Reliance, were completely out of cough syrup and nasal decongestants, yet neither item appeared on the overnight reorders. Unfortunately the DC was almost depleted of these SKUs as well, indicating that wrong inventory information had passed up the supply chain to the vendors. Wade had Big John ration the last two cases among the needier stores, then add the items to the emergency drop shipments from third-party distributors.

Wade and John had also been coordinating warehouse inventory with company DCs in neighboring regions, arranging to shift stock rapidly to where it was most desperately needed.

"You know," Wade said during a short break in the early morning madness, "since we're all going to have to second-guess the computer anyway on every order, why the hell don't they just switch over to manual operation until the system gets fixed?"

"Awfully damn drastic," John said.

"Exactly what Ocean Plains told me." Wade had made the suggestion on the phone yesterday to one of Roy's assistants. And he intended to make it again today, with more details and urgency and, he hoped, persuasiveness. This time he'd e-mail all of top management. "Granted, John, going manual would cripple us. But look what happened in Plainfield. Sometimes you *have* to shut down a system. It's the old garbage-in, garbage-out principle. If our computer system isn't serving our customers' needs—and it damn well isn't—we have to find another way to do it. *Now.*"

"If Royal was still running things, I think he'd agree with you. But down there now, seems like *nobody's* ready to make a tough call."

John was right. Wade had talked to a couple of his Ocean Plains contacts and gathered that with headquarters chaos escalating hourly, Roy was overwhelmed. Panicky calls were coming in from the regions, from individual stores, and from Akers's trading partners—including some from major overseas suppliers. But the home-front stores were hurting the most—especially the shortages in small towns dependent on Akers for everything from bread to Band-Aids. Wade had seen several segments like that on local TV, heard mentioned too many others. Where real shortages didn't exist, Wade visualized zealous news directors—was there any other kind?—telling their camera crews to go out and find them anyway. Even if the system damage could somehow be repaired quickly, repairing the damage done to the Akers name was going to take a lot longer.

But that, thank God, wasn't his headache. He had more than enough to do with his dozen stores. By the time the trucks were gone and the sun was up, Wade was ready to take ten someplace quiet.

That was when his belt pager warbled. He unclipped it, checked the readout—his office extension. He'd left his cellphone on his desk. He dialed his own number from a wall-phone, got his assistant, Patrick DePalma.

"Ocean Plains is on the other line, Wade. They want you."

"Want me to what?"

"They want you down there, on their crisis team."

"Shit!"

"Hold on, I'm transferring. Go ahead, sir."

Silence. Then a resonant baritone. "Wade, you there? This is Don Landsman."

"Don? No kidding! Shouldn't you be teeing off in the Florida sunshine about now. What *are* you doing?"

"I'll tell you when you get here. We need you, buddy—and fast. Aren't you supposed to be working out of here anyway?"

"Thought I could accomplish more in Milwaukee."

"Heard about that. Fact is, that's why I'm calling. We're getting a new crisis team in place, trying to do what you suggested—company-wide. Operate without the system—keeping the stores stocked whatever way we can—till we get all the damn computer bugs exterminated."

"Somebody must have had an overnight conversion down there. My idea got shot down real fast yesterday. Obviously I'm all for it—especially if I'm not having a breakdown and imagining this whole conversation. You really back, heading up this team?"

"Just helping out. And only temporarily."

"Then who *is* running it?"

"Royal's daughter. Goes by the name of Jane Malcolm now. Turns out she's a logistics consultant and—"

"I know who she is. She ask for me, Don? Nah, you must've."

Landsman chuckled. "I'd like to take the credit, but so far she's been one jump ahead of me. How soon can you get here?"

Wade glanced around the warehouse, his mind racing.

"I'm on my way."

He hung up, then redialed his own line. "Pat, how'd you like to be in charge of our stores for a while?"

"Sure thing. You'll be back when? An hour? Two?"

Wade laughed. "Not hardly. Don't know exactly, but I figure a week, at least. Anyway, you *are*. In charge. As of right now."

THE WAR ROOM
9:13 A.M.

"Your guy in Milwaukee is definitely on to something, Chip," Jane said to the information technology director. "Isn't it obvious? We need to take down the new software, disable inbound and outbound communication, notify all EDI partners that they may be infected and should do likewise. That includes EDI links to all company computers in stores and distribution centers."

"I'm afraid the ramifications of all that would be pretty apocalyptic," Bragan said carefully.

"Well, what do you say we lay them all out and compare them with the apocalypse we're up against right now."

"Jane, think," Roy said. "The simple fact is, we can't do what we do without computers."

Jane bit her tongue, recalling her promise to Aunt Louise. But she was getting awfully tired of Roy's patronizing tone.

"Sad, but true," Chip Bragan said. "There's no way—"

"Hold on," Jane said. "What is it we do, Roy? Are we in the computer business? Retailing is what we do—not data processing, right? Computers are just tools we use to help provide our customers with what they want—the right selection of quality merchandise at low prices. So our computers are incapacitated. *We* still need to do our job. Get the goods to our customers.

"Seems to me it would help to see our problem for what it is. It's not essentially a computer problem. It's a logistical problem." She took a sip of cold coffee, giving her words time to sink in.

"That means we need a logistical solution, or solutions. Computers help us move information, just like trucks help us move goods. If our computers break down and can't move information, or move wrong information, we have to take them off-line and find another way to move the stuff. At this point, your new system, what do you call it again, integrated . . . "

"Integrated logistics system," Chip Bragan said. "ILS."

"Well, it's clearly not living up to its name. We've got *disinte-grating* logistics."

"You're talking about, what, using networked PCs?" Bragan asked.

"That's among the backup options. *If* our networked PCs can be quarantined."

"There would still be risk," Don Landsman said. "A lot of folks here in Ocean Plains download database files from the mainframe with terminal emulation, then manipulate the numbers in the PC environment. Couldn't a virus come right along, Chip?"

"I wouldn't rule it out."

"Okay," Jane said. She stood and went over to the window, looked out. Her father's domain. Tumbling like so many dominoes. She turned back toward the table.

"The entire computer system has to be dismembered," she said. "We don't put it back together again, we certainly don't link it to the outside world, until we've found who's been feeding it viruses and can guarantee it won't happen again."

"What would you say to installing the previous version of our inventory software?" Landsman suggested. "Minus EDI?"

"I'd say run tests on anything and everything," Jane said. "But for the time being you can't expect to operate the company safely on any networked software. A few loose strains of virus are one thing, a plague is another."

"When a plague invaded a house," Elie Grey said, "half measures didn't save you. You had to burn every bedsheet. Am I right, Jane?"

"That is definitely the idea. It may seem like I'm suggesting we go back to the start of this century, or even the Stone Age. But the backup technology isn't really that ancient. I'm talking about using stand-alone PCs, faxes, telephones. Forget electronic documents. Print up old-fashioned reorder forms, have store managers fill 'em out and fax 'em in to the DCs and headquarters."

"How about poor man's EDI?" Khalif suggested politely.

Jane turned to the young Saudi prince. "What's that?"

"I'm sure they already have it." The young man gestured vaguely around the table. "IDCS is the technical acronym. Integrated data collection servers. Fancy term for faxing and scanning."

"Khalif is right," Don said. "Quite a few of our smaller suppliers that can't afford EDI use IDCS. Fill us in, Chip?"

"Sure. Basically, with IDCS, you fax in your order forms to a scanner-equipped PC, which reads the information fields, using OCR software—optical character recognition—or OMR, which is optical mark recognition, like filling in the circles on a school test. The results feed into the database. We've got a slightly more sophisticated program for reading faxed bar codes."

"That sounds like something we could definitely use on an interim basis," Jane said. "Not only as a bridge to our disconnected EDI partners, but in-house. Chip, could you get some people working on that right off, studying how much data we could move that way?"

Chip Bragan nodded—a little grudgingly, Jane thought, but at least without a questioning look in Roy Jr.'s direction.

"Look, if this all seems unnecessarily drastic . . . " Jane hesitated.

"Hell, you folks just told us you lost your local dry-grocery warehouse yesterday," Elie Grey jumped in. "Inventory totally bollixed up. Far as I can tell, you could lose the rest of 'em in the next ten minutes. I'd call that drastic. And as far as that warehouse goes, you've already pulled the plug."

And your boss, Jane thought, is ready to pull the plug on the whole company.

"Here's our proposal," she said, moving on. "Wade Crain up

in Milwaukee has already drafted an emergency plan for disconnecting the mainframe connection in his region. Don and I have asked Wade to fly down here—he should be here in an hour or so—and join the team. We'd like to use his region, those dozen stores, for a pilot program. Run it twenty-four hours, monitor it, review the results. If it's panning out, we take it nationwide, then international. Here's a copy of Wade's latest e-mail on his plan."

She dealt Xerox copies around the table. "I like Wade's emphasis on basics. Look at his third and fourth bullet points. 'Focus on items most critical to the customer. Medicine, staple food items, fast movers. List those, get accurate physical inventories. When you've got control of those SKUs, add more, a few at a time.'"

"The bottom line is this," Don said. "We've got to move fast, not just sit around debating." He stood.

Roy stayed put. "We all appreciate the need for urgency," he said. "But I think a little more discussion is in order."

"Go ahead, Roy," Jane said to her brother, knotting her hands in her lap to stop herself from gathering her papers. "What's your idea?" She noticed that Don sat again, but only on the arm of his chair.

"I don't want to be preemptive here." Roy swung toward Wallace Conover. "Wally, what's your thinking?"

Conover steepled his fingers, glanced down at Wade's just-distributed proposal. "A pilot program seems worth a try. If we can come up with a way to operate temporarily on PCs and faxes—whatever—I'd certainly look at that. But we have a lot of communicating to do to get this little experiment under way."

"Great!" Jane said, ignoring the put-down. "Let's start right now. Is that thing plugged in?" She pointed to a small teleconferencing unit midtable.

"We used it earlier this morning," said Reid Smith.

"Then I suggest we start dialing out, linking up as many

regional managers, warehouse managers, and store managers as it can handle, tell 'em what we're doing, and ask for their input. Maybe we can head off some of those 'apocalyptic ramifications' Chip Bragan was mentioning. Don, will you run the meeting?"

Landsman gave her a thumbs-up. "Sure. Let's get a fresh pot of coffee."

AKERS STORE #28, LAKE RELIANCE, WISCONSIN 10:41 P.M.

"So we're all getting ready to hit the beach, eh?" quipped Tony, the wiry ex-marine who ran the sporting goods department. "What are they calling it? Operation Unplug?"

The wisecrack got general laughter from the assembled department managers, and Phil Leavitt joined in, because that's what a good boss did. But there was little about the situation he found amusing. Almost two hours after closing time, and five hours after his normal going-home time, Phil was still furiously busy. Like the eleven other Akers store managers in the Milwaukee region "pilot program," he felt himself very much in the front wave of risk taking.

If it wasn't amusing, it *was* funny in a way. For days he'd been leading the plaintive chorus over the computer snafus and thinking they should scrap the entire snake bit system. Now that it was actually happening, his stomach was revolving like a damn washing machine.

He surveyed his department heads—electronics and appliances, toys and sporting goods, hardware and housewares, baby store, men's, women's and children's apparel, grocery, lawn and garden, tires and auto parts, groceries and liquor. "Getting down to it, troops, one of the obvious dangers is that when we split up in a minute and you double-check your stock, you'll basically be reordering blind, because we'll be getting our replenishment later tonight. Our last computerized reorder for a while. Obviously, at six A.M. after restock, you'll have a more accurate picture of what you need. Trouble is, Wade says we can't wait for six A.M., and he's right."

A moment later his department heads fanned out, each to make sure his or her area was adequately stocked, especially on critical and high-velocity items. Lower-priority items, slow movers, would get cursory attention. Priority SKUs were to be physically counted in-store, then matched against supplies in the distribution center.

All this would be done low-tech. Throughout the facility all system workstations were dark, network plugs pulled. Phil's office staff, also double-shifting, was moving its critical number crunching to off-the-shelf PC spreadsheet and database programs, along guidelines provided by Ocean Plains. Aileen McMasters, head of #28's Accounting Department, had promised Phil she'd transfer daily inventory data to a notebook computer he could carry around, enabling him to call up rough store stocks of key items—by SKU, department, and vendor.

We can do this, Phil told himself, just as he'd told his people. But he couldn't escape a queasy sensation. An always demanding job had suddenly escalated into a high-wire act—without a net. In this case, a computer net.

His mind reeled as he tallied just how many daily store operations depended on the system that was no longer going to be there. First and foremost, streamlining of inventories and replenishment would be out the window. In fact, where the DC couldn't supply a product in a timely fashion, alternate sources were already being lined up and deals made—costs be damned. Which, of course, was a violation of Akers commandments numbers two and three, decreed by Royal Akers two decades earlier: Leverage buying power to reduce the cost of goods; and Trim operational costs to the bone, so stores could charge customers the lowest possible price for quality merchandise.

But not today or tomorrow. And who knew when? The cost of goods and moving them, usually scrupulously contained, would all rise; some would skyrocket. Tomorrow morning, for instance, there would be extravagant drop shipments of dry grocery items from a Chicago wholesaler and housewares from a Milwaukee distributor. Deliveries would be made not only to the DC, but direct to several stores.

It didn't exactly square with the company management seminar Phil had completed only two months earlier. The lecturer had emphasized key Japanese management principles, such as eliminating *muda*, or waste, from the supply chain. Examples had included excess inventory, unnecessary transport of goods, redundant processes. The goal was to convert *muda* into what the lecturer termed *kaizen*: "continuous incremental improvement."

Well, maybe next week. Or next month. If Akers was still in business.

As he walked the aisles one last time, checking in with his department heads, Phil paused beside the infant formula shelf. There was now an abundance of powdered soy milk. More *muda,* but at least little Amy—would he ever forget that little girl's name?—wouldn't be going without. Another baby food line seemed spotty. The missing items, Phil was assured, were covered by another brand.

He checked the dairy cases up front. Their last delivery from the company dairy, deli, and frozen food distribution center was half what it should have been. Before having its plug pulled, the system query screen had shown the correct refill numbers in transit, but God only knew what would actually show up on the dock. A local dairy was standing by, more than happy to fill any gaps left by erratic deliveries from the DC. For this essential favor, Phil would pay more per carton than he would charge his customers. He recalled the old joke: Sure, we lose a little on each sale, but we make up for it in volume.

Phil ducked into the break room, heading for the coffee machine. With any luck he could be horizontal before midnight. Definitely decaf, then. As he reached for the steaming pot, his assistant manager, Dave Daley, edged alongside, rolling his eyes.

"Phil, I could have gone to law school, you know that?"

"What, and miss all this fun? So, how are the clothing lines? Are we gonna get in all our sale items?"

Daley whipped out a pocket notebook. "How many problems do you have time for?"

Phil glanced up at the wall clock. Eleven-nineteen. "All of 'em," he said, putting down the decaf and grabbing the high-octane handle.

DAY 3

WEDNESDAY, MAY 12

AKERS HQ
10:41 A.M.

The Akers guard couldn't understand the small, dark-skinned man. His chirrupy speech was as peculiar as his all white dress—slacks, golf shirt, and turban. After several vain repetitions, the newcomer took out his business card and carefully printed on its back the name of the person he wished to see. The guard examined both sides of the card, consulted his clipboard, then a list inside the gatehouse, finally shook his head.

"Sorry, but you're not on our visitor list. There's no Malcolm in our directory, anyway."

"A moment, please. I will obtain Ms. Malcolm's current telephone extension from her associate in San Francisco." Calmly, the strangely attired man unzipped a leather pouch, pulled out a flip-phone and palmtop computer.

A red Mazda Miata convertible had just nosed to the gate. The guard waved it through, but the car stayed put. The driver was staring out the window.

"Hey," Lenny called out. "You're R. K. Singh."

"That is so," said the small man, lifting his fingers from the phone's keypad. "How did you know?"

"Your picture's on your home page. But I've also read your books and Internet postings. Heck, I use some of your software. Whatcha doing here?"

"Failing to penetrate your perimeter security," Mr. Singh said with a faint smile. "Do you know Jane Malcolm?"

"Of course."

"I have an appointment with her, as it happens, but this gentleman doesn't recognize her name."

"He's new. Dave, it's okay. Jane Malcolm is Royal's daughter. You know, like in Royal Akers? She's a consultant. And this guy is strictly legit. In fact, he's world famous. Sign him in, then call extension 7796. I'll take him in."

R. K. Singh was there to take a look at their computer problems—exactly what Lenny Cristofaro figured. "We'll be seeing a
lot of each other, then."

"How is that?"

"I'm Lenny Cristofaro, systems administrator here. So it's my
baby that's in trouble. And boy, can I use your help."

"And I can use a guide. Seems like an admirable arrangement."

Lenny parked his Miata behind the general offices and
walked the newcomer into the building. "Jane actually has an
office, but she's rarely in it. She's usually down here in our War
Room. That's her talking to the old gent. Before you go in, you'll
want to know. We had Unix consultants poking around here all
week. Nobody in your league, of course."

"And what have they found?"

"What you might expect. What's your first name, R. K.?"

"It's Rikki, but I prefer Mr. Singh, if you don't mind."

"Sure. Well, they found plenty of digital footprints through
our file systems. But so far nobody can trace 'em."

"I need to see their notes."

"No problem. Really, though, you shoulda been here yesterday."

"Ideally, Mr. Cristofaro, I should have been here the day the
damage was discovered, when the digital footprints, as you call
them, were fresh and not muddied over."

"Sure, right. The thing is, we pulled the network plugs early
this morning," Lenny said. "So how'd Jane Malcolm track you
down, Mr. Singh?"

Singh explained that he'd been in New Orleans to deliver a
paper on Internet crime at a National Security Agency conference, when he'd gotten a call from a former Berkeley classmate
who now worked for Ms. Malcolm. In response, Singh had flown
to Champaign via Chicago, taken a taxi to Ocean Plains.

"I assume you're going to stay till you get us healthy, right?"

"Unfortunately, I have other commitments. I am only here
today to take a look."

"That's a bummer. Unless you come up with an instant cure." Lenny smiled ingratiatingly.

"Unlikely. Perhaps, however, I can give some useful advice."

"Lotsa luck," Lenny said. "After you see Jane Malcolm, if you want me to show you around our turf, I'd be glad to."

"Thank you. Where will I find you?"

"In the next building over, second door on the right. Oh, and I'd wait till I got to know you better to ask, but since you're leaving soon, I'll just ask up front. Would you autograph a couple of your books for me?"

AKERS HEADQUARTERS, OCEAN PLAINS 10:45 A.M.

Jane was trying to reconcile the sun-bronzed, businesslike-looking man standing before her with the freckle-faced teen of her memory.

"How many years has it been, Wade?" she asked. "On second thought, don't answer that."

Wade grinned. "Our class did have a reunion last year—fifteenth or fiftieth, I forget. I was kinda hoping you'd come."

"I left OPHS in the tenth grade. You probably didn't notice."

"I noticed. I even have a pretty good idea of your doings over the years. Why do you look puzzled? Are you that surprised your dad liked to brag about you?"

Jane instructed her mouth to smile. "So, you've stayed with the company all these years?"

"My one and only job. From stock boy to whatever it is you want me to do now."

"Help us expand on what you wrote here." She pointed to his e-mail memo. "What you proposed for your region? We want to do it nationwide. Better make that worldwide. We're just getting the first wave of problems from overseas, some on a scale that seems almost diaboli—"

"I know. The Milwaukee DC got hit Monday with those two truckloads of wrong PC monitors, most from Korea, some from Mexico. What have I missed?"

"Containers full of scrambled fall fashion orders," Don Landsman interjected. "From as far away as Pakistan. Parkas for the Sun Belt, tank tops for the upper Midwest, you name it. When Jane says diabolical—our saboteur has a twisted sense of humor."

"Where are these garments in the pipeline?" Wade asked.

"They moved through our West Coast import hub on Monday," Jane said, glancing at her notes. "The paperwork checked out, just like the Hyon electronics, so nobody caught it. We're catching it now"—she winced—"in some RDCs."

"Can't turn any of the trucks around?"

Jane shook her head. "They're carrying too much stuff we need—and badly. The scariest thing is, we didn't see this coming. We didn't have a clue—there wasn't one. Which means we can't predict what else may be headed our way. Don's already got a separate team working on what to do with the wrong monitors and the wrong fashion. Basically it'll cost us a whole lot of time and money—in interfacility transfers, inventory carrying charges, inventory warehousing, warehousing labor charges, reselling at cost or at a loss. But if we keep getting blindsided like this, the hit is going to be deadly. We're going to go under, plain and simple."

Wade shook his head. "Let's see. You've got one group trying to keep the wrong stuff *out* of the stores, another one figuring how to get the right stuff *into* the stores. Which group am I in?"

"Definitely a 'right stuff' guy." Jane smiled. "You'll be helping us stock shelves, Wade, just like we used to as kids in Daddy's Store Number One. Meet your teammates."

She gestured at the three men seated around the small round conference table in her office. "Don and Elie you know. The gentleman on Don's left is Prince Khalif al-Marzouki. Khalif is a teaching assistant at MSU—"

"At Eli Broad?" Wade asked, impressed. Michigan State's business college was known for its supply-chain management concentration.

"I have been a guest lecturer there on logistics strategy, but mainly I am working on my doctorate in economics. Or was until Jane rang."

"Khalif helped our logistics effort in the Gulf with food procurement," she said. "Thanks to him, our troops didn't have to survive on prepackaged food—MREs—but got fresh pita, fruit, and some other foods more interesting than they probably got back home. He'll be working on 'right stuff' distribution, too."

"Sounds like we can sure use you," Wade said, shaking hands and taking a seat at the table.

"I'll do my best, I assure you," Khalif said, displaying his dazzling teeth.

"And, Wade, we're not done recruiting," Jane said. "In case you think of someone we ought to have."

"What we need are a few folks used to stocking stores the old way," Elie said. "Before computers took over."

"You may be right," Jane said. "I was just talking to one of my Penn State logistics professors, Dan Sammons. Dan's in his late sixties, and he made the point that a lot of young Turks, you know, leading-edge types—"

"This is not me," Khalif said, grinning. "I may be young in years, but I am ancient in logistical lore."

"Yeah, we know," Jane said. "But a lot of young computer whizbangs may be hard-pressed to invent work-arounds. They don't begin to know how things were done before—in the store, in the warehouse, so on. So we're going to need—"

"Folks like me," Elie said. "I'll browse through my mental Rolodex. But most people with that level of expertise are *retired.*"

The phone rang. Landsman picked up, nodded, handed the phone to Jane. "Front gate calling. Says some guy in a turban is on his way in."

"Another Unix guru?" Wade said when Jane told them she wanted them to wait a minute and meet Mr. Singh. "Two days ago when I stopped by, Lenny Cristofaro told me we were already up to our ears in computer expertise, with zero results."

"Then one more won't hurt, will he?" Jane said pleasantly. "Mark Gibian, who works with me, says this guy is absolute tops in the field. He's based at Berkeley's Experimental Computing Facility."

"I hope he's a miracle worker," Don said. "But why don't we just turn him over to Bragan? We've got plenty of problems of our own to plow through."

"Because we've got to fight this war on two fronts at the same time," Jane said. "While we're improvising full-speed ahead with traditional logistics, we also need to be attacking our computer problems." She paused. "Look, I have complete confidence in this team—with the five of us as a nucleus. But based on our meeting this morning, I have some misgivings about the people trying to get our computer network restored to full operational health. And I'll feel a whole lot better about what *we're* doing, if I can feel better about what *they're* doing."

"Point taken," Landsman said.

"What I hope Mr. Singh can do is, basically, track all the electronic dragons to their lair, drive a stake through the heart of each one, and construct an impregnable moat around the castle."

Wade, while attentive to Jane's words, was taking a certain pleasure in watching her run this meeting. Here was this special person he had eyed from afar in high school suddenly up close, looking like, talking like—*being*—a really smart top executive. But was this woman really so different from Jane as a girl? In school, hadn't she spearheaded their class activities, from staging pep rallies to decorating the gym for dances? Wade also couldn't help noticing nonbusiness details of his new team leader—the wayward curls at her neck; the pearls in her earlobes; the absence of a ring on her tapered fingers as they steepled or splayed, closed into emphatic fists or flew up in vivid gestures.

The arrival of the computer expert cut short his inventory. *Just as well. Got to keep my eye on the ball, not the umpire.*

R. K. Singh was a study in contrasts—skin the rich color of roasted espresso beans, and garments half Western, half Indian, entirely white. The only break in the white was the ebony pager on Singh's white woven belt.

Jane made introductions all around, then got right to the point. "Mark Gibian tells me you've been able to solve some major breaches of computer security, break-ins, and so forth."

"My actual field is artificial intelligence and machine learning. But it only takes one or two successes of the sort you mention to gain a certain notoriety." He sighed without a bit of sadness. "As Mark must have told you, I now do systems and network security consulting."

"I'd like your card," Khalif said. "I might have some work for you sometime."

Singh obliged; then, getting a nod from Wade and Don, dealt cards around the table.

Jane recapped the earlier meeting, describing the kinds of viruses that had been isolated so far and the varieties of disaster they had wrought upon the company's inventory and replenishment systems. "You'll get your specific questions answered by our information technology director, Chip Bragan," she told Singh. "Of course, the entire technical staff working with him will be available to you as well. What I'm hoping for—I can't tell you how sorry I was to hear you'll have to leave at the end of the day for other commitments—is some kind of general diagnosis or recommendation for action. How *you* think we should proceed. How long you think it will take to get us healthy. What resources, if any, we need to go out and get. And, frankly, your evaluation of the personnel here."

Mr. Singh nodded. "You suspect that the sabotage may have originated within, rather than without?"

"Your call," Jane answered. "But I'd like to hear your thoughts as to when or whether we should bring in law enforcement."

"You would certainly qualify for a great deal of attention! I'm familiar with many of the people at the FBI Computer Crimes Division—I've lectured at Quantico—and I can tell you that they allocate the highest priority to cases involving the highest dollar loss. From what you've just told me, it's plain that your company's recent losses are extremely high. Some of that will be in intangibles, of course. But one way to calculate it might be to multiply the computer-related drop in your stock price by the number of shares outstand—"

"We've run those numbers," Don Landsman said. "And you're right. It's like having all our cash receipts siphoned off for several weeks running. Can the feds really help us?"

"To track down someone hacking into your system? Of course. The FBI has the most sophisticated phone-monitoring equipment, and they can get trap-and-trace orders for local phone companies. And ultimately, to be sure, you will want the person or persons apprehended, arrested, and prosecuted. But for pure digital detective work—finding the predator—well, let's just say you'd be better off with my eleven-year-old nephew, Manjit, or the average high school computer science student."

"What's your nephew's per diem?" Wade said, eliciting a chuckle from Elie and Khalif.

But not from Don and not from Jane. "That would be funny, Wade," she said levelly. "Except nothing about this is funny."

AKERS PR DEPARTMENT
1:33 P.M.

"She's expecting you," the receptionist said, leading Jane across a small bullpen in which a half dozen people hunched over PCs or Macs. They seemed to be ignoring a wall-mounted TV on which spidery models pranced down a runway, flapping red vinyl raincoats.

How many offices kept CNN on all day? Jane wondered. It was getting to be the new Muzak. Of course, here it made some sense. The PR department needed to maintain a media vigil, monitoring all the negative reports on the company. Which had to be pretty depressing for these troops, whose mission was to put a happy face on everything.

"Jane Malcolm to see you," said the receptionist, pushing open the door of a small, windowless office. "This is Doreen Heckler, our assistant PR director."

"Ms. Malcolm, please come in." A tall and very pregnant young blond woman half stood, taking off her glasses and extending her hand. "I'm delighted that you came by. Just before she left to get a quick bite, my boss, Esther Tice, and I were talking about you—your coming aboard, I mean. Everyone's behind you, you must know that. It would be so great if you can help us out of this mess. If there's anything we can do to help, just say the word. Esther should be back any minute."

"It's you I really wanted to talk to."

"Really? Well, I have drafted a status report on our recent efforts. Esther hasn't seen it yet, though." Doreen hesitated, then swiveled to her desk. "I guess it'd be all right to show it to you . . . so long as you understand Esther hasn't—"

"I do, and I'll be sure to let her know you made that clear. I'll look it over as soon as I get back to my desk. What I came in to ask you, well, it's sort of personal. I understand that you're the granddaughter of Sid Skoggs. Is that right?"

"You know my grandfather?"

"Everyone who grew up in Ocean Plains knew him. I once worked two whole days—well, almost two whole days—in his grocery store downtown. I was fifteen, and I had this elaborate idea for decorating my dad's store window for Halloween, only he wouldn't give me the job. Said my ideas were too fanciful or too fancy or something. So I went down the street and talked your grandfather into giving me a chance. Dad didn't hear about it till I was almost finished. Then he came by and yanked me right out of there." Jane picked a scrunched paper off the floor and tossed it in the wastebasket. "I'm sure Mr. Skoggs has forgotten the whole thing, but I never have."

"That's a nice story," Doreen said.

"Well, he was a nice man. He seemed different from most grown-ups." She smiled. "He didn't seem to like saying no to us kids. But I bet that was true when it came to his adult customers, too."

"Sounds like Grandpa. But I'm afraid he hasn't been quite so positive since he lost the—Sorry."

"Don't you be sorry," Jane said. "I'm the one who should feel sorry—and I do. Whenever I've driven by, since I've been back, I have to think about . . . what happened. The way Mr. Skoggs really took *care* of his customers. Then to have Daddy lure them away. It's no wonder if he didn't keep his positive outlook on life."

Skoggs Market had been in Ocean Plains for twenty-five years, the last five losing a war against Akers Discount. Mr. Skoggs had closed his doors a dozen years ago, unable even to sell the business. The windows were boarded up now, the graffiti-marred bricks crumbling.

"He won't set foot in our stores, you know, Ms. Malcolm. He pretends he doesn't know where I work, and believe me, I don't mention it anymore. He didn't come to the funeral, but my mom—Grandpa lives with her—she told me he kept turning on the local radio station all day. Didn't say a word, just sat there, listening to the same reports of what was said by whom at the funeral, over and over."

"Well, I'd very much like to talk to him, Doreen. Could you tell him that? And that I've never forgotten the favor he did for me that Halloween?"

Doreen looked dubious. "I could try."

JANE'S OFFICE, AKERS
HEADQUARTERS
3:40 P.M.

The all-hands meeting was scheduled for four. Jane, who was to give the main presentation, glanced at her watch and realized she had almost no time to prepare.

She spent ten minutes making some notes, and then, with five to go, she shut the door and ran her main points by Don and Wade. They had a few suggestions—and she jotted down additions on her three-by-five notes. "Thanks," she said.

"In my case," Wade said, "it's not the first time. Helping you with a phrase or two, I mean."

Jane, in no mood for games, frowned. "Oh?"

"When you ran for freshman class president?"

"I don't remember that," she said, walking very quickly down the corridor, Wade barely keeping up, Don not even trying to.

"You won. Remember *that*?"

Jane, regretting her curt tone, tossed Wade a half smile. "Some things don't change, do they? I was always running for something. Apparently still am."

"I was one of your volunteers," Wade said. "In fact, I came up with your campaign slogan. It was pretty corny. 'Jane Reigns at Ocean Plains.'"

"Dear God! But we won!"

"Of course you won."

Jane opened the door and strode toward the building the auditorium was in. Wade stopped to hold the door for Don. "Hey, Wade," Jane called back. "How could you remember all that?"

"How could I not?" he hollered after her.

On the way to the auditorium building, she ran into her brothers. Roy Jr., she knew, was going to preside at the meeting. But she was surprised, and delighted, to see Alex.

"I was afraid I'd missed you somehow," she said. "That you

might have already gone back to Hilton Head."

"Without saying good-bye? I wouldn't do that, Janie. Anyway, Roy's asked me to stay on for a while in Dad's house. Give Anne someone to pick up after, I guess."

"That's very thoughtful," she said. "On your part, and on Roy's, for thinking of it. I know Tyler must really need you, Alex, missing his granddad and all."

"What about you, Janie? Maybe you could use me some, too?"

"How do you mean?"

"Give it some thought. I'll be around."

Alex saluted her and moved off to say hello to Don. What on earth did Alex have in mind? she wondered. Schmoozing investment bankers?

His staying on did make a kind of sense, though. Anne Akers probably had a support system—friends, family—and was smart enough to tap into it. Still, it couldn't hurt to have a step-son underfoot as adept at people pleasing as Alex was. As for Tyler, she wished she had some time so they could get to know each other, but she lacked Alex's natural ease with children. Better for Tyler to have the real thing.

That was the moment it occurred to Jane to wonder if Roy had another reason for asking Alex to stay on. For instance, wanting his brother on hand to sign proxies on short notice, without the necessity of FedExing documents to and from Hilton Head.

That was pretty thin reasoning. Besides, she shouldn't be wasting time trying to read Roy's mind. She had enough on her own mind right now—like how much she could accomplish with one talk.

At the door of the auditorium, she rejoined Don and Wade, and they became part of the employee flow inward. Most of them were strangers to her, of course. But she noticed how many greeted Roy at the door and how he responded by shaking hands, slapping backs, then playfully introducing his kid sister.

Not her all-time favorite introduction.

Of course you won.

Thanks, Wade. Suddenly Jane's jitters vanished, and she moved on through the foyer with renewed resolve.

She could do it again.

AKERS REGIONAL DISTRIBUTION CENTER, MILWAUKEE
3:55 P.M.

John Brinkema leaned a haunch against the IBM AS/400 midrange computer, now off-line. Across the room, Henry Torres, the DC systems operator, was sitting in front of a workstation, fingers dancing across the keys. Patrick DePalma was watching the PC screen over Henry's shoulder. John drew out his flip-phone and punched a number.

"Wade Crain here."

"Big John, Wade. The word for today is '*shit*.' Make that '*horseshit*.' As in up to our armpits and rising fast."

"John, make it quick. I was just switching off my cell-phone when you got me. I'm on my way into the Akers Auditorium to hear Jane Malcolm. She's gonna be talking to the employees in a minute."

"Having champagne and caviar afterward, are we?"

"John, what's happening?"

"We pulled the plugs, just like you said to. So it's real quiet here in the computer room—the label printer especially. As in dead silent."

"Shit!" Wade said.

"You finally smell it."

The label printer operated with very little noise, producing merely a continuous whir slap as three across, bar-scannable, laser labels fan-folded into its output hopper. These labels, keyed to order and store number, drove the entire distribution center. At the moment, however, there was *no* sound because no labels were being produced.

As a result, the entire warehouse floor, viewable through the computer room windows, was also still. The warehouse, like the printer, was not operating.

"What about the other printers?"

"Just sitting there, Wade."

The other laser printers generated pick lists for each order, including PO and ticket number, sold-to and ship-to info, along with the number of the cases, cartons, and pallets as well as the total weight per order. Essential outbound documentation—shipping and packing lists, bills of lading—was also printed.

But not at the moment. And since these printouts constituted the DC's basic work orders, Big John's labor force was mostly just standing around.

"John, is Pat there?"

"Hold on." John handed the cell-phone to DePalma.

"Boss?"

"Make sure I've got this right, will you, Pat?" Wade said. "The backup plan was for a stand-alone workstation to drive the printers, once we got the replenishment orders in via faxing and scanning. Is any of that happening?"

"Some of it. We're getting the orders into the PC database okay. Henry's *massaging* the data now—his word, not mine. He's printed some labels, but some fields are apparently scrambled. He says the program needs more tweaking. Henry says we shoulda been gearing up for this weeks ago. That we just can't cut over on a dime."

"Dammit, Pat, we *are* cutting over on a dime. Today. The rest of the DCs tomorrow. That's how it's gotta be."

"Ask him about the AS/400!" Henry called out, still typing.

"Wade, Henry wants to know are we *sure* the AS/400 can't be used? He thinks maybe he can clean up the database."

"Tell him to e-mail his proposal to me. Right now, John, we keep it off-line and the whole network stays down. Pat, you gotta get Henry to generate some kind of pick lists fast for John's people. We gotta get moving."

"We're working on it. Thing is, how will the pickers know what they're doing? I mean, without the computer, we'll just be releasing orders, without knowing which we can fill and which we can't. Plus, without our warehouse management system working, *we're* working without our routing programs. We'll have guys stumbling over each other, picking stuff in the wrong sequence, loading the trucks backward, stacking heavy pallets on top of light ones, crushing them—you name it."

Wade swore again. "Hang on, Pat."

Pat found some encouragement in the silence on Wade's end.

The man finally realized what they were up against and he was doing something about it!

He had to be. Because, the way Pat saw it, the situation wasn't just bad, it was a bad situation someone covered in foil and then put in the microwave to heat up. The routing software not only calculated the quickest pick path through the warehouse, but also the correct piece of equipment to use— walkie-riders for building pallets, forklifts for moving full pallet loads, man-aboard order pickers to pull individual cartons from the racks, reach trucks to put away and pull full pallets from racks. In fact, the program juggled all kinds of tricky variables, ensuring that items arrived at the right dock door in sequence, so Kleenex got loaded on top of bowling balls and not vice versa.

The transportation routing software kicked in next, calculating cubic dimension and weight for each order, number of orders or stops per truckload, how many trucks were needed. If orders for several stores went in one truck, the software plotted the optimal driving routes, even down to secondary roads, then made sure the stuff destined for the final stop went into the trailer first, the first-stop cartons last. It all sounded obvious, but with two hundred people wandering around down there, Pat knew Murphy's law was the only thing you could really count on.

"Pat, you there?"

"Right here. Got something for me?"

"I was just talking to this guy . . . " Wade sounded out of breath. "He's a Saudi . . . never mind . . . he teaches logistics up at MSU. He had mentioned this routing program—it's got most of the features we need, runs on a Pentium. Right after Jane's speech, he'll arrange to get copies FedExed to every one of our DCs tomorrow morning by ten-thirty."

"And by eleven, we find out you *don't* just put the program in your A drive and, abracadabra, it works? The program would have to be adapted to each DC—but how could the sysops, the systems operators, find out quickly how to do that? Thanks, but no thanks."

"I have it," Wade said. "Once they're in-house, we'll put all the systems operators on a conference call, or maybe some kind of BBS, a bulletin board setup. Khalif—that's the guy—he'll work with them on making modifications for each site. Tell Henry to be ready, okay?"

"I will, but—"

"Look, Pat, I know it's gonna be a real son of a bitch up there for a while, and I'm sorry I'm not there to help you and John. But we can make it work. Got to. Now I gotta go. Call you later."

Pat slapped the phone shut, started to hand it back to Big John. But the burly night-shift manager wasn't paying attention. His gaze was riveted on the label printer, which was again purring and extruding fan-fold sheets. John pulled one up and peered at it closely.

"Henry, these look pretty good."

The computer operator turned, still typing, and flashed a victorious smile. "I did it! I massaged the damn thing till it loved me. For the record, pal, these labels aren't just good," he said. "They're fucking fabulous."

AKERS AUDITORIUM
3:57 P.M.

What was now the Akers Auditorium had been the Siler Air Force Base movie theater—and still looked it. Two hundred red plush seats sloped down toward a high stage, while the walls retained a faded mural commemorating the history of aviation—Icarus with his melting wings; the Montgolfier brothers shaking hands with the Wright Brothers across the centuries; John Glenn tilting *Gemini* in tribute to the ghost of Lucky Lindy.

The house was pretty well packed when Jane filed onto the stage with her brothers, her team members (including a spiffy-looking Eliezer Grey), and top Akers executives, all taking seats behind a long table. Esther Tice had caught her on the way in, introduced herself, and said that a camera in the back of the hall would be feeding the event live, not only to an expected overflow crowd in the nearby cafeteria, but, via satellite, to Akers stores and distribution centers worldwide.

"Thought you'd want to know a lot of people are going to be listening to what you have to say." Tice smiled. "Good luck."

The feeds made sense. With rumors and facts vying for the title "Worst News of the Day," *something* had to be done to revive employee confidence. Apparently that task, too, was hers.

She stared out at the faces—senior management up front, regional and district managers behind. Farther back was the contingent she figured to be most receptive to her message—the matrix management team. Their assignments and responsibilities crossed classic organizational lines. They used logistics to support and coordinate main operating areas like merchandising sales and marketing, purchasing and accounting, transportation and warehousing, information technologies. Working with them were the customer support teams for each of the company's main vendors. All in all, Jane thought, a sophisticated and skeptical audience, seeking reason to hope.

Exactly at four, Roy Jr. tapped his microphone to make sure it was on, and that was enough to make the hubbub subside.

"Good afternoon," he began. "As many of you know, we've put off this all-hands meeting for several days, until the executive committee and the board could agree on a definite plan of action. I'm happy to announce we have done that. Though our problems remain formidable, I think you'll all be encouraged by the steps we have already taken and others we will be taking in the days ahead. I trust you will all continue your extraordinary efforts to pull together in this difficult time for Akers. . . ."

Not exactly the announcement Roy had hoped to make, Jane knew. He had anticipated announcing the sale of the company to Tom Soverel and the ingestion of the Akers empire by its chief rival, 4-Mor Stores.

I changed that, Jane thought, and a wave of apprehension was heading her way when, suddenly, she remembered a story her father had told her in one of his undelivered letters.

He said it happened in 1973, when she was seven. He remembered because 1973 was the year he opened the first Akers store in Ocean Plains. Jane, he wrote, had organized her Girl Scout troop to sell cookies. She'd written a cookie song and organized a chorus to walk up and down Main Street, singing their way from one shop to the next. They'd racked up more sales that day than the store, he said. Jane remembered exactly what he wrote next. "I told your mother then, 'Someday Jane will put this company statewide . . .'"

Jane felt her lower lip tremble.

I'm here now, Dad.

I want you to know, that makes me proud. You make me proud. I never did tell you that often enough, but right now you better listen to what your brother's saying. It's almost your turn.

Jane glanced at her hands and watched them stop trembling. She turned toward her brother.

Roy Jr. *was* finishing up.

"My sister, who I know is new to most of you, worked in Dad's first store, learned the business from the ground up, from rounding up shopping carts to cleaning the popcorn machine. Dad often expressed his confidence in her. Since her résumé is included in the handout we're making available, I won't rehash it here. But Jane has been a logistics consultant to several multinational corporations, including TranSonic Airex.

"She'd be the first to tell you—I don't mean to steal your thunder here, sis—that she isn't here to solve all our problems. What she *wants* to do, what I and the board hope she *can* do, is to look at those problems from a fresh perspective, and perhaps be able to formulate a new strategic plan to put Akers back on a sound operational basis."

He twisted around to smile at her. "Sis, it's all yours."

Jane stood. She hadn't missed the way he'd covered himself at the end. Nor could she recall Roy ever calling her "sis" before—another kind of diminishment.

As the microphone was being passed down to her, someone pinched her left elbow. She turned to see Elie giving her a thumbs-up.

The gesture helped Jane dispossess her anger; it could only get in her way. "Thank you, bro," she began, winning a few chuckles. "Seriously, thank you for those generous words and for the confidence you and the Akers board have seen fit to place in me during this rough time for the entire Akers family. I'm talking about all of you"—Jane looked slowly over the audience, stopping several times to focus on a particularly attentive face—"the way my father thought of you, you who have contributed so much to the company's long and great success.

"As some of you may know, I've been involved in some

pretty challenging situations before—most notably in Saudi Arabia, during Operation Desert Shield. During the first few months there, in addition to solving all the daily emergencies, the first duty of our logistics team was to figure out what we'd do if Saddam Hussein decided to attack us that day.

"I guess that qualifies as Crisis Management 101."

A handful or two of friendly laughs, quite a few smiles. More important, not a cough in the house.

They're ready, Jane. Are you?

You still here, Dad? Good, listen up.

"I have to tell you, I look at this challenge as the greatest I've ever had. And I'm prepared to devote my very best effort to it. My guess is, all of you feel the same way. Of course, you beat me to it. You've already been looking this challenge in the face without flinching. The sluggishness of Akers's recovery has not been due to any lack of effort."

Jane paused, looked at the people seated behind the table, and then, taking her time, scanned the audience again. Lifting her arms, she tried to embrace the entire room. "We've had the talent and resources all along to right this ship. Now we finally have a *plan* in place to do it." She paused again, so the scattered applause may have been in response to what they read as a hint. Immediately she stopped it, holding up a firm hand.

"What we don't have on our side is time. But I don't have to tell you that, do I? You've seen our dilemma firsthand in our stores, and second- and thirdhand, often unfairly magnified, on TV, in newspapers and magazines. You've seen it, perhaps most distressingly, on the stock ticker."

Akers common had opened at 35⅛ that morning, the lowest price in nine years, according to Don Landsman. It would have been even lower, he added, except for rumors that the company was in play and about to be gobbled up.

Jane glanced down at her notes, then ignored them. "According to many Wall Street wizards, the odds are against us.

But that's because those folks *don't know us.*"

There was no applause, but Jane saw a number of heads nodding—which was better.

"Our crisis recovery plan is straightforward," she went on, satisfied because they were listening with such attentiveness. "Our objective is the same one that's always been the company's mission statement, my dad's mission statement: 'Akers will provide its customers with the highest-quality products at the lowest possible prices—when and where they need them.'

"How can we restore full faith in that creed, and once again do the job our customers count on us to do?" She took a sip of water. "*We just start doing it.* We must not—cannot—wait for a complete restoration of our electronic systems. If the autopilot conks out temporarily, you fly by the seat of your pants. But you keep flying.

"Today, as some of you know, we disconnected the main integrated logistics software for the Milwaukee region. And tomorrow we'll be doing it worldwide. Which means losing all the functions we've come to rely on, including EDI and VMI. How soon will we get the software back? I can't tell you at this moment. Right after this meeting, my team is going to be sitting down with Chip Bragan and his IT people, plus our consultant teams. By the end of the day, I expect to have a better fix on how soon we can plug back in.

"But we don't need that software to do our job. Here's how we're going to do that."

She briefed them on the main points of the plan she'd worked out with her team. A lot of functions that had become centralized in Ocean Plains would be temporarily regionalized. For the duration of the crisis, regional managers would relocate in their own distribution centers and begin operating semiautonomously, assuming responsibility for their own inventory and replenishment, concentrating on crucial SKUs, cutting their own deals where necessary, especially for perishables. Wade Crain, whom many of them knew, would coordinate those efforts.

Don Landsman, whom *all* of them knew, would work with sales and with purchasing to smooth over relations with the most critical accounts, the company's first-tier suppliers. Most of these household-name giants were major trading partners. A very few had escaped the viruses. The rest were still trying to sort out the mess. Akers's mission would be to help these accounts on a daily, even hourly, basis, get accurate snapshots of all their product throughout the Akers supply chain—on store shelves, in warehouses, in transit—and track down any missing shipments. Then, using historical numbers where necessary, Don's group would help suppliers determine what quantities needed to be relocated, returned, or reordered and what new production cycles triggered.

"If we have to use semaphores and smoke signals, then we'll do that," Jane emphasized. "As long as we can provide our vendors accurate and up-to-date information by whatever means, we can continue to pull products from them, and they can generate production cycles in response."

Jane sipped water again, to let that hefty dose of information sink in. She had been surprised that there was only a smattering of applause when she'd mentioned Don's name, but now that she had a moment to think about it, it made sense. Word that he was gone had probably not made it throughout the company before he was back. If Roy and not Don was running the meeting, well, he was an Akers, and what had happened was a family responsibility. Wasn't that why she was at the podium?

She straightened and, seeing her audience was also ready, went on. Don's group would also take charge of making the initial investigation into all supplier complaints and financial losses linked to inaccurate electronic reorders. Processing of these cases would be expedited by a separate adjustment group being set up in Accounts Payable.

The War Room in the data-processing center would remain in operation, she told them, but two more War Rooms would be added in the general office building, using existing conference rooms. The main one would be for dealing with immediate crises, the second to handle contingencies farther out and work on strategic planning.

"A lot of the work," she said, "will be low-tech but with proven reliability. White-boards, colored markers, three-by-five cards. We will have fax machines, a lot of them, and PCs, some with scanners, for fax-and-scan order entry. But that's about as close to network as we'll get till we get the all-clear from our software team. We'll use stand-alone software, network with floppies—and even be very careful there. And we're not talking about doing all this tomorrow or the day after. Both those new War Rooms will be up and running tonight. Personnel is already working on the staffing plan."

She wound up her presentation in just under thirty minutes. Q&A took another half hour—longer than she'd expected, but she was determined to give the audience the same respect they'd shown her. At some point, Jane realized, there had been a subtle mood shift. She felt the audience pulling together, taking on the solidarity of a genuine community. Or a fighting contingent that knew how vital the sense of being in it together could be to winning the battle at hand.

When the last question had been answered, she said, "I think most all of us have a good idea of which team we're going to be working on, and even what our immediate tasks might be. If not, you know who to ask."

Nods. More nods. A sea of nods.

"Okay, then. If I may quote my father one more time, 'There's work that needs doing—let's do it!'"

There was a silence, during which most of the audience got

up. But then the clapping began, and as the rest of them stood and joined in, it grew. Jane breathed it in for one precious moment, then stopped herself—and them.

"Thanks," she said, "but let's not congratulate ourselves on anything yet. We've got a lot of work ahead of us."

AT A COMPUTER
5:29 P.M.

```
She's a good talker, I'll give her that.
    And . . . I think I'll give her something else.
Something really nifty. Big and loud. Rhymes
with cloud. Mushroom cloud.
    Got it.
```

Fingers tapped away rapidly, backed by a happy tune.

```
Okay. Hit Enter. Done.
Rhymes with fun.
```

WAR ROOM,
AKERS HQ
7:41 P.M.

While the overtime crew worked furiously toward the dinner break, Jane dispatched a foraging party to local fast-food franchises. Her promise to have both new War Rooms up and running tonight was taking more effort by the crew than she had anticipated, but she hadn't heard much muttering, and both rooms *would* be ready before midnight struck and this third day of the ten she'd been given was officially over.

All four walls of the large, commandeered meeting room were now paneled with white-boards, with sections for each operational region. Down the middle of the room, banks of computers had been set up—but not networked or modemed. Overhead, extra fluorescent fixtures already provided a surgical-theater blaze. Flanking the PC stations, phone-conferencing areas were being slammed together, partitioned, and wired; these, like the white-boards, would be divided into regions, to be staffed by territory-specific project teams.

At one end of the big room, another long, partitioned phone bank was designed for the customer reps handling the largest accounts. At the opposite end, a smaller, adjoining room with far less equipment was designated for longer-range planning. This Strategy Room would also house Jane's team, though it seemed likely that she, Don, Khalif, and especially Wade would spend much of their time in the main War Room, monitoring and troubleshooting.

The layout, Jane thought, roughly approximated the now dormant computer system. The tools they would be using for information management were relatively rudimentary, but the data flow was still from order taking through detailed scheduling to fulfillment and delivery, with a continuous conference of all concerned parties.

The question was, would the ersatz system work?

As she played out all the elements in her mind, Jane felt suddenly overwhelmed. The task that lay before her and her teammates was nothing less than the rebuilding of the company's logistics operations, component by component.

I need to walk them all through it—and myself—to make damn sure we've got every base covered.

Customer service, order entry, purchasing, inventory management, transportation, distribution—all were elements of logistics. Akers already had topflight logistics managers toiling in each of these areas, Jane had no doubt. She would therefore have to be diplomat as well as taskmaster, in order to mesh the talents of her teammates with existing staff.

As long as it's understood who's running the operation.

They do know that, Dad. And, yes, I know it, too.

Jane shivered. It was done. It was fact now. The locus of power had shifted from the corporate office suites into this bustling and inelegant room.

Looks to me like those who don't like it better get out of the way. Do me a favor, Dad. Take a rest.

"Lord-a-mercy, child, you should see yourself, standing there, staring off into space something fierce and muttering to yourself!"

"Oh, hi, Uncle Elie. It's like chess, you know? Staring at the board, all the pieces and positions, playing all the moves out in your mind."

"I thought so. Now you come with me." He took hold of her hand.

"Where're we going?"

"Never you mind, just come." He dragged her off to a corner table and directed her attention to a box of takeout chicken, baked beans, and potato salad.

"You win," Jane said with a sigh, unwrapping a plastic fork. She took an obedient bite and even attempted to take her brain out of gear for a few minutes. Elie was right, she thought. She did need a break—and some fuel. But an instant later she was fully engaged again.

Rikki Singh was coming through the War Room door. She'd nearly forgotten about him—and his report was crucial! Unfortunately Singh was not alone, as she would have preferred. His escort included Chip Bragan and Chip's IT assistant, Norma Klein; Sammy Shah of Bowerman Consulting; Dick Lai of Positek; and—Robert Grooms? The new security man had approached Jane earlier in the day, wanting to run a background check on Mr. Singh—"on general principles," Grooms had said, "but especially given the man's Berkeley connection." Jane had turned the request down flat. Did the man think this was still the sixties? And why was he here now, anyway? As far as she was concerned, just because the others had briefed Singh did not mean that *any* of them needed to be in on his debriefing.

She closed her dinner box and directed the group into the Strategy Room. Wade Crain, Khalif al-Marzouki, and Don Landsman were fully engaged in the larger room; she'd handle this alone.

She asked Chip and the others to enumerate what they'd shown Singh on the tour. When they finished she said, "Sounds to me as though you've been pretty thorough. Good. Now. Mr. Singh has already told me that he is going to withhold comment on our situation, pending evaluation of his own resources and scheduling. But there are matters I do need to discuss with him, so I thank you all very much for your time and expertise."

A slight raising of Chip Bragan's eyebrows was the only visible hint of discontent at their dismissal. Hands were shaken and the escort group retired to the buffet in the next room.

"What about you, Mr. Singh?" she asked, indicating her boxed chicken. "Are you hungry?"

"Some fruit, some salad, would be most welcome. I am a Sikh, you see, and a vegetarian." On the way to her office, Jane made a pit stop at the buffet and picked up a fruit salad and some cookies for Singh.

She shut the door to her small office and removed a carton from the only seat besides hers. Singh took out his palmtop computer and hooked it to a wireless modem. "Permit me to check my messages before we proceed," he said. By the time she'd finished her chicken, he had downloaded and perused his mail from the Internet, responded to several messages, and put away his equipment. He'd even managed to take a few bites of fruit.

"Take another minute and finish your dinner," she suggested. "It's meager enough."

"Thank you. I am finished."

"Well then," she said. "What can you tell me, Mr. Singh?"

"As I told Mr. Cristofaro, it is very difficult to solve a crime so long after the fact. The crime scene has been walked upon by too many people."

He had been shown samples of several strains of virus, decompiled back into an approximation of the original C-language source code. Many other samples had self-destructed after doing their dirty work and could not be recovered or unerased. "The programs I saw were all well written," Singh said. "I didn't have time to study examples of all their tricks, but I was told some viruses have altered calculation algorithms, so that's how replenishment numbers were thrown off. Other viruses have changed product-volume factors in your company database upward or downward, so that when an order was sent for, say, 200 units, then multiplied by the cubic dimensions in the database to plan the amount of transportation required, or optimize vehicle capacity, there would be either too few or too many trucks dispatched to move the product. Even more insidious, it seems to be a self-governing epidemic. In other words, the viruses have apparently been very good at lying dormant, killing themselves off, only affecting random files."

But was all this industrial espionage financed by a competitor? Or maybe some disgruntled employee, or ex-employee, on an electronic vendetta? Singh couldn't tell her that, but—

"Excuse me, Mr. Singh, but what I'm really most anxious to learn is the *origin* of the sabotage. Are we being attacked from inside the company or outside? By some ingenious hacker dialing in or someone with on-site access?"

"Based on what I've seen so far, I can't tell you whether the saboteur was remote or local. No one would log in under his own ID, commit sabotage, and then log off without cleaning up after himself, whether he came in via modem line or via the main system console."

"What's your hunch?" Jane asked. "Local or remote?"

"Perhaps both," he said. "You've definitely been attacked from without. As to whether some of the viruses have also been introduced locally, I'd need to do further investigation before venturing an opinion. This is a clever person."

"And someone who obviously knows a great deal about our business and how it operates."

"There could be more than one person involved."

"Great." Jane sighed.

"Either way, please understand that a knowledge of your business doesn't necessarily indicate an inside job. You've heard of 'phone phreaks' taking over a company's digital central office switchboard by dialing in from their computers? These naughty chaps often know more about the manipulation of telephone company hardware and software than many communication engineers."

Suddenly Jane understood that R. K. Singh was not going to give her more than generalities, unless . . .

She leaned forward. "What do you need? To take this on full-time."

Singh did not mask his lack of surprise at her question. He steepled his fingers. "I have a client in Texas I am supposed to call upon tomorrow. I would have to postpone, or cancel. I have other work pending that would have to be delegated to assistants."

"What would it take to clear the decks?" Jane asked. "And how soon could you do it?"

"I might be able to arrange everything tonight. I'll need to fly in my main assistant from Berkeley, along with an investigative tool kit—a special portable computer, certain diagnostic programs I've written, and so on."

"How much?"

Singh did not look taken aback by the directness of her question. Indeed, he seemed entirely prepared for it—and with an answer. He quoted a per diem—one with performance incentives that would pay him more the faster he could solve their problems and be gone.

His expertise came high. But going on her own instincts and

Mark Gibian's strong recommendation, having Rikki Singh aboard seemed essential.

"Agreed," she said. "Now, do you have any definite ideas about whom you want to work with here—your primary contact?"

"It's best that I work with everyone," Singh said. "And trust no one." He offered a small smile no one would mistake for deferential. "With the exception of yourself, of course."

"Don't rule me out," Jane said, returning his smile with eighteen percent interest. "Hercule Poirot wouldn't."

"I'm afraid I don't know him."

"You'll find him in Agatha Christie mysteries. What do you think of Mr. Grooms, the security man? He wishes to install his own surveillance procedures in connection with the sabotage. Do you think that would be useful?"

Singh shrugged. "Two weeks ago, a hidden camera might have been useful in the machine room, monitoring the console and various workstations. We might have caught someone in the act. Really, I know of such cases. But now, especially with your system taken down, I suspect it is too late."

"Right." She stood and offered him a hand, which he shook vigorously. "Welcome to the team, Mr. Singh," she said.

"Now here is a key to a condo in the Akers complex. I'll have a rental car arranged for you."

"Thank you. But, alas, I do not drive. I would appreciate having a bicycle. The repetitive motion stimulates my thinking."

"A bicycle will be outside your condo door in the morning."

Singh nodded and headed for the door.

Jane hefted her stack of phone messages. Each one had to be taken care of before she left the office; it was one of her personal rules, begun the day she'd opened her own office.

"Ms. Malcolm?"

Jane looked up. Singh was turned at the door.

"Yes? Did you forget something, Mr. Singh?"

"I have an excellent memory, thank you, Ms. Malcolm. It

occurred to me that as you embark on your evening's work, you might like to know something."

About a thousand things. She nodded encouragingly.

"I am not much of a gambler, but occasionally . . . " He lifted his hands in a gesture indicating his participation in the weaknesses of man. "After my tour, I would have been reluctant to bet on your success, and would have done so only if I were given two-to-one odds. However, as a result of our conversation, I would feel it incumbent to revise my wager." He paused. "I'd accept even money." He bowed slightly. "Good evening, Ms. Malcolm."

9:06 P.M.

Jane was on her knees, sorting through wastebasket debris for a scribbled phone number she hadn't meant to discard. There were three sharp knocks against her open door. She jerked her head up. Alex stood in the doorway, grinning.

"Need help?"

"It's hopeless," she said, scrambling up. She hit her head with her hand. "Stupid! I bet Don knows that number—my brain must've shut down about half an hour ago."

"Whatever problem you've just solved, can we come in anyway?"

"Sure, c'mon in," she said, toeing scrap paper out of sight under her desk. "But who's 'we'?"

Anne Akers appeared in the doorway behind Alex, a tote bag over her shoulder, a tentative smile on her face. "Me," she said.

"Why, hello, Anne. What a surprise." Jane pushed her tousled hair back from her face.

"I hope you don't mind us popping in like this," Anne said.

Alex patted his stepmother on the shoulder. Jane didn't miss either the familiarity or the obvious affection in the gesture. "You were right about her still being here," he said.

Anne shrugged. "An old family habit when there's work to be done."

"Maybe it *is* in the genes," Jane said. "Anyway, I may not be due for a break, but I can sure use one."

Anne opened her tote. "We brought you some tea and a lemon bread I baked. You're not one of those young women who diet all the time, are you?"

"Nope. Especially not when I end up skipping more meals than I eat. Tea and lemon bread sound wonderful. Please, both of you, have a seat." She laughed. The room's third chair was piled high with folders. She picked up a batch, looked around for a place to stash them.

Alex took the pile in her hands and put them back on the chair. "I'm fine here." He perched on a corner of Jane's desk.

"Maybe we shouldn't stay, just leave these for you," Anne said, reaching into the tote. She produced foil-wrapped slices of dessert bread, which she handed round, and a tall, two-quart, dented gunmetal cylinder. "We saw Don Landsman outside. He told us it's pure bedlam here tonight. We obviously interrupted you."

"All true, but stay. Please? I really do need a break, and I couldn't imagine a pleasanter one."

"Well, then, we will," Anne said. She unscrewed the stainless cap of the thermos and poured a steaming liquid into three hot cups she'd brought.

"That's Dad's old thermos!" Jane said. She hadn't laid eyes on it in years and years, but seeing it brought back those rare times when her father had taken her along on one of his fishing trips, leaving before dawn, the hot metal cup passing between them in the front seat, watching for the sunrise.

"Spearmint," she said, inhaling the tea. "How did you know?"

"I remembered," Alex said. "The truth is, Anne made it for me, and I hate the stuff. So we thought, Hey, let's go bug Jane! And here we are."

The phone on Jane's desk rang. "Don't move!" she said.

Khalif sounded tense. "Bad news. More problems with those

fax-and-scan forms. We have to redesign them. Fast."

"Can you handle it, Khalif?"

A beat. "Not without your input, Jane."

"Okay. Give me a minute, though."

Alex and Anne were already on their feet. "It's okay, dear," Anne was saying. "Do whatever it is you have to do. We both have an idea of the pressures you're up against. I do want to say one thing before we leave. Thank you."

Jane raised an eyebrow.

"For what you're trying to do," Anne said. "It means a lot to me. It would—"

"I hope so, and thank *you*," Jane said quickly. Then, "If you don't mind, I'll share the treat you brought with my associate. He's Saudi but adores American sweets."

"Well, when—if—you have time for a genuine break, you know where we are," Anne said with a good-bye wave. "The door is always open."

"I'll try, I really will. Wait, you forgot Daddy's thermos."

"No, I didn't." Anne smiled and was gone.

But Alex still stood in the doorway. "Janie?"

"Yes?"

"I told you before, I'd like to help. And I mean with more than foraging through wastebaskets—though I'm up for that, too."

Jane searched his face. Not a twitch of a tease in sight. "Thanks, Alex. I'll think of something. Promise."

JANE'S CONDO
11:58 P.M.

Jane yanked off her running shoes and went to fill the tub.

She poured in a capful of bubble bath she'd bought on impulse this morning during her quick daily check of customer attitudes at Store Number One and pulled off her clothes. If the weather got any warmer, she'd have to pick up some lighter-weight clothes. Given how few things she had with her, she'd probably have to do that anyway—God only knew when.

She stepped into the warm tub and eased her aching body down until only her head and knees were out of the water. She pushed large mounds of bubbles about and began to laugh. She flattened a few peaks and laughed some more. Then she settled into the warmth and let the waters of remembrance bathe her, too. Anne's gift of the old thermos and her thank-you had touched Jane, and Alex really did seem serious about lending a hand. Then she remembered what Rikki Singh had said on his way out of her office earlier in the evening. He had meant it as a compliment. From his expression, a sizable one.

Even odds!

Jane sat up straight in the tub.

I'll bet you, Mr. Singh, we improve those odds to two to one in our favor by the end of the day Friday.

Janie?

Okay, Dad. Saturday.

Now they were laughing together. The way they used to when she was little and he would sometimes stop by during her nighttime bath and make himself a mustache and beard out of bubbles. He looked so funny, she'd laugh till her tummy hurt.

So why was she crying?

DAY 4

OCEAN MEADOWS CONDOMINIUMS, OCEAN PLAINS
7:35 A.M.

"Sounds like it's going really well, Mother."

"Janie, I've nearly sold out! Did I tell you what Prentiss Jones said in the *Times*?"

"I take it he was complimentary?"

"He was downright gushy! It's full of the usual gobbledygook, of course, but I have to read it you anyway. Hold on, it's here somewhere . . . Okay, ready? 'Joycelyn Malcolm's current series of Andean tribal motif acrylics (at the Mid-Madison Gallery) does not disappoint. A cautionary note: The faint of eye may find the artist's neoprimitive, primary-clashing impastos approaching the optical pain threshold. And, truth be told, her works *do* tend to overwhelm in the wide-angled aggregate. But contemplate a single painting—the *Cordillera Negra at Sunrise,* say—against whitewashed gallery walls, and you're vortexed into untrammeled exoticism.' Now, Janie, don't ask me what it means. But it's a rave—tons better than my own self-promotion. I'm having it reprinted—it'll do wonders with collectors and curators and dealers."

"I'm sorry I couldn't have been there, Mother. Things have been pretty hectic here."

"So I gathered from Roy Jr."

"He called again?"

"I told him it wasn't necessary. I trust his judgment."

"Mother! You promised! You can't sign any proxies yet. We're making progress."

"Janie, I haven't signed anything. But I have to tell you, Roy makes a very good case for selling now, not next week. He's afraid our window of opportunity will be closed. My broker agrees, by the way. I mean, are you aware how much money we could all lose?"

"Yes, Mother, I'm aware. You *won't* lose."

"Well, I don't know how you can be so sure, Janie. And I really don't understand what possessed you to get involved in . . . *mass merchandising* . . . after all these years. I'll feel much better when we're all well out of it. Some very bright people have been advising me for years to sell my Akers stock, and I would have, if Roy hadn't made such a fuss about losing control."

"Mother, listen to me. I'll make ten times more of a fuss than Roy ever did if you back out on your promise to me. I know Roy can be persuasive, but this time he's just plain wrong. Please don't cave in. If you think you're about to, call me first. I need you to promise me, Mother."

"Of course, dear. When have I ever failed you?"

How about when you kept Daddy's letters from me?

But, of course, this wasn't the time—for several reasons. Not first but also not last was that Jane desperately needed Joycelyn's cooperation—and her proxy.

"I'm counting on you, Mother."

Like I did when I was younger. Even if I couldn't count on you for much in the way of hugs and kisses and affirmation, I could always rely on your honesty. You never held back the truth when I got a bad haircut or decided to try purple lipstick. Who dreamt you'd lie by omission—a terrible omission the size of the chasm that grew between me and Dad?

Jane took a serrated breath. The hurt, the anger, over what her mother had done all those years ago would have to come out . . . sometime.

"Of course you are, dear," her mother was saying.

"And Mother, please don't worry about me—I'm almost enjoying myself. Mass merchandising is not some kind of contemptible, boorish activity. And it so happens that what I do, logistics, makes the real world *work*. Most people can't shop designer boutiques and phone for gourmet takeout."

"There's no need for that tone, dear."

"Sorry. Guess I could have used fifteen minutes more sleep."

"I forgive you. Just don't go totally midwestern on me. I couldn't bear it."

"Glad about your show. Now, I've got to get mine on the road. Bye."

The amazing thing, Jane thought, was not that her mother and father had split after twenty years of marriage, but that they had gotten together in the first place—let alone stayed together so long. Joycelyn had been eighteen, Royal twenty-one, during the summer of 1960 when they'd met at a lakeside dance in Mecosta County, Michigan, and discovered they were both camp counselors, working opposite ends of the lake. "Your father was as handsome as the Arrow Shirt man," Joycelyn had told Jane more than once when she was little, as though that explained everything.

Later, after they'd left Ocean Plains, she'd said, "He talked about a career in politics, maybe getting a law degree. He never mentioned retailing." The way she'd pronounced "retailing," you'd have been tempted to check the Ten Commandments to find if it was on the forbidden list.

"JUST BILL'S," OCEAN PLAINS 8:15 A.M.

The diner was a block behind Main. Breakfast and lunch specials were hand lettered on the plate glass. Jane's entry rattled sleigh bells. A hefty young waitress at the cash register tossed Jane a "come on in" smile. The short counter was empty; three men in overalls and farm caps occupied the first booth in silence. Maybe they were listening to the lively chatter from the five women squeezed into the middle booth. In the last booth a lone man, facing the door, slowly lifted his hand elbow height toward Jane, like a less-than-enthusiastic schoolboy.

The waitress arrived the same moment Jane did, bringing scrambled eggs, hash browns, and toast. Mr. Skoggs already had coffee and the *Tribune* sports page in front of him. He got up slowly, tugging at the wrinkles in his seersucker coat. He had to be seventy-five, Jane thought, but his back was straight, his white hair thick, his brown eyes alert under shaggy brows. He motioned her to join him.

"Just coffee," Jane told the waitress.

"Hope they never change the name to that," Sid Skoggs said. "Best hash browns in four counties. Sure you're gonna abstain?"

"Maybe next time. Thank you for agreeing to talk to me, Mr. Skoggs. I'll only take a few minutes of your time."

"No hurry. I'm here every morning, maybe Doreen told you. Bill's is my office."

"I remember it as the Snack Attack. We used to hang out here after school."

"It's gone through half a dozen names since then. Closed down for a year before Bill Myers bought it and saved it. I chipped in a little myself. It barely breaks even, but the town needs a place like this."

Jane nodded, absorbing the point. Nowadays kids hung out at the Akers fast-food stalls, and their mothers had their second cup of morning coffee there. Many families parked their shopping carts long enough to grab lunch or supper. But Mr. Skoggs obviously didn't consider the giant emporium part of *his* town.

"I have three things I'd like to tell you, Mr. Skoggs."

"You're very organized," he said.

"I guess I am. Sorry."

"It's not a fault. Go on."

"Well, first I want to thank you for hiring me, all those years ago, and letting me paint pumpkins and goblins all over your window glass. It meant a lot."

"You're most welcome, but I ought to have checked with your dad. He didn't like it. Maybe made him even tougher on me."

"I never thought of that."

"Not worth a whole lot of thought. Wouldn'ta mattered either way. I didn't have the determination, or the stomach, for a thirty years' war."

"That's the second thing, Mr. Skoggs. I want to say that I'm sorry. For what happened—to you and all the other Ocean Plains merchants I remember."

"Buggywhip makers, all of us," he said, finishing a forkful of hash browns. "But I appreciate the thought, Jane. Like a lot of old folks, I suppose, I spend too many hours living in the past, missing things long gone . . . never coming back. The past wasn't perfect, like somebody said. Go-getters in every age come along, eager to tear down the old and build everything brand new. But when I look around . . . " Skoggs gestured out the window at the abandoned storefronts and shook his head heavily. "You'd think I'd get over it, but it's kinda like grieving: doesn't end just 'cause time's up. But I'm not bitter. Not anymore." His eyes were a plain medium brown, but now they seemed flecked with gold, kind of like there were little lights inside blinking on and off. "No, not even at your dad."

"But Doreen said—"

"That I don't like his name mentioned in the house, and that I don't shop there? All true." Skoggs bit off some toast, chewed it attentively with teeth he must have considered over priced, swallowed. "One or two items, we shop down at the gas station minimart, pay through the nose. Once a week we take our list over to the 4-Mor in Red Knoll. All that shows is I'm stubborn and stupid, not bitter."

"Well, if some people have their way, you may not have to leave town to shop at 4-Mor," Jane said. "There's a lot of talk we may sell out." She took a sip of coffee, aware that Mr. Skoggs was *not* saying how sorry he was to hear it. "I'm heading up a team that's trying to stop that from happening, trying to rescue the company."

"That bad?"

"We've got a week left to do it." She took another sip. "And I'd like to ask your advice."

"That sounds like maybe the third thing you came for. Unless it's actually number one."

"Maybe it is." Jane grinned and thought she saw the shadow of one around Mr. Skoggs's mouth.

"Now, what makes a smart woman like you think *I* could tell you anything that would help Akers? I'm not saying *would*— that's another question—but *could*. Remember, you're sitting across from the fuddy-duddy who got left in the dust a dozen years back."

"Mr. Skoggs, what I remember is how you used to greet your customers by name. How welcome you used to make me feel when I walked in—when you had every reason not to. Dad used to copy you, you know? Giveaways, games for kids, those silly costumes you and your clerks would wear."

"I'd as soon *not* remember."

"The point is, we desperately need to reach out to our customers. You've heard about the shortages?"

"'Course I've heard. Lately, I been seeing a lot of locals over at the 4-Mor in Red Knoll. They tell me you've been out of all kinds of stuff they need, from rat poison to baby food. Is it true all those fancy computers of yours have gone on the fritz?"

"It's true. Right now we're doing everything we can to bypass them and get the empty shelves filled in, all around the world. We will, too. But I've been thinking a lot about you, Mr. Skoggs, the way you always interacted with your customers, and I was hoping that maybe you could think of some way we could let our customers know what we're doing to try to make things better. I've got Doreen—oh, shoot—"

"It's okay, I know where she works."

"Well, then, I've got her working on a series of big newspaper ads, to tell our customers how we're fighting back to solve our problems and earn their loyalty again, but I have a feeling they—"

"What about the radio?"

"I suppose we could do that, too. Make a few changes to the copy, put it on radio stations in Akers towns."

"Not what I meant. Why not put a daily bulletin on the air—'course you'll have to buy the time. Report what's in store and what isn't. For anything missing that day, you tell folks the nearest place they can get it, especially critical items—food, medicines, so forth. Make it like a public service announcement, the way they broadcast school closings when there's heavy snow."

Jane thought the idea was charming but impractical. Then she realized that their crisis plan charged each Akers store manager with determining this precise information— the most critical stock-outs—on a nightly basis. So they'd *have* the list. It wouldn't take much extra work for the managers to tape a thirty-second spot to air on the local station the following morning—maybe even air it every fifteen minutes from sunup to door opening. She'd have to check her budget. Every penny she allocated was being watched, she knew.

Hell, running spots like that could well be genuinely helpful all around—she'd find a way to fit them in. Her father always said there was no better way to generate customer loyalty than to put yourself in the customer's shoes.

"You know what, Mr. Skoggs? That's exactly the kind of thing I was searching for."

"No charge," he said, spooning more jam onto his toast.

"Thank you, thank you very much, Mr. Skoggs." Jane stood. "If you'll excuse my not finishing my coffee, I want to get on it right away."

Skoggs took a bite of his toast, looked up at her. "Got to admit one thing, young lady. I've only met one other person as gung ho as you." Mr. Skoggs emitted the driest chuckle Jane had ever heard. "And he got what he was after."

WAR ROOM,
AKERS HQ
9:08 A.M.

Don Landsman swore, accidentally knocked over his Styrofoam cup, spilling coffee across a yellow legal pad filled with scribbling, and swore even louder.

He'd just hung up from a thoroughly miserable conversation with the vice president of sales at GroMor, the outfit that supplied a wide range of products to Akers Garden Centers—seeds and fertilizers, pesticides and houseplants. Or *had* supplied these items—until this phone call.

The new VP was stopping all sales to Akers. It was his first act in his new job, he said, since his predecessor had been reassigned for basically doing nothing about GroMor's supply-chain problems with Akers during the last three weeks.

And GroMor's problems were pretty impressive, though far from unique. Because of high, and totally false, replenishment signals fed into their computers, they had ratcheted up their production nationwide, investing in more material, more staffing, shifting garden chemicals into areas where Akers's demand had suddenly skyrocketed.

When the pipeline suddenly backed up, GroMor found itself with a horrendous inventory of unwanted product. They were, as the executive had graphically put it, "up to their rafters in steer shit," with nowhere to move it.

A third-party logistics company had been hired to help sort out the mess, and GroMor was selling those garden chemicals at a huge discount to all of Akers's competitors—4-Mor buying the biggest share.

Don Landsman had tried every inducement he could think of to persuade the new VP to give them a second chance. "Tell you what I'll do," the GroMor man said finally. "I read the *Journal.* If they tell me you folks are healthy, then we'll talk. Until then, I'm not going to risk one more dime with you."

Five minutes after he had hung up, Don was still frustrated, still trying to think of some way to salvage the situation. In their first few hours under the new War Room setup, his procurement specialists had done pretty well overall in soothing and satisfying first-tier suppliers. They were prepared to hand-hold on an hourly basis, if necessary. But as the loss of GroMor proved, promises couldn't substitute for performance.

Don told himself to swallow his anger, even if it went down like bones, because it wasn't getting him anywhere. A minute later he had his procurement specialists dialing and dealing with alternate garden products suppliers and paying a premium due to all the negative press. As soon as that was up and running, he contacted an intermediary he trusted and instructed him to start bidding on GroMor's bargain-basement backlog. Don knew he had to move swiftly on this. What with the planting season in full swing, and GroMor cutting off supplies, the wildly inaccurate replenishment numbers of the past three weeks might very soon become accurate—while Akers's fifteen hundred garden centers would have nothing to sell. Which would mean a whole lot of their customers would have to go elsewhere for supplies—or have nothing to grow.

"And GroMor isn't an isolated case," he explained to Jane when she walked in. "I can think of at least four executive casualties directly traceable to our screwups. I'm talking about some of our biggest vendors. And we've got other key people in real trouble—guys and gals hanging on by their fingernails, taking the heat for production cycles we've screwed up, for overstocked warehouses or expensive, last-minute interfacility transfers. People at this very moment staking their whole careers on my personal promise to deliver. Trust is a heavy load, Jane. Their trust gives me the chance to deliver. But what if I can't? What've I done?"

Jane rested a hand on his shoulder. "Will it make it better or worse if I tell you I think their trust is well placed?"

He looked up at her, considering. "A mite better, thanks."

"Good. Got to admit," she said, "I came by hoping for some *good* news."

"Well, I guess we got some of that, too," he said. "We're still hemorrhaging, but at least our heart is beating again. Will that hold you?"

"Bet I can go sixty full minutes on it."

AKERS STORE #28, LAKE RELIANCE, WISCONSIN 12:12 P.M.

Phil Leavitt did a wondrous thing. He went home for lunch—walked, in fact, taking a leisurely ten minutes to do it, hardly able to believe himself. En route he daydreamed about taking a short nap after lunch. God knew he deserved it. He was exhausted from working double shifts four out of the last five days. But battle fatigue wasn't all Phil was feeling; the less familiar feeling was a sort of tentative optimism.

The emergency plan seemed to be working.

Wade's plan, as Phil thought of it—though he could imagine any number of headquarters types claiming credit if the nationwide implementation proved successful—was definitely having good results here in southeastern Wisconsin. His store was a prime example. Of course, again thanks to Wade, they'd had a head start on the rest of the field.

In many ways basic operations had not changed. Checkout lanes were moving at a normal rate. Bar-code scanners were still plugged in, capturing item numbers and prices. The registers and receipt printers were working as usual, as were the manual-swipe magnetic card readers and credit verification systems. Finally—thank God!—the bar-code scans were still able to update an uncorrupted accounting system.

Unplugged and therefore unavailable, of course, were all the invaluable point-of-sale data downstream—inventory activity and stock alerts, department and vendor statistics, hot-and-cold sellers, hundreds of other information printouts.

Which meant that Phil, and his people, instead of relying on a perpetual electronic inventory, had to take physical inventory at the end of each day, just like a decade or so ago, and hand calculate the quantities needed to restore normal stocks, adjusting for any variable demand factors.

And instead of reorders being satellite dished direct to the Ocean Plains computer—and then sent on by EDI to vendor computers—they were being filled out by hand on special forms, then faxed to the regional distribution center to be scanned into a PC database. In Phil's case, last night he had done the faxing himself, to the RDC in Milwaukee, where Pat DePalma was filling in for Wade Crain. And this morning, several hours before store opening, that shipment had arrived on the Lake Reliance receiving dock. Just before the doors opened at eight A.M., Phil Leavitt had walked his store and found most of his high-volume SKUs at or near optimal levels. The daily just-in-time replenishment cycle had been achieved—never mind the extra sweat.

Oh, there were some stock-outs, but no vital ones. Some shelf displays were scanty. The worst cases were where first-tier suppliers had been so bollixed up with bogus reorders, they had depleted or lost track of inventory items and couldn't respond. As for the few suppliers who had actually stopped doing business with Akers, Ocean Plains was being pretty successful at cutting deals with alternate vendors.

Basically, Phil thought, they had gotten the fallen giant back on its feet and plodding forward, though pretty damn slowly. When Phil had checked out for lunch, his assistant manager, Dave Daley, was on the phone with Wade in Ocean Plains, going over the problem product lines.

Phil turned into his own front yard. Sheba, their orange

tabby, stirred from her chaise longue nap to note his coming. Sunny met him just inside the screen door, looking more distressed than pleased. He could smell soup.

"Oh, Phil."

Not the enthusiastic greeting he was expecting.

Oh, God. Dave must have called. There's a crisis at the store. I knew it was too good to last!

"You forgot to bring home a quart of milk from the store," Sunny was saying. "I asked you last night, remember?"

Phil couldn't help smiling. "Sorry, Sunny. Put it down to supply-chain foul-up."

AKERS REGIONAL DISTRIBUTION CENTER
1:41 P.M.

For Patrick DePalma in the Milwaukee DC, lunch was still a sandwich-on-the-desktop affair, with bites grabbed between phone calls. He had just given an update to Wade Crain, mostly favorable.

By midnight Henry Torres had scanned reorder faxes from the dozen stores in their region. These reorders were then combined, broken into product lines and/or supplier batches. So far so good. It was during the next steps that complications set in, where they really missed the systems.

The AS/400, using downloads from the mainframe in Ocean Plains, could instantly review all reorders against warehouse inventory before generating billing documents, then automatically deplete that inventory by the number of items billed, and, finally, transmit replenishment orders to vendors in order to bring the DC stock back to acceptable levels.

All those electronic processes now had to be hand hammered, including keeping tabs on the current inventory. Like computer room staff at all Akers distribution centers, Henry Torres had been conferring yesterday and today with Khalif al-Marzouki in Ocean Plains to transfer those operations to

PC databases, along with basic warehouse management programs. But they just weren't there yet. So Henry, with continuous input from Pat and John Brinkema, had been guesstimating inventory levels, then calculating replenishment numbers for faxing on to suppliers, with copies to the Akers Accounting Department.

Despite the retro tech methodology, the operations were essentially the same. As before, each supplier was allocated a precise time and loading dock number for his goods to reach the DC. Because the Milwaukee region had started the emergency system a day ahead of everybody else, Pat was able to tell Wade that the goods reordered the previous day had arrived in those specified delivery windows—with only minor variances. Inbound trailers had been directed to the dock doors closest to put-away locations in the towering steel shelves. Meanwhile, all night long other pallet loads had been rolling from those racks to areas on the warehouse floor assigned to each store, then consolidated into truckloads. In the hours before dawn, trucks had rolled out from the DC to each of the dozen satellite stores in Wade's area, to replenish stock before doors opened.

"The pipeline's flowing again," Pat assured his boss. "The thing is, I just don't know how long we can keep it up. We're all working like sons a bitches."

"Ain't that the truth," Wade said, chuckling. He thought a moment. Wouldn't do to have his people so overworked they came down sick. "Would it make a difference if you had some more college flextimers?"

"It'd help, sure."

"Then find 'em and hire 'em."

WAR ROOM, AKERS HQ 1:47 P.M.

Wade had a grin on his face as he hung up the phone. He

couldn't wait to tell Don Landsman how his old region, just one jump ahead of the others, was back on track. Don thought the good news warranted sharing with Jane, so they tracked her down to her office, where she was finishing a call.

"Both of you smiling at once, must be terrific news."

Don naturally deferred to Wade, who began his report. Jane beamed but cut it short.

"That *is* terrific. I was actually about to come find the two of you. You'll need your jackets."

"What for?"

"We're wanted in that meeting with the Hyon execs. Roy thinks they may pull their whole line and sue us for millions in damages. He wants us to help calm 'em down, convince them we've got our problems pretty much behind us."

"Have we?" Don asked.

CORPORATE CONFERENCE ROOM, AKERS HQ
2:00 P.M.

The three dark-suited men from Hyon Group offered only stiff handshakes—unaccompanied by even perfunctory smiles—before taking their chairs. Jane had been in some extremely tense meetings, especially during the first deployments of Desert Shield. She would never forget the SITREPS, daily situation reports, conducted against the paralyzing prospect of a hundred thousand Iraqi soldiers just over the horizon. Yet even those meetings did not begin in an atmosphere more strained than this.

She was not surprised.

"These Hyon guys have a lot to worry about back home," she'd said to Don and Wade on their quick walk to the meeting. "Unless you two have seen different articles from the ones I've been reading."

"I *used* to read the paper," Wade had said. "Haven't had time to even glance at it the last few days."

"I've seen those pieces." Don shook his head. "Although you doubtless know more about how the area's economics and politics interact than me."

"I've got to admit, working with TranSonic Airex—that's been a cram course. The way we do things in this country is pretty damn straightforward compared to Asia. There, everything and everyone seems interdependent."

Jane took very seriously the need to keep a close watch on Asia, where any and every change had unpredictable ripples. Korea's export-driven economy had slipped seriously from its peak growth rates of the mid-1990s, as she'd reminded Don and Wade, and was now running record trade deficits. Trade union demonstrators had taken to the streets of Seoul again, protesting government-industry antilabor collusion. And these days, few references to Hyon Group in the Korean press failed to mention the corruption and bribery charges leveled against its former chairman.

"In other words," Wade had said as they'd approached the conference room, "these troubles we're causing them couldn't have come at a worse time."

"Exactly," Jane had said.

As soon as the meeting got under way, the delegation's silver-haired, senior man, Lee Sung Whan, lost no time in confirming Jane's point. "Due to our special relationship with Akers," he announced, "all of Hyon Group is now in peril. With your permission, I will take a few minutes to explain exactly what I mean. Mr. Akers, Mr. Conover, you have apologized for your computer mistakes. I expect you will do so again. But in Korea these mistakes are considered *our* mistakes. We cannot point to our American trading partners and complain of your computers transmitting incorrect numbers. It is our business judgment now on trial. There will be resignations required."

"We sincerely hope not," Wallace Conover said gravely.

Lee shook his head. "Our judgment *was* at fault. We placed too much reliance on Akers, resulting in far too much fiscal exposure."

Until the last couple of weeks, Jane knew, Hyon had been happy with its "special relationship" as primary electronics supplier to the Akers chain. And with good reason. The relationship had given them access to massive sales. In exchange, Hyon had agreed to extremely slim margins on all their products. This was standard Akers policy—giving their purchasing agents maximum bargaining leverage by picking a single firm in each main retail category. Given their pared-down profit margins, it was no surprise that the Korean firm was absorbing heavy losses as a result of scrambled orders and reorders.

"The gravity of what's happened is plain to all of us here," Roy said. "Our regret is profound. But our meeting together has to go beyond charges and apologies. It must be focused on the realities of your situation and ours. We're here to examine those together, and see what we can all do to improve matters. Now, about the computer monitors, we can assure you that—"

"Excuse me, Mr. Akers," said Lee. "But this last shipload of monitors is only one critical issue between us." He asked his frail-looking, bespectacled colleague, Roh Jang Koo, to summarize several more. It was an alarming list.

In the past month, according to Roh, there had been a serious decline in Akers reorders throughout their product lines—PCs, TVs, VCRs, camcorders, washers, dryers, microwaves, cordless phones, stereos, and answering and fax machines. Different items were affected in different Akers regions, but the overall result worldwide was very serious.

As Roh began to dip into statistics, Lee Sung Whan cut him short. "During this period we made numerous inquiries, demanding to know *why* our numbers were down. We were given many explanations by our account executive in your Purchasing Department, Mr. John Flowers, whom I do not see at this meeting."

"He is no longer with the company," Wallace Conover said.

"Ah! Well, just two weeks ago, when he *was* with your company, this man spoke of reduced seasonal demand, economic turndown, and so forth. Only in the last week have we been informed that these reduced numbers were due to errors in your computer system, and that you urgently need *more* product from us—electronics, appliances, everything—to make up for what you should have ordered previously.

"But Hyon can fill only a fraction of these urgent reorders. Inventories are very low now in our U.S. warehouses. And, as you know, we have no inventory of finished goods in Korea. So Hyon must generate new production cycles. It will take our Mexican assembly plants at least a month to make and ship the color TVs. 'White goods' will require longer, because they must be manufactured in Korea. Those time frames begin, gentlemen, *after* we give the order to commence production. But we do not dare give this order. Let me tell you why.

"First, Hyon has no assurance that these new numbers are more correct than the previous ones. Second, as a result of reduced sales to Akers, our cash flow has been seriously reduced. This limits our ability to invest in new production. In fact, we are now being forced to close one of our manufacturing plants in Suwŏn, a step which will not only put several thousand men and women out of work, but may also cause trade union leaders to retaliate by striking *all* our plants. The hard truth is that we have no choice but to risk this. Finally . . ." He paused and, uncharacteristically, looked directly at Roy. "Hyon Group now has very serious concerns about the future of your company."

"Well, we clearly have a lot to answer here—"

Lee held up a palm, silencing Roy. "I have come *almost* to an end. We do have *some* product in the pipeline, although not as much as you say you need. Tomorrow morning, a container ship owned by our partners, Kimyung, will arrive at the Port of Los Angeles. This ship will unload many containers filled with Hyon product consigned to Akers. But we have no confidence that these will be the correct products in the correct numbers for your stores. We have no confidence in your purchasing systems, or your financial systems, which have held up payment on certain received orders. And we at Hyon cannot continue our relationship under such conditions." His expression made clear that in his opinion he had said the final word on the matter.

If the mood around the table had been somber when the meeting began, it was now positively funereal. Jane glanced at Roy to find he was looking at her expectantly. He had asked earlier if she felt capable of organizing the response to Hyon. The implication seemed to be that if she really thought she could save the company, here was a critical juncture. She had agreed with the premise—and to the challenge—with the sole proviso that Don Landsman play a role as well.

Jane nodded at Roy. Roy bent his head ever so slightly in acknowledgment, then made some appropriate noises—how vital it was for all concerned to keep in mind their mutual interests, how Akers had taken decisive steps to restore the integrity of its electronic systems, how important it was to reach consensus and avoid a breach between their companies.

Unstated but understood by both sides, Jane knew, was the fact that competing electronics firms—several in Korea, more in Japan and Taiwan—would be eager to fill the Akers electronic supply channel. If, that is, Akers had a future.

Suddenly—it felt sudden—Roy reintroduced Jane, and every head in the room swiveled toward her.

In any other circumstance, Jane would have smiled.

So she did. Modestly, briefly, and—*God help me*—confidently at each of the three Koreans in turn. Then she said, "Since my brother has already expressed our company's—our family's—regrets, I hope you will forgive me if I go directly to specifics. First, the display monitors. Akers will absorb all the Hyon product that has been shipped in error or caused to be mistakenly manufactured."

It was a commitment that was certainly owed, yet one that would entail little financial risk to Akers, or so Jane had been advised. During a discussion with Don Landsman of potential short-term solutions to the Hyon problem, Jane had speculated that a massive loss-leader promotion could blow those excess electronics out the front doors. "Not necessary," Don had assured her. "Other retail chains or warehouse clubs will be tickled to buy those high-res monitors off our hands." At the very worst, Akers could always move them—at cost—to discount electronic warehouses, the Madman Marvins and Crazy Archies that bought up liquidated stock and returned merchandise.

She did not blink at the open skepticism in the Korean faces. "The funds for these monitors were released by our bankers in Seoul as of today," she said.

She paused only a moment to let that sink in before she continued. "Now let me address some of your other critical issues, all of which are linked directly to our computer problems."

She explained their efforts to quarantine their EDI systems from further damage by computer viruses. "Until we have achieved that," she said, "we are operating on backup technology. For instance, we are handling order entry by use of faxes and scanners."

"Scanners are notorious for making errors," interrupted Roh Jang Koo, and continued to polish his spectacles.

"Absolutely correct, Mr. Roh," Jane said. "Therefore, to make the process as fail-safe as possible, we've incorporated numerous order-accuracy checks. These somewhat slower procedures require every transaction to be monitored closely. This close scrutiny, in turn, assures that we will notice any unusual transactional activity, such as the recent problems you have experienced with erratic reorders—sudden decline followed by spiked demand."

Jane looked from Roh to his fellow countrymen, without either the haste of nervousness or the stare of arrogance. "All of us at Akers," she said, "hope that these steps will go a long way toward restoring your confidence in our current and future orders."

"Don, do *you* have anything to add?" Roy asked, his expression unreadable, except that delight wasn't among the possibilities.

"One thing," Don said, managing to address all three men personally at once. "I would like to assure you that, from now on, you will be dealing only at the highest level of our company, no matter what the issue. For the time being, that will be with Ms. Malcolm and myself. As soon as we have your go-ahead, we will be setting up a task team to investigate all Hyon order irregularities and to monitor the flow of Hyon goods through our supply chain."

Lee Sung Whan nodded, but without expression. Roh Jang Koo stared at his sheet of statistical data. The third Korean, a young man introduced only as Mr. Choi, had yet to utter a word.

"These are steps in the right direction," Lee said. "However, we will need to see them detailed in a memorandum, so we can discuss them among ourselves and with our colleagues in Korea."

"You'll have it by the end of the day," Jane said.

"And we will require other assurances. You have mentioned emergency procedures. I think we can all agree that they are very much justified. But you do not tell us when you intend to return to full electronic commerce. I assume that is your intention?"

"Of course," Jane said.

"When?"

"We can't give you a definite date," Don said. "We will resume full EDI as quickly as we can, but not before we are absolutely certain there will be no recurrence of our problems."

"Ah, yes, recurrence. Recurrence, Mr. Landsman, is a very big issue so far as we are concerned as well. Can you tell us exactly what you are doing to prevent this?"

Don glanced at Jane. He had suggested how they might handle this question if it came up, and she had expressed reservations. But hell, why not? After all, the three Koreans were probably the only humans on the premises today above suspicion.

"I suggest we take a brief tour of our data-processing center, and you can see for yourselves what we are doing to make sure this never happens again."

DATA-PROCESSING UNIT, AKERS HQ 3:41 P.M.

Jane funneled the Hyon delegation into Chip Bragan's designer-appointed office. Letting them see R. K. Singh's workshop was out of the question—but only because it was such an unholy mess. The security specialist and his ponytailed graduate assistant, Ron Hagstrum, had managed to turn their assigned work quarters into what looked more like an electronic salvage shop than a computer lab. CPUs and disk drives with exposed circuitry and cabling occupied tables and workbenches, even the floor. Atop this high-tech jumble were various keyboards and laptops and blinking monitors. Jane left the Koreans in Chip's hands for the moment it took to slip inside and ask Rikki to lock up and join the meeting in Bragan's office.

She was glad she had. Singh's credentials obviously were not lost on the men from Hyon, and his quiet manner seemed to reassure them. So far, they told him, their own software sleuths had been unable to find any viruses in their own systems or any logic bombs they might have acquired from Akers.

"Nor will they find any of these things," Singh said cheerfully. Hyon had gotten infected with some very bad numbers, he went on, which would have to be rectified in their databases. But their own programs and database structures were quite unaffected. As for logic bombs, Singh said he had now verified that their EDI link had moved no destructive code of any kind.

"Excuse me, Mr. Singh," said Mr. Roh Jang Koo, who had previously questioned the reliability of scanned documents. "How can you be so sure that the infected data which passed into our computers are themselves not contagious?"

"Because of a test I myself conducted," Mr. Singh said. Overnight he and his assistant had taken a version of the company's corrupted operating system and file system to a duplicate site in Chicago arranged by their hardware vendors. There they had reproduced a miniature of the Akers network configuration, with EDI bridges to various platforms, including the type used by Hyon Group. They had run the test system, activating every strain of virus so far isolated and generating sample replenishment orders.

Using an experimental laptop of his own design, Singh had then been able to monitor what was happening in real time, as well as to capture it for later analysis. It was, he said, like observing a controlled riot in a rat maze. He could see exactly how the virus attached and concealed itself, how it mutated and traveled, how it affected executables.

"And you are writing a vaccine?" asked Roh.

"Not actually a vaccine," Singh said, smiling at the good question. "More of a countervirus. Or a counterworm, perhaps. I am writing a search-and-destroy program, which will travel, on its own, throughout our system and linked network, sniffing out all varieties of this nasty little code and killing them."

"Even dormant code?" Roh asked.

"Yes. My program has a very good nose. But it will not affect anything else."

"Ah," said Lee Sung Whan. "But you will only be able to activate your search-and-destroy program when you go live again with EDI, isn't that right?"

"That is correct. I will set it loose throughout the network for several days, to be sure."

"So first you must track down the source of these viruses, to make sure there will be no reinfection?"

"Yes, Mr. Lee, that is also correct."

"And have you made any progress on that?"

"Not yet," said Mr. Singh in his sunny way. "But I am very confident."

AKERS HQ
4:11 P.M.

"We will walk," said Lee Sung Whan when the Akers minivan arrived to take the Hyon delegation to the adjoining airfield.

Jane had to quicken her own stride to keep pace with the Koreans, but it felt good to stretch her legs; she'd been getting precious little exercise. Roy chugged along a few steps behind. Wade and Don had gone back to the War Room, while Wally Conover said his good-byes at the general offices.

Despite the ten o'clock meeting scheduled for the following morning, the Koreans had also decided not to stay at the Akers company condos. Instead, they announced, they would commute in their business jet to and from O'Hare, overnighting at the airport Hyatt Regency. Could they possibly think the condos were bugged? A twinge of anger shot through Jane.

As they approached the main gate, Jane saw Alex standing beside the guard shack, waving good-bye to a vanload of TV people from Champaign. Jane had had the station's tour request on her desk, meaning to give it to Esther Tice in PR; but after Alex asked for something to do, she'd decided to toss the job his way.

"Try to orchestrate what they're filming, Alex," she'd counseled. "Aside from that, just be your irrepressible self. Tell them how we're coming back stronger than ever."

"Don't worry," he'd said. "By the time I'm finished, they'll be buying our stock."

That flip remark gave Jane second thoughts about her decision, but to call Alex off and Esther Tice in at this point would be a clear insult to Alex. Luckily the TV crew wasn't national.

During the day, Jane had caught glimpses of Alex's group. From a distance, at least, the tour seemed to be running smoothly. She was glad she hadn't acted on her fear that he would overplay the confidence hand. She was not in the least surprised, however, to observe, even at a fair distance, that her handsome brother seemed to have enthralled the on-air reporter, a long-stemmed, elaborately coiffed brunette.

As the media van drove off, Alex met her group at the gate. Roy made the introductions.

"What's your handicap?" Alex asked after he had shaken hands with the strangely silent Mr. Choi.

Good God, Alex! What on earth are you thinking?

Mr. Choi's persistent silence had certainly aroused Jane's curiosity. For all she knew, it was the result not of deference to colleagues of higher rank, but of a speech impediment. She thought it prudent to determine the reason for his silence if she could and had assigned a company researcher the task of finding out a little more about Mr. Choi. Only minutes earlier the researcher had passed Jane a note informing her that the man her brother had just insulted was Choi Hong Soo, nephew of Hyon's current chairman. That said a mouthful! Jane cringed inwardly but stepped forward quickly to try to limit the damage.

"Mr. Choi—"

But Mr. Choi ignored her, so busy was he staring at Alex. *Oh, Alex, when will you grow up?*

Then he smiled—Mr. Choi *smiled*. Shyly, to be sure, but unmistakably. "I play to a five, Mr. Akers."

He could speak!

"Bring your clubs?"

"I regret that there will be no time to play."

"You could make the time, couldn't you? Roy said you gentlemen are staying out at the Hyatt near O'Hare, right? That puts you next door to one of the great courses in the country. Medinah Number Three."

"*Medinah Number Three.*" Mr. Choi repeated the words with unmistakable respect. He turned to his colleagues. A guttural burst of Korean ensued.

"How'd you know he was a golfer?" Jane whispered to her brother.

"Look at his tie tack."

Mr. Choi turned back, beaming. His tie tack, Jane saw, was a tiny gold putter. "You could get us on tomorrow morning?"

"No problem, Mr. Choi. The pro and I went to school together. What time would you like to tee off?"

Another guttural conference, this one briefer, after which Mr. Lee Sung Whan said, "Seven o'clock would be best. Even so, we would have to reschedule our meeting here."

"I'll have you back right after lunch. What about it, Jane?"

"I think we can arrange that," Jane said evenly. "Say, for two o'clock?"

Alex fell in beside her as they resumed walking toward what was left of Siler Field. In a few moments they had a view of Hyon's sleek Gulfstream V.

"Say," Alex said. "Would you gentlemen mind if I hitched a ride with you? It'll save me flying over later, and the truth is, I've never been in a G-5."

"We'd be honored to have you join us, Mr. Akers," Lee Sung Whan said.

"Don't you need to pack a bag?" Jane asked.

"I'll pick up what I need at the hotel. Do me a favor, though. Call Anne for me, and tell her I won't be there for dinner and not to wait up. Better yet, why don't you have dinner with her?"

Moments later the sleek bizjet had roared off the runway. Roy turned to Jane. "What was *that* all about?"

"I think our brother may just have taken a giant step toward solving our Korean problem."

**JANE'S OFFICE,
AKERS HQ
4:34 P.M.**

Jane was startled to see Robert Grooms seated on a corner of her desk. He was in profile, leafing through a copy of *Logistics Management* magazine, only his blue eye visible. Then he turned to her, bringing the gray one into view, and flashed an amiable smile.

"What are you doing in my office, Mr. Grooms?"

"I stopped by to apologize."

"Oh?"

"Obviously I've done something that's gotten us off to a bad start. I want us to be able to work together."

"That would probably be in everyone's best interest. All right, I'll tell you what you did that I didn't like. You had no reason to target Mr. Singh for investigation aside from the fact that you don't like the politics his university was notorious for three decades ago."

"Let me apologize for that, then. And you're right. That's when I attended Berkeley—during the psychedelic age. I was a real straight-arrow type, and I'm afraid that bias turned out to have a long shelf life. But guilt by association *is* a bad rap. I might as well confess I went ahead and did a little checking on your Mr. Singh, and it turns out he's damn good at what he does. He's very well thought of at the NSA. That's the National Security Agency. He does work for them now and again."

"Yes, I know." Jane smiled. "Now, I have a great deal to do, Mr. Grooms. Is there anything else?"

"Well, I do have kind of an offbeat idea, a way you could maybe guarantee there'll be no new system sabotage. Then, since Singh seems confident he can purge all existing viruses, you could just plug back in."

"There's no way we're going to do that till we're a hundred percent sure. But let's hear your offbeat idea."

"This is definitely not going to improve your opinion of me, but here goes. You suspend everybody with computer access during the last month. Pay them, but keep them away from the computer. Get Singh or your friend Khalif to train backups. Then you plug back in. If you don't get any new viruses, you've probably isolated the culprit."

"Like isolating an allergen, is that it? I suppose we reintroduce them, one by one, till we get more damage?"

"You got it. Too Draconian?"

"Just a tad. And not my style. By the way, the suspend list would include a lot of top management, I'm afraid."

"Well, then you've got another issue," Grooms said. "System access."

He stood. "Before you do get up and running again, maybe you should sit down with Mr. Singh and Chip Bragan and whoever else you designate, and revamp the computer room security policy. Looks like too many folks have those coded-access cards."

"That's a good suggestion," Jane said. "I accept your apology, Mr. Grooms, and I'm glad you dropped by."

"Mission accomplished, then."

"And you're welcome to borrow that magazine. It's always nice when a layman shows an interest in learning a little about logistics."

DAY 5

FRIDAY, MAY 14

Wade Crain was definitely ready to crash. He'd been working eighteen hours straight, much of it in the windowless War Room. He'd just wrapped up a forty-minute conference call with Patrick DePalma, John Brinkema, and Henry Torres at the Milwaukee distribution center. They were still having trouble adapting Khalif's PC routing software to their site.

One thing they'd discovered was that the older workers in the DC did better in the current makeshift environment than recent hires. The warehouse veterans were used to making basic logistical decisions on their own. They knew how to plot a pick path through the racks; how to build a pallet with heavy stuff on the bottom, light on top; how many pallets filled a twenty-eight-foot trailer and a forty-eight-; which crushable products required deck trailers that folded down for double pallet stacking. Old-fashioned skills like these were having to be dusted off and put to work in Akers distribution centers worldwide.

And Wade was responsible for all of them now, not just his familiar Milwaukee stomping grounds. Which was what he finally had to remind Pat DePalma, who continued to crave a chunk of Wade's attention, which now had to be subdivided.

There was, for instance, the situation in the Plainfield, Illinois, dry grocery warehouse. Thank God it was due to come back on-line today, having completed its inventory—sixty thousand pallets pulled down and rescanned into a separate computer database, with no link to Ocean Plains. Even counting the chlorine gas spill in Milwaukee, so far Plainfield had been the biggest DC disaster. But Akers distribution centers across the country were all having problems getting product in from suppliers and out to their stores without total electronic oversight.

Shortages were still a plague. On this day several Akers stores in Missouri had gotten scrambled shipments, not because of computer sabotage, but from misrouting by their regional DC. Yesterday an Akers driver in western Pennsylvania had apparently gotten so disgusted with snarled-up paperwork, he'd walked off the job without notifying anybody, leaving his big reefer truck full of perishables in a rest stop beside Interstate 80. It hadn't been located till late today. One community affected by the driver's novel way of quitting happened to be the governor's hometown. A panicky phone call to his office in Harrisburg had triggered an emergency food shipment from the nearest National Guard armory. Wade had been warned to expect the story in *USA Today*.

And that was far from the day's only fiasco or the only emergency shipment that had to be gotten out. Wade had arranged almost a dozen himself.

Now he needed to arrange some sleep. Urgently.

He walked down the carpeted corridor of executive offices, thinking about the unlikely foursome teeing off in less than six hours at Medinah Country Club. Would the men of Hyon be sleeping now or figuring how best to implement their latest instructions from Seoul, where it was already four o'clock tomorrow afternoon? Somebody on the Hyon customer service team had told Wade that Korea's labor problems had escalated again, shutting down a major shipbuilder.

Not Akers's fault, not this time. But that wouldn't help their situation. Pray for low scores on the golf course, Wade thought.

He reached the turnoff to the lobby, stopped abruptly, and glanced back along the cross-passage. Halfway down, about where Jane's office was, an oblong of light spilled across the carpet. Curiosity drew him toward the light. As he neared the door, he heard the clickety staccato of a computer keyboard. He glanced in. She had her back to the door and was hunched over her notebook PC, typing away.

"I forget," Wade said. "Just which shift are you on?"

Jane spun around in her chair, found her smile. "Same as you, I guess. Think we need a union?"

"I wondered where you'd gone to. You been here all this time?"

"Actually, no. About six, I felt so stiff I took off for the condo, took a long hot shower to loosen up my joints, and got back here, feeling a whole lot better—until I remembered Alex had asked if I'd sub for him at dinner with Anne Akers. Luckily—I guess—I hadn't called her yet to see how that sat with her. By the time I did remember to call and tell her *he* wouldn't be home, I couldn't spare another minute, so I guess I'll have to get to know my stepmother better another time."

Suddenly she colored. "Why am I telling you all this? Don't really know you any better than I know her."

"Maybe I could get on the rain check list, too?"

"It's not that long, believe me."

"All the better," Wade said.

Jane picked up a paper, glanced at it, and shook her head. "Speaking of lists, look at the length of this printout of my 'to do today' list. I'm only about half-done—those are the items highlighted in yellow. I just hate it when I have to carry over more items than I cross off. That's why I absolutely had to come back to my cell."

Wade grinned, leaning against the doorjamb. "My 'to do' list is in my brain somewhere—in the part that doesn't seem to be working anymore. I'm going home. How about you?"

She glanced at him a moment, then realized he hadn't even asked to give her a lift to the condos. He was just being polite.

Good. That leaves only one of us who said something too personal tonight.

"You go on," she said. "I'm good for another few pages, I think. Believe it or not, I'm writing a radio script. Actually just the 'lead-in'. The local store managers have to fill in the body. Wade, you're talking to the RDCs. How does it feel out there?

What's the morale like?"

"Pretty good. A lot of Akers folks are doing what we're doing, busting their tails. But there seems to be a sense of relief out in the field, because we've managed to turn off the spigot on all the wacko replenishment orders—most of our guys don't know about those crazy Pac-Rim imports already in the pipeline. And we are getting stuff out on the shelves, admittedly with a hell of an effort. My take on it is that that's boosting morale some, and if things keep moving the way they are, the morale will stay up there. But I just don't know how long we *can* keep—"

"As long as we have to."

Wade nodded. No need for her to know that he was surprised by her response, so immediate and uncompromising. He'd expected some hesitancy, perhaps a little self-doubt. Maybe she felt that certain, maybe she was just putting up a good front. Either way, she suddenly looked to him very much like a leader.

She also looked quite lovely, sitting there with her face lit only by the desk lamp. She obviously hadn't bothered to put on makeup again after her shower and he could see the tiny lavender veins in her eyelids. And the pink in her cheeks that didn't come out of a bottle.

"I'm off," he said. "Don't you stay all night. Remember . . . " He glanced at his watch. "Today is another day."

OCEAN MEADOWS CONDOMINIUMS
2:04 A.M.

Jane switched off the engine and was enveloped in sudden silence—the first quiet to come her way since her alarm clock had started chiming twenty hours ago. It felt . . . as if she'd been let out of a drum.

Someday she ought to trek to a mountaintop, maybe into the foothills of the Himalayas, where R. K. Singh had mentioned growing up. Just sit there, cross-legged, staring at a tree.

Not necessarily to achieve illumination. Just to savor the quiet.

Hell, she wouldn't last a New York minute. She liked things hectic!

Hectic? How about insane? Jane scooted out of her car. As she hefted her briefcase, something spilled from her lap to the asphalt, crumpling underfoot. It was the empty bag from the vending machine chips she'd devoured during the ninety-second drive home.

She picked up the bag and tossed it back into the car. Then, on impulse, she stuffed her briefcase back inside, too, pulled her Nikes out of the book bag in her backseat, and laced them on. That one minute of blessed silence was enough to reactivate Jane's normal appetite for self-indulgence, in park since Sunday. She felt impelled to do something for herself before she crawled under the covers and stared at a magazine or the boob tube or the condo's ugly, stippled ceiling until sleep overcame the jagged motion of her mind.

Across from the Ocean Meadows complex was a square expanse of land, lined by maples on three sides and by the Akers perimeter fence on the fourth. When Siler AFB was operational, Jane recalled, that entire block had been covered with slapdash, mustard-colored apartment units. After the base closure, the air force housing must have been bulldozed. A large, weatherbeaten sign proclaimed the wasteland a commercially desirable industrial park. Jane figured a jog along the three-sided perimeter at maybe a half mile.

She set off. Since she was wearing a skirt, she settled for more of a bouncy walk than a jog but found it exhilarating nonetheless—inhaling clean night air and gently purging herself of the day's accumulated toxins, both mental and physical. The vastness of sky and prairie put all her feverish concerns in perspective—at least for a few recuperative moments.

She trotted across the narrow street and onto the tree-shadowed, uneven sidewalk of the old housing block. Then she stepped up the pace a bit, despite her confining clothes, find-ing release in the physical challenge. Exactly what she needed!

Her exhilaration lingered to the lot's farthest corner, leav-ing a last outbound leg to the Akers property, then three legs back to the condos. But Jane opted to turn around instead. No point in overdoing. If she did, she knew, the exercise might have a rebound effect and end up keeping her awake till all hours. Getting too little sleep was one thing; getting hardly any might put her judgment at risk. Couldn't afford that. She headed back toward the condo, the distant lumines-cence of the Akers compound flickering on her right between the passing maples.

Halfway down the long block, Jane sensed she was being watched.

She tried to shake off the feeling. This was, after all, Ocean Plains. There were no voyeurs here. Certainly not . . . a stalker?

Still, the feeling intensified. She felt very exposed, seeing her-self a moving target against the prairie night, hearing her own panting like some fleeing animal, her shoes slapping the rough pavement. What if she twisted an ankle and couldn't keep run-ning? She wished herself instantly home, with the door locked. Then she wished she'd accepted Wade's offer to accompany her. She could at least see her street now, and the silhouetted geom-etry of the condo complex, still a few hundred yards away.

Jane moved sideways, off the sidewalk and away from the flanking maples, where someone could—*dear God, easily!*—be hiding. She darted into the weed-infested street. Now she began to really run, heedless of her skirt being hiked up by the sheer motion of her legs. By the time she glimpsed her own car-port, she was almost sprinting.

She didn't slow down as she ran up to her door, key in hand, not even stopping to retrieve her briefcase. She fumbled twice before fitting the key into the lock. Inside, she bolted the door and leaned against it. Then, slowly, she slid to the floor. Some minutes after, she could not have said how many, she got up and switched on the lights—all the lights in every room. She looked around every corner, opened every closet, actually looked under the bed.

Then she sat in the living room and listened to her heart beat. She could not recall ever having felt so scared before. Not since childhood, anyway. After seeing *Jaws*, it had been hard *not* to imagine a shark coming after you underwater, even in Porter Lake. The memory didn't make her smile. *That* made her scowl. This was probably no more than the result of having worked herself into a highly susceptible state. Her terror had been self-induced.

What else could it be?

Don't you mean who"

But who would know she'd be coming home this late all alone?

One person knew.

It wasn't possible. She'd known him since high school.

You don't know him at all. You said as much to him only tonight.

Jane looked at her hands. She ordered them to stop shaking. They weren't taking orders.

He wouldn't have said he was on his way home.

Maybe watching her was a sudden inspiration. Or an irresistible impulse.

Jane huddled in the chair, knees to her chest, hands hugging her knees, and began to rock back and forth, hardly feeling the tears on her face.

OCEAN MEADOWS CONDOMINIUMS
7:55 A.M.

Jane was at the door when the phone rang. *He*—whoever he was—knew she was about to leave the condo! She stared at the phone, but it didn't stop. On the fifth ring she walked back and picked it up. It was only when she recognized Mark's voice that she realized she hadn't taken a breath since it rang. She breathed out, in. "Mark, you're in the office awfully early."

"I'm calling from home, Jane. But I've been trying to get through to you since five-fifteen my time."

"Sorry about that—no way for you to know that by a quarter past seven here I'd already made four calls. It's been crazy, Mark. It's been crazy every day. I should have been at work an hour ago. I'm just going out the door, okay? I'll call you in an hour, I promise."

"Bridey's sister died," Mark said.

Jane held her breath, reached for something she knew. Wasn't that sister Bridey's only family? She let the breath out. "I'm so sorry. I—I don't think I ever met her. She lived somewhere in Canada, didn't she?"

"Nova Scotia. Bridey left in a hurry last night. Said she had to arrange the funeral and something about having to stay on afterward to go through her sister's things. She left a number there, but she'll be out all next week. Jane, how am I supposed to handle everything here without her? Deal with clients and run the office at the same time? I never was able to rub my stomach and pat my head simultaneously. And frankly, that trick wasn't part of my job description."

"Read the small print in our verbal contract, Mark," she said lightly. Then, firmly: "It says you handle what—whatever—needs to be handled. And don't you go bugging Bridey while she's up there, calling to ask where she keeps the Coffee-mate, for God's sake."

"Thanks for the vote of confidence in my judgment. Jane, when are you coming back?"

"You *know* when I'm coming back. I've got five more days here. Then I'll be finished, one way or the other."

"You know, our client's *problems* don't know this isn't a good time to happen just because Bridey's sister died—or, for that matter, because you decided to try to resurrect your father's company. How about *this* company? *Your* company?"

"Mark, I know with Bridey having to be away too, that puts more on your shoulders. But get a grip. I hope to God you don't sound like this when clients call."

"There you go showing extravagant confidence in my judgment again."

"Well, what am I to think when you sound like that?"

A noise. A chuckle?

"That I need to vent and you're my designated ventee. I wouldn't dream of using anything even resembling a desperate tone with one of our clients. Don't worry, Jane. I'll survive without you for five more days, and so will Malcolm and Associates."

"That's better. A whole lot better. And, Mark, hire a temp."

"I already looked up the number of two agencies."

"See? I told you you're terrific."

"Not lately you haven't."

"When I call you back, if I forget to mention it then, remind me. Bye, I'm late."

Jane hurried out of her condo, thinking how absurd it was to get into a car to drive all of five hundred yards over to the adjoining Akers parking lot.

Bullshit. It would be absurd not to!

Daylight, after barely two hours' sleep, hadn't obliterated the sense that someone was watching her last night. She didn't want to be walking any distance on her own when she came back tonight—or, more likely, at some ungodly hour tomorrow morning. And she certainly didn't want to have to depend on being seen safely home by someone.

Definitely not by Wade Crain.

Feeling funny about walking alone at *any* hour felt . . . funny. After all, this was Ocean Plains, the town where, when she was growing up, people used to keep their houses unlocked. Where the only threat to life and limb usually involved fast cars, dark two-lane roads, and six-packs of beer.

But this didn't feel like that town anymore. Besides, for all she knew, these days everyone locked their door.

**PARKING LOT,
AKERS HQ
8:03 A.M.**

As Jane stepped out of her car, a dozen spaces to her right a blue-green cargo van pulled in. Robert Grooms got out and headed for the main entrance, which was where she was headed. He got to the door ahead of her and held it open.

"Good morning, Mr. Grooms."

"Morning, boss."

"I'm not your boss, Mr. Grooms."

He flashed a smile. Used to his mismatched eyes, Jane now became conscious of their luminous intensity.

"By the way," he said, following her inside, "I enjoyed that logistics magazine, but I could have used a glossary of all the buzzwords, like 'virtual response' and 'just-in-time.' Do they really mean something?"

"We like to think so."

"My first impression of logistics was back in Vietnam, watching supply officers and loadmasters move mountains of stuff in and out of Da Nang. You know, bundling up all those gas cans and mortar rounds and candy bars and getting them choppered up to our firebases."

Jane halted in midstride. "You were at Da Nang?"

"Among other garden spots. I was with the First Force Recon Marines."

"My grandfather flew in and out of there during the war with the Blackjacks—the 53rd Airlift Squadron. Colonel James Malcolm."

"Colonel Jimmy was your uncle? Quite a man. I saw him maybe a dozen times. Real knack for telling a story. Your grandpa fought in three wars, right?"

"No, he skipped Korea."

"Okay, wait a minute, what it was, Jimmy was the guy who served in three different air forces during World War II—all on our side, of course. Did I get it right this time?"

"You sure did," Jane said, pleased that this stranger remembered her grandfather's favorite exploit. When war broke out in Europe in 1939, James Malcolm was a high school senior in Toledo. Desperate to get into the fray, he'd quit school, hitchhiked north, and signed up with the Royal Canadian Air Force. In 1940 he'd wangled a transfer to the RAF, where he flew Spitfires in the Battle of Britain. Finally, when the Americans came in, Captain James Malcolm ended up in the Eighth Air Force, flying bombing missions over Germany.

"I lost track of so many men whose paths I crossed over in 'Nam," Grooms said. "What's he doing these days?"

"My grandfather died in 1975, Mr. Grooms. During the evacuation of Tan Son Nhut Air Base. He was taking off in a C-130 that was hit by a North Vietnamese SA-3 rocket."

"I'm really sorry to hear that. He was a great guy."

"Yes, he was."

"But then you must have been just a toddler when he died."

"Not quite. I was nine. I remember him very well. When I was younger, I would sit on his lap and play with his medals—two Purple Hearts, a Distinguished Flying Cross, even a *Croix de Guerre*. He's the reason I got into the military, and into logistics."

Jane ended up walking Grooms to the cafeteria and getting herself a coffee to go while telling Grooms about her grandfather's work during the Berlin airlift. The security man was a good listener, asking perceptive questions.

Her cell-phone cut them short.

She listened for a minute. "Right away."

Then to Grooms, "Speaking of war, I've got to go man the barricades. Seems we've got incoming." She smiled. "I enjoyed our little reminiscence, Mr. Grooms."

"Call me Bobby. Your grandpa did."

"All right. Guess that makes me Jane."

OCEANOS TRANSPORT SERVICE
PORT OF LOS ANGELES
7:51 A.M.

"One more time, then," said George Korchek. He was on the phone to the manager of the Akers import distribution center in Chino, only an hour's truck drive away. "I've got your ninety Hyon containers sitting here on my pier—I can see 'em through the venetian blinds. But I can't move 'em off. I've been on the phone since fucking sunrise—talking to your customhouse broker, to your trucking company, and to my good pals over at U.S. Customs. Your import documentation never showed up, my friend. Your broker has no ocean bill of lading for these goods, no commercial invoice. Which means no way to calculate duties. Which means Customs can't clear it. Which means it can't enter U.S. commerce. Am I being clear enough now?"

Korchek tilted way back in his office chair, nodding patiently as he listened to another set of arguments, followed by another set of pleas, wishing like hell he hadn't quit smoking. He thumbed a wider gap in his blinds, peered through the fog-shrouded tracery of towering, insectile cranes. Three of his fifty-tonners were currently extended out over the nine-hundred-foot length of the *Kimyung Empress*, yanking off containers and plunking 'em down onto wheeled chassis, to be hauled away by waiting trucks. The unloading of the huge Korean ship had started at six A.M. and would finish by five or maybe six tonight. Thousands of containers off-loaded and hauled away, for either intermodal rail transfer or truck delivery.

All, that is, except for the ninety Hyon containers consigned to the Akers Chino distribution center. They were stacked ten long, three wide, and three high, a big blue barricade taking up terminal space Korchek damn well couldn't spare, blocking traffic, giving him a massive headache.

"Look, pal," Korchek cut in, "I don't know where the hell your import documents are. But I do know this. You better generate new originals fast. You want your merchandise? I want those damn cans out of my way even more! And your shipping company, Kimyung, which pays my salary, wants 'em back empty to put goods in. And the folks at Hyon want to get paid. Let's make everybody happy, that's my motto.

"Otherwise, you got ten days to get that shit off my dock, or Customs is gonna seize it, stash it in their general order warehouse, and auction it all off a year from now."

WAR ROOM
11:06 A.M.

The first call came into Wade from Chris Azarian, manager of Akers's West Coast import hub. A minute after he hung up, Wade buttonholed Jane in the hall just outside the War Room.

"We got brand-new trouble with Hyon. You remember that shipment Mr. Lee What's-his-face mentioned?"

"Lee Sung Whan."

"Right, the shipment he said was docking in L.A. today? TVs, VCRs, stereo equipment, and small appliances? Well, it docked all right. But our containers are just sitting there, not going anywhere. The terminal operator can't release them. There's no paperwork."

Jane shook her head. "There has to be."

"Well, the Korean steamship company—Kimyung—has their bill of lading, which lets them get paid and release their cargo. But Chris Azarian's been talking to everybody in the loop—the terminal guy, our Customs broker, the trucking company. They've got no delivery documents. And without that paper, the merchandise doesn't move. Period."

Jane took a breath, just as she'd advised Mark to do earlier, rehearsing the import sequence in her mind. Several copies of the commercial invoice would have been transmitted by EDI from Hyon to Akers's International Department, which would have transmitted it to their customhouse broker in L.A. The network shutdown should not have affected that process; all import documentation had been backed up onto PC workstations. Or *should* have been.

"Let's go talk to Jessica Zhuo," Jane said, taking off toward the office of the International Department's director. "Whatever went wrong, at least she can see that they retransmit whatever documents are missing."

"I sure hope so," Wade said, falling in step beside her. "And fast. Chris says the terminal operator wants our stuff out of his way now. And God knows our stores need it."

"What about Hyon?" Jane asked.

"That's the worst part, really. Until we actually receive the merchandise, there's no way Purchasing can let Accounts Payable release the funds to Hyon. Right now we don't know what the hell's actually in those containers or what we're being billed."

That's just dandy.

Jane quickened her pace even more. She recalled Mr. Lee's recital of how Akers had already caused the Korean company serious cash-flow problems, enough to call for the closure of a key manufacturing site in Suwŏn, putting people out of work during a time of nationwide labor unrest.

Mr. Lee and his two colleagues were probably finishing up the back nine at Medinah now. In less than three hours they'd be back for their meeting in Ocean Plains. Jane had been looking forward to that meeting. Not anymore.

She glanced at the man keeping pace beside her. The last few days, she'd found herself thinking how good it was to have him in her corner.

Now she had to wonder if he *was* in her corner.

And she remembered how everyone had praised him for

being the first regional manager to get around the computer problems.

My God, could it be he had time to prepare?

**INTERNATIONAL DEPARTMENT,
AKERS HQ
11:13 A.M.**

As Jane and Wade strode in on a scene of a department in turmoil, she forced herself to put her suspicions out of her mind. There was no way to examine them right now. Later, when she had a chance, she'd tell Elie, talk it through with him. Everyone in International seemed to be moving at fast-forward speed— employees pawing through file drawers, others scrolling down computer screens, all back-and-forth talk shrill with anxiety.

"Where's Jessica?" Jane addressed the room at large, loudly enough to be heard above the din.

"Right here." Jessica Zhuo appeared in the doorway to an adjoining office, holding a ledger. Jane recalled, from their only previous meeting, that the young woman seemed to know her business and didn't waste words trying to sound reticent about that, a virtue in Jane's book. "I assume you're here about the missing import documents," she said now. "The ones for this morning's shipment on the *Kimyung Empress*?"

"Right. How long have you been looking?"

"Five minutes, if that. Don't worry, we'll find them. Before the ILS system was taken down, we made backups of all those documents, including hard copies. So they're definitely here. What I don't understand is, according to our import schedules, all our documentation went out. That paperwork should be in place."

"It isn't," Wade said. "Believe me."

"I know that. But it doesn't make sense, Mr. Crain. I immediately faxed Hyon's Export Department to retransmit copies of all relevant invoices, but it's one in the morning there, you know, so it'll be a few hours at best. Meanwhile, we're retrieving our backup files, but they're on streaming tape, and I'm afraid it's a sequential search with rather a lot to go through. We imported more than forty thousand containers last year, and we're well ahead of that pace this year. You can be sure the instant we find them, I'll have them faxed to our customhouse broker."

"And notify the War Room."

"That goes without saying. The urgency of the situation is clear."

"We'll leave you to it, then," Jane said. She led the way out of the room as she'd led the way in.

Don Landsman was hustling down the hall toward them.

"What's the latest word on that shipment?" he called out. When Jane had briefed him, he shook his head. "We can't just stand around, dammit. There has to be something we can do *now*."

Practically the same words had been banging around in Jane's head. Suddenly they came to a stop. There *was* one thing they could try. She stepped back into the International Department office.

"Jessica?"

"Something else, Jane?"

"While you're searching, and waiting for Hyon to fax you back, how about contracting with a foreign trade zone operator out there and getting those containers picked up and moved into a bonded warehouse?"

As Jane understood it, a foreign trade zone, or FTZ, was a sort of mythical entity—an area within a country that the local Customs service treated as a foreign territory. Shippers could land, store, and process goods within an FTZ without incurring any import duties or domestic taxes. Facilities within FTZs could receive goods on which duty had not been paid and hold those goods in legal limbo, as if they were still in a foreign country. In the case of the Hyon shipment, when the paperwork arrived, duties could be paid and the merchandise entered into U.S. commerce.

"That's going to require two moves and cost more money, isn't it?" Jessica said. "Hauling the containers off the pier to an FTZ, where they'll have to be unloaded. Then, after duty is paid, reloading everything into containers for the move to our Chino import hub."

"Maybe not," said Don Landsman. "Why don't you check with our trucking company out there. Since they provide assembly services for their import customers, maybe they also operate within a foreign trade zone—"

"It's Harbor Cartage," came a woman's voice from behind a partition. "And they *do* operate a bonded warehouse. A large one."

"Right, Gwen," Jessica said. "Thanks."

"In which case," said Don, "they could haul those containers off the dock right now, maybe stall a little bit in transit—"

"Long enough for us to get those documents retransmitted and in the hands of our customhouse broker," Jane picked up. "That way, Jessica, we could get the merchandise cleared through Customs and back on the road to our DC, with Hyon paid off— maybe without having to unload and reload those containers."

"So we really wouldn't lose that much time," Wade said. "Sounds good to me."

"It's definitely worth a try," Jessica said. "Glad you came by— and I don't just mean Ocean Plains."

"Thanks. Now, why don't you make that call and get it started?"

JANE'S OFFICE
1:30 P.M.

With the documents still missing, Jane took a call from someone called Stickley Shears, whose secretary, before she put him on, identified him as undersecretary of state for East Asian and Pacific affairs.

"My secretary has been on the phone with various people there for nearly half an hour, Ms . . . Malcolm," he said testily. "She has other things to attend to. Finally, she was assured that you're a person of some authority there. Is that correct?"

"Yes, it is, Mr. Shears. How may I help you?"

"Expeditiously, I hope. I'm calling because *I* had an urgent call within the last hour from Barbara Ganahey, our ambassador to Korea. Ambassador Ganahey expressed deep concern about Hyon Group in particular, and the Korean situation in general." Mr. Shears swung into a lecture on Korea's current afflictions—increasing labor strife; an escalating trade war with America, with each side slapping on countervailing duties; serious political instability, with opposition party leaders making common cause with labor protesters and radical elements supporting the North—

"Excuse me, Mr. Shears," Jane said, "but I follow these events pretty closely myself. They are indeed disturbing. May I assume you're calling about our recent problems with Hyon Group?"

"Indeed you may, Ms. Malcolm. To the highly unstable political and economic equation I just outlined, which threatens not only Korea, but all of East Asian security—and therefore the whole Pacific Rim—we must now factor in the troubles of Hyon Group. Hyon is one of the five pillars of Korean industrial power—a pillar that may be in danger of imminent collapse. And according to what Ambassador Ganahey was told this morning, almost all of Hyon's problems have their origin in corrupted electronic data transmitted by your company. As a first step, Ms. Malcolm, can you comment on these allegations?"

"Inflammatory rhetoric is seldom helpful, Mr. Shears. I'll just say this. We *have* experienced computer sabotage here, and the damage *has* affected some of our trading partners, Hyon Group perhaps foremost. We've taken radical and, I might add, work-intensive as well as expensive measures to assure that no further damage can occur. We're working now to repair the harm that's been done. In fact, there's a high-level delegation from Hyon here at the moment, as determined as we are to find the best way to resolve all the issues between us."

"But just this morning, Ms. Malcolm, the ambassador was told—actually awakened from her sleep to be told—that a sizable export shipment from Hyon was refused entry into U.S. commerce by your representatives. An investment to Hyon, Ms. Malcolm, of millions and millions—"

"*Hold on a minute!* We did not refuse it, Mr. Shears. The paperwork was lost."

"Lost? Is that what I'm to tell the ambassador? One of the largest retailers in the United States of America has mislaid *some paperwork* and so cannot accept a crucial shipload of goods?"

"I think we really ought to stick to the facts, Mr. Shears. An entire shipload is not involved. Nonetheless, the problem does concern, as you pointed out, a critical amount of goods—ninety containers out of several thousand. And we *are* accepting it. We just can't clear Customs, so we can't pay Hyon yet. But we will, as soon as we are able. Believe me, Mr. Shears, and please assure the ambassador, we're moving as swiftly as we can on this. And we're conferring with Hyon Group every step of the way."

There was a beat, during which Jane wondered what Mrs. Stickley Shears called her husband in their intimate moments—if they had any.

"Very well, Ms. Malcolm," Shears said finally. "I will convey that information. But if I need to call you back, I'd prefer to do it on your direct line."

Jane gave him her cell-phone number and hung up.

Then, for a long moment, she sat very still. She'd handled the undersecretary's call well enough, but only now did she allow its full significance to hit her. If anyone needed proof that logistics was a worldwide web unto itself, involving domino effects on a global scale, here it was. For want of an invoice, it seemed, an entire government was being shaken up—and conceivably might fall.

Jane stood and shrugged her shoulders vehemently to rid herself of the sudden extra weight she felt on them. There were some things it was best not to dwell on. As she hurried to find Don and Wade, she began to smile.

There wasn't anything the least bit funny about the situation. Still . . .

Who's my next call going to be from? Madame Ambassador? Or maybe the First Lady herself, desolate because the nearest Akers is out of her daughter's size in a pair of black jeans everyone on campus except her already owns?

THE WAR ROOM
3:28 P.M.

Wade noticed the roar of the Gulfstream overhead first and signaled Jane and Don. Jane buzzed Roy in his office, and several minutes later Jane, Roy, Don, and Wade were striding onto the apron of Siler Field as the sleek business jet taxied to a stop. The bad news was that the original documents were still missing. The good news was that all the Hyon product was off the dock and en route to a bonded warehouse.

Only five minutes earlier Jessica Zhuo had reported to Jane that duplicate "original" invoices were finally coming across her fax machine from Hyon's Export Department. By nightfall, she said, if certain financial incentives were offered the broker and trucking company, those goods could have cleared Customs and be back on the road to Akers's import hub in Chino.

"Use maximum leverage," Jane called over her shoulder as she hurried out toward the plane.

The Gulfstream's stairs lowered and locked. The door opened, and the Korean trio descended in Technicolor golf finery, squinting into the afternoon sun. Alex ambled down behind them, grinning. No one would describe what the Koreans were doing as grinning, but they definitely looked pleased.

Either they'd all switched off their cell-phones, Jane thought, or her brother was so darned charming, the Hyon guys had called a hiatus on caring *what* happened off the links.

In which case, maybe *Alex* should be ambassador to Korea.

JANE'S OFFICE
10:03 P.M.

"Did I wake you, Uncle Elie?"

"Of course not. Why are you whispering?"

Paranoia spreads its wings wide, Elie, that's why. She'd already closed the door to her office.

"Sorry," she said in her normal voice. "I need to ask you something. It may sound downright stupid to you, but it's really important to me to have your opinion about something. Some*one*, actually."

"I'd never think something important to you was stupid, Jane. Who's this about?"

"Wade. How well do you know him?"

There was a pause. "Better than I know you right now, to put it bluntly. Wade's always made time—even since he's been responsible for twelve stores—to come out to the farm and visit with the kids. Once I had to have some surgery in Chicago and he took a week of his vacation to stay with them for me."

"You trusted him with your kids."

"Sure did. Really knows how to handle them. His younger sister, DeeDee, had cerebral palsy, remember?"

Jane inhaled. "I don't. Where is she?"

"She died—long time ago, when she was thirteen. But what is this about Wade? Did something happen that has you worried about him?"

"No."

"There's something you're not telling me, Jane."

"I was just confused. I'm not anymore. Thanks, Elie. I've

gotta go. Wade and I are going over to Don's for a late pizza and half an hour's R and R."

"Sounds good to me. Think you can relax and have a good time—for a whole half hour?"

"I just might be able to."

Now.

**DON LANDSMAN'S HOUSE,
PORTER LAKE, ILLINOIS
10:33 P.M.**

All in all, Don and Wade and Jane agreed, they felt encouraged. Certainly not jubilant, but encouraged. The Koreans were no longer pretending that finding a solution was solely in Akers's interest, and some small yet palpable progress had been made by the time Alex suggested flying to Chicago for a farewell dinner in a little Korean restaurant he wanted their opinion on.

Alex was doing so well with them on his own that Jane thought it could only cramp his style if she went along, so she said she had to work and the Koreans nodded as though that were a fine idea.

Now she and Wade were at Don's house, just across Porter Lake from Roy's. No one mentioned Roy, who had stayed only long enough at the afternoon meeting to prove he'd been there and then had himself called away. Just as well, because that left seven people in the room who seemed equally determined to make some progress.

Because they *had*, Don went scrounging among the moving company cartons that crowded his living room for a case of California zinfandel he evidently prized. As he poured the wine into paper cups, he said, "I've got some money in this vineyard—but more pride. The founder and I were roommates at Morehouse, where I spent my first two years."

Jane thought the dry red zinfandel was worthy of Don's pride in it—and, she guessed, in his entrepreneurial roommate.

Wade praised the wine, too, but Jane had a hunch he'd have preferred a brew with his pizza. She raised her cup. "Congratulations to my fellow Californian, and cheers to us."

Don smiled. "Smitty, that's my friend, was born in White Plains and you were born in Ocean Plains. And you've both found success west of all the plains. I guess 'Go West' is still a good piece of advice for the enterprising young. Speaking of which, Jane, how is your business holding up without you?"

"Mark Gibian, who's my associate, is holding down the fort. Some days he feels the fort is being attacked from too many directions at once for him to be able to keep the enemy at bay, but in fact he's very able. Besides, in less than a week I'll be back there. And you'll be free to get off on that cruise of yours."

There were boxes all over the place, marked "STORAGE" or "CONDO." Don lifted the lid from the most recent addition, a giant box of "with everything" from the Akers Pizza Kitchen.

Don refilled their cups. "To victory," he proposed.

"Confusion to our enemies," Wade said, grinning as he lifted his cup. "I think that's what Napoleon said to his generals before some big battle."

"I just hope we can find out who our enemies are," Jane said, feeling guilty. Thank goodness the man had no idea what craziness her imagination had dreamed up.

"Rikki Singh may be the best sleuth out there at what he calls digital forensics," she said. "But if he can't track back from the crime to the culprit, we're still nowhere."

"I don't know," Don said, serving out the pizza. "But I thought we just had a meeting so that *this* wouldn't be one. At least through the first slice?"

Jane managed a smile. "Sounds to me like an excellent executive decision," she said. "How about those Cubs?"

"Oops," Don said. "I'm strictly a White Sox fan."

ON THE ROAD
11:10 P.M.

Jane got into her rental car and followed the taillights of Wade's Blazer out. Off to her right, the dark lake flashed quicksilver, and she glimpsed a bright moon drifting along behind the lakeside willows. The lovely sight, so unexpected, conjured that island cruise she'd never gotten to take. Which reminded her, when she'd hurried in and out of her office after final farewells to Messrs. Lee, Roh, and Choi, there'd been two phone slips on her desk from Chevy, which she'd neglected to bring along. She'd definitely remember to call him back in the morning.

The gravel road ended, and they swung in tandem onto the two-lane blacktop back to Ocean Plains. Jane still felt bad about what happened—what *didn't* happen, except in her mind. All the way home, a guilty feeling hovered over her.

There were only a few spaces between their carports at the condo complex.

Jane stepped out and closed her car door. Wade, standing beside his own car, didn't move, a broad-shouldered silhouette against the overarching emptiness of sky and plains, his features not quite discernible. Jane felt a quiver run through her, the tension she felt quite different from the fear that had overtaken her nearly twenty hours earlier.

"Walk you to your door?" came his voice.

"Of course not," she said, keeping her voice light. "It's just a step away."

"Mind if I *watch* you home, then?"

"Night, Wade. See you tomorrow."

She let herself into the condo and leaned against the door.

She waited. Slowly, slowly, the desire ebbed. Then she went to bed.

And dreamed a wild, logistically impossible dream.

DAY 6

SATURDAY, MAY 15

**WAR ROOM,
AKERS HQ
7:49 A.M.**

The War Room was actually beginning to function the way Jane had hoped. Complaints, however, kept flooding in from every Akers region. Most of these were about the fatigue—and extra staffing cost—of nightly physical inventories. One question was asked over and over: How soon before they could switch back to computers?

That was exactly the wrong question, in Jane's estimation. The real question—the real goal—was: How soon could they achieve full servicing of their stores *without* computers?

She was, of course, concerned that no one had yet discovered the source of the contamination; but solving that mystery was not her prime focus. Jane had mandated Mr. Singh to throw every available resource at their system problem and to keep her informed of his progress—or lack thereof. And when she had questions, as she did this morning, she sought him out.

"What about all that nifty stuff you let Chip Bragan tell the Koreans?" she asked, tracking Mr. Singh to the computer room, where he was talking to Lenny Cristofaro and Mimi Takuda. The weariness in Singh's features was unmistakable. According to Bragan, the consultant had been getting less sleep than any of them. "Something about 'firewalling all the inputs,' wasn't it, Mr. Singh? That won't really work, will it?"

Mimi shook her head. "Lenny and I would've already tried it."

"I appreciate that, Ms. Takuda, but I'd like to hear Mr. Singh's answer."

"Sorry."

"You and Ms. Takuda are both correct, Ms. Malcolm," Singh said. "There is no black box sophisticated enough to do what you wish. A virus could still enter from any store, disguised as sales data. What I have done is write a filter to check every incoming data packet for viral evidence. It works. However, it can't keep up with the speed of your datastreams. My filter would basically cripple your operation."

"So we're back to trying to find out 'whodunit,'" Jane said grimly.

"But thanks to Mr. Singh here, we're starting to get somewhere," said Lenny Cristofaro. "We've already found certain log-ins used to break into—mostly dormant—accounts, which we've now eliminated. We've analyzed whole sessions, keystroke by keystroke. Most were by modem, but some were done right here." He pointed to the system console.

"Can't you pull personnel records for those times and narrow it down?" Jane asked, recalling her chat on computer room security with Robert Grooms. "Assuming nobody could get in here who wasn't authorized."

"It's kind of a long list," Mimi said. "In fact, Lenny and I are both on it. So is practically half the IT Department, as well as the Bowerman and Positek people. What I think we should do is hire a lie-detector expert and put everybody through a test."

Not a bad idea, Jane thought. But, dammit, there was too much urgency for that lengthy a screening now.

"Thanks, Mimi. You too, Lenny. Mr. Singh, would you walk me back to my office?"

Moments later, when she was assured of their privacy, she pushed Mr. Singh. "If only someone had thought of polygraphing when this began! Don't you have *anything* that might work faster?"

"I do," Singh said.

"What is it?"

"Linguistic analysis."

"I'm listening."

"It will take more than a moment to explain."

Jane sat behind her desk and indicated an empty chair. But Mr. Singh preferred to stand. In a minute Jane understood: he had become teacher and she student.

She found herself trying to follow a far more extended and technical discourse than she'd bargained for and actually began taking notes. Basically, it seemed, Mr. Singh intended to analyze all the samples of decompiled virus source code and compare them with the unique C-language syntax and structures of the various programmers employed on the project.

When she'd got that much, Jane broke in. "Mr. Singh, I don't want to know any more right now. If what you're describing will take long-term study, forget it. There's no time. I need you to track down this criminal in the next three days. If your syntactical analysis or whatever it is can help you accomplish that *within that time frame*, then go to it. Do you understand?"

Singh dipped his turban toward her ever so slightly. The gesture seemed one of deference. But it didn't answer her question.

"*Can* you do that, Mr. Singh?"

"I will not willingly disappoint you, Ms. Malcolm."

God, I hope that was a yes.

8:19 A.M.

Flipping through her stack of phone messages, Jane discovered two new ones from Chevy. Damn, it was rude not to have called him back before this. Just as she reached for the phone, her pager went off. The display showed Don Landsman's extension. She hurried down the hall to his temporary office.

"What's up?"

"Sorry, Jane. I thought you were still over in the data-processing center with Singh. Could you close the door?"

"Uh-oh. What now?"

Don waited till the door was shut and she was seated. "Tom Soverel has sweetened his offer to the family with a hefty infusion of cash and given them until tomorrow afternoon to accept—before it's withdrawn."

Jane expelled an exceedingly nasty wish regarding Tom Soverel's personal longevity, slamming her fist down on Don's desk with such force that it toppled a picture of his wife. She straightened the photo.

"Sorry."

"Don't be," Don said. "Just because we're trying to accomplish a superhuman goal doesn't take away our right to a human response when someone changes the rules in the middle of the game."

"I was given ten days!"

"Somebody now thinks that's too long," Don said quietly.

"Just when we're making progress!"

"Exactly."

They were, too. Real progress. Despite the wave upon wave of horrendous problems. Jane's three-day ultimatum to Singh was based on the fact that she actually had four days remaining in her deal with the Akers board. Now this!

"Have you been able to find out the family reaction so far?"

Don reached over and touched Jane's hand. "I'm afraid they're buckling."

JANE'S OFFICE, AKERS HQ 8:50 A.M.

"I didn't promise never to sell, Janie," Joycelyn insisted. "I promised not to act hastily."

"But you *are* acting hastily."

"Roy said we must act now—or we could lose everything. Don't forget that applies to you, too, dear, now that you've got your trust fund. Everything your father built up, the entire fortune, could be lost. Aren't you paying attention? The stock dropped another three points just today—*after* you told me the corner had been turned."

"It *has* been turned," Jane said. Joycelyn had to be made to understand. "Mother, Wall Street only measures the ripples downstream, and they're still negative. It will take a few more days for the positive effects to work their way out. The company isn't lost, unless you and the others help Roy lose it."

"That's not fair, Janie. Roy has only the best interests of all of us at heart."

"Okay, Mother, I won't argue. I'm going to beg. *Please* give me four more days."

A deep indecipherable sigh was the only response.

"All right, three. I absolutely have to have three. You wouldn't cut me off when I'm that close, would you? You know I wouldn't ask you to keep on backing me if I didn't know I can do this. I can, Mother. With just three days, I can have us back on track. Let me have the three days. In my whole life, I've never asked you for anything I wanted more."

She couldn't go lower, so she used the silence to pray.

"I can't, Janie," Joycelyn finally said in an uncharacteristically quiet voice. "Roy had a courier pick up my signed proxy an hour ago."

"Mother!"

"Janie, I don't see why you're even there. You escaped from Ocean Plains!"

Jane took a shallow breath and spoke very carefully. Maybe there was still a small chance her mother would revoke the proxy. "I'm here, Mother, because I discovered I do care what happens to the company. The company and our family—it's all connected. That's why I don't want to sell Akers. I want to save it . . . to try to save what's left of us as a family."

"Janie, if you're suddenly so invested in family, why don't you listen to your brother? Roy doesn't want to sell the company either. But he's not letting his emotions get in the way of his business judgment and his responsibility to all of us."

She'd been wrong: there was no chance Joycelyn would change her mind. "All right, Mother, you've made your position very clear." Jane said it levelly, determined not to feed into her mother's current view of her as overemotional. There'd be another time to tell her that she wished she'd paid more attention to all Jane's emotions about her father—not just the angry ones.

"I've got to go now, Mother. My pager is beeping."

"Bye, darling. You'll see, this is all for the best."

Suddenly Janie understood something.

Even now, if her mother knew that Jane knew what she'd done, Joycelyn would expect her to see that keeping her father's letters from her was all for the best.

Now, what was she supposed to do with that nifty insight? Whatever it was, she couldn't do it now. *Now*, she had to focus.

Her mother's vote was gone. The thing to do was move on to the other family partners—as quickly as possible.

Who next? Alex? Anne? Aunt Lou? There was no obvious choice. She was trying to weigh the odds when the phone rang.

Let it be good news.

When she heard the voice of Mark Gibian she knew it could go either way.

"How's that guy in the turban I sent you doing?" Mark asked.

"If anybody can track down our phantom hacker, I'm convinced Singh can do it."

"But he hasn't?"

"Not yet. But, Mark, we are putting the Akers supply chain back together, using a lot of low-tech solutions and old-fashioned teamwork. To tell you the truth, I was feeling pretty damn good about things until about five minutes ago, when I discovered my little family coalition may be falling apart."

"Families! Judging by mine, what they do *best* is fall apart. You know, Jane, how much I'd like to have you back—this minute, if possible—but not that way."

"Thanks. That's really nice."

"You won't think this is. I think you better call Dave Ellsmere."

"What's happened?"

"Apparently somebody at TranSonic Airex wants to pull the plug on your deal."

"They can't!"

"Not on your consulting contract, on the South China Air agreement. If they do that, they probably won't renew us."

"Who told you they're thinking of dropping South China Air?"

"I know this financial planner in TSA's New York office, Hank DeWitt. We actually went to Brooklyn Tech together. Anyway, Hank heard some high-level scuttlebutt that the Macao air hub project was dead and the company was now looking for a, quote, cheaper, safer way to go—like handing off all their Asian cargo traffic to a Taiwanese air freight company."

"They can't really be thinking that way again! How many times did we talk them through the downside of the Taiwan option? After all that, somebody is spreading the word that my plan is too costly and too risky?"

"Worse than that, Jane. The rumor references your full-time involvement with Akers—like maybe you're in no position to deliver full value."

"Talk about sabotage! Who's doing this?"

"Hank said he'd try to find out. But even if he does, we'll have some major damage control to do."

"Look, stay on your friend. And get back to me the minute you find out any more, all right? And, Mark, whatever it takes, we'll handle it. Together. That's a promise."

"I'll hold you to that, boss."

"You won't have to."

Jane hung up and stared into the middle distance. After a couple of minutes she shook herself out of that passive mode and made a list of the family members she had to call—then tore it up. She didn't need a list; what she needed was some hope that calling them might accomplish something, and she didn't have that. If Roy had already persuaded the rest of them as he had Joycelyn, it was all but over. Akers was doomed. And now her own business was being threatened. TranSonic Airex had a six-month out on her consulting contract. If it became known they'd discounted her advice, then dropped her . . .

What next? she wondered. What part of the sky *hadn't* fallen?

9:22 A.M.

Anne Akers confirmed what Don Landsman had told Jane—there was a new takeover offer from Tom Soverel, with a deadline attached.

"I was just talking to Louise about it," Anne said. "I was going to call you next, to see what you thought of it."

"I wasn't informed," Jane said. "I only heard about it through Don, who heard from Elie, who was told by Aunt Lou."

"That's not right, Jane! You should have been told!"

"Sure should have."

"I'm awfully confused, I have to tell you. I thought this was all settled. Now Roy tells me that circumstances have completely changed, that Mr. Soverel will have to withdraw his offer if we don't accept. He wants to make his final presentation to the family in his office in Chicago tomorrow afternoon at three. He's putting us up at the Drake. He's even sending a limousine for us."

"Is that supposed to impress us? Never mind. Is Alex going along?"

"I don't know. He's up at Roy's today, playing with Tyler."

Jane fell silent, trying to absorb all the blows. She suddenly had a queasy feeling that she could keep on throwing punches, but the fight was already over, the decision rendered.

She glanced at her watch and realized she had just two minutes to reach the War Room, where the Hyon customer logistics team was about to meet.

She made a lunch date with Anne, then hung up and hurried out of her office.

"Whoa there!" called a resonant voice behind her.

Jane turned.

"Why don't you let Wade and me take over the Hyon meeting?" Don Landsman said as he caught up with her. "You've obviously got other problems on your mind."

Jane hesitated. If the reps from Hyon were going to be present, she wouldn't dare skip the meeting. But with the latest import crisis now defused, relations between the two companies seemed headed toward normalization. Meanwhile, the Akers customer logistics team needed to stay on top of the account, service it like all get-out—and pray that the trade union protesters in the streets of Seoul would stop shouting and go back to work.

"Don't worry," Don persisted. "I'll give you a complete report."

"Okay," Jane said. "Thanks. I *am* swamped."

JANE'S OFFICE, AKERS HQ 9:36 A.M.

"Oh, Ms. Malcolm, you were next on my list," Dave Ellsmere's secretary said brightly. "Let me just check, I think he's free."

Jane nodded to no one in particular and found herself listening to a cocktail piano rendition of "Don't Cry for Me, Argentina." She held the receiver away from her ear, staring down at a list headed "To do: Urgent"—made less than three hours ago and already out-of-date.

Dave Ellsmere's commanding drawl silenced the Muzak. "Jane, perfect timing. I need you."

"What's happening down there, Dave? Mark Gibian has been passing along some pretty disturbing rumors."

"Not just rumors anymore, I'm afraid. We've got a full-scale mutiny on our hands. Some of our alarm-prone bean counters got panicky over their long-term cost projections on the Macao-South China Air deal, and they've succeeded in spreading their panic to our board—to three or four members for sure."

"Aren't we pretty far down the road for this kind of second-guessing?"

"You kidding me, Jane? Our engineers have been on-site ever since Hawaii. And day after tomorrow six of us were due to fly to Macao for the ground breaking—or water breaking, since some of it's going to be landfill. I'm talking a big-time photo op, with honchos from South China Air and all the local Macanese politicos. So, yeah, I'd say we're pretty far down the road."

"But you're telling me you could still walk away?"

"Sure we can! By paying big-time penalties to get out of the contract and, in the process, screwing up our relations with half of China. That means not only Macao, but Beijing. That's how."

"I'm sorry, Dave. I thought we made a persuasive case."

"You did. That's why I need you to do it again. I'm going before a special session of the board tomorrow afternoon. Frankly, I'm putting my job on the line for this, that's how deeply I'm committed to your idea. Jane, I know you're in deep up there, but I'd really like you here tomorrow to make your presentation to the board again. I'll fax you the 'Mutineers' Mad Memorandum'—my name for it, not theirs—with the points marked I want you to demolish." He paused.

She waited for the punch line.

"I hate to spring this on you," he said. "But the board meeting's at nine. If you can somehow get here by eight, we could have some breakfast and talk things over in *my* office first."

It wasn't a request. Sure, he said "I'd really like you here tomorrow . . . " but what he meant was that he *expected* her to be there.

Even some nonquestions required an answer. "I'll make it by eight," she said.

"Great, that's just great. I knew I could count on you."

It was becoming almost comical, Jane thought after giving Dave her fax number and hanging up. She took her pen and scratched, above the first item on her "to do" list:

```
May 16 - Tomorrow!!
   v. early fly Dallas
  8 a.m. Ellsmere's office
  9 a.m. TSA board
```

What on earth was she going to wear? She'd been basically living out of her Hawaiian-trip suitcase for two weeks, with some minor additions, like the black outfit she'd worn to her father's funeral and a navy blue skirt and three washable white blouses she'd picked up at Akers Number One. Ocean Plains was definitely not one of the great shopping cities of the Western world. There was one dress shop left on Main Street, with fashions for the septuagenarian set, and a thrift shop run by the Presbyterian church. Beyond that there was only Akers.

AKERS PARKING LOT
9:56 A.M.

Jane was angling toward her car when an out-of-breath Lenny Cristofaro caught up to her. "Ms. Malcolm, can I talk to you?"

"I thought we settled on first names, Lenny. God knows we've been in enough meetings together."

"Yeah, I forgot. Okay. Jane, can I talk to *you*?" His face bunched into a cheeky smile that made him look about twelve, a few tufted gray hairs notwithstanding. Lenny always gave her the impression, as did certain former music prodigies, of someone whose maturation had ceased right around puberty, thus rendering his boyishness eternal.

"How can I help you, Lenny?" she asked, feeling slightly schoolmarmish.

"It's more I wish I could help you, Jane. That goes for Mimi as well. We both feel kind of superfluous, if you know what I mean."

"Aren't we scheduling enough meetings for you?"

He chuckled. "Seriously, with the network down, and Singh doing all the cyber sleuthing . . . I keep volunteering to help him, like maybe running some simulations for him, but he never gets back to me."

"That's kind of a basic policy decision, Lenny. We want Mr. Singh insulated as much as possible."

"Sure. The bad guy could be any one of us. The thing is, Mimi and I are left with just sys-admin junk, redrawing wiring schematics, updating user manuals. I don't feel like I'm earning my pay here."

"Have you talked with Chip Bragan?"

"Yeah, he's given us some piddling little tasks. But we could really help with some of the PC stuff you got going, maybe take some of the workload off Prince Khalif."

"You don't have to call him Prince, Lenny. And I'm afraid that's up to Khalif, not me. I can tell you this. One way or another, it won't go on too much longer."

"You mean the system will have to restart?"

"No, I *don't* mean that. I mean this operational phase will come to an end in a few days. Be ready for something, I just don't know what." She smiled. "Sorry I couldn't do better."

"That's okay." He turned and began trudging back in the direction of the data-processing center. With his head drooping that way, he reminded Jane of a disconsolate boy walking home from school, kicking at rocks and cans. Of course he wanted to be more involved. How else would he get praised?

**AKERS STORE #1,
OCEAN PLAINS
10:10 A.M.**

Jane was pleasantly surprised by the stock levels in all the departments she walked through. The store manager and his assistant both confirmed her hasty findings. While there were still shortages, critical and high-velocity SKUs were all above safety levels. To get this done on a daily basis, they were all working harder and longer, even volunteering to work through lunches and breaks. Yet employee morale was surprisingly high.

"You know what was really demoralizing, Ms. Malcolm?" Glenn Hawkes asked. "Seeing customers unable to find what they needed, and being unable to help them."

"By the way, those radio spots turned out to be terrific PR," volunteered Glenn's assistant, Pattie Daran, the redhead Jane had connected with on her store tour five days ago. "While focusing on what brands or items we might be out of each day, we're also able to highlight special sale items."

"You'll have to credit Mr. Skoggs for that idea," Jane said. "He told me it might generate extra sales."

"Who's Mr. Skoggs?" asked Glenn. Pattie seemed equally in the dark.

"He ran the local supermarket here on Main Street for twenty years. He went out of business about a dozen years ago."

"Oh," Pattie said. Glenn merely nodded.

Jane chalked up the underwhelming response to either a shortage or excess of sensitivity. "If you'll excuse me, I've got to do some shopping. Let's see how you're doing in women's clothes."

Quantity wise fairly good, was Jane's verdict on the department called "The Akers Woman." Racks were full, being browsed by a respectable Saturday morning turnout of customers. But the clothes didn't strike her as right for her upcoming meetings. Still, she had no option; she *had* to find something suitable for tomorrow.

She clicked her way through rack after rack of dresses and suits, her cell-phone clamped under her chin, on prolonged hold. Finally she came across a black-and-white-checked blend suit in her size. She held it up to herself, trying to visualize wearing it as she faced TranSonic Airex's conservative board.

Worth a try-on.

Suddenly the receptionist with the stilted accent was back on the line. "I'm putting you through now, Ms. Malcolm."

"Tom Soverel," said a mellow baritone. Soverel's craggy features remained fresh in her memory from their brief meeting at her father's memorial service.

"How nice of you to return my call so promptly," he went on. "It's not two minutes since I left the message."

"I don't want to disappoint you, Mr. Soverel, but I didn't get your message yet. Were you calling to finally notify me of your latest offer to the Akers board, and of our accelerated deadline to accept it?"

"Among other things, yes."

"Why wasn't I notified last night, along with the other family partners?"

"I want to apologize for that, Ms. Malcolm. I only just learned of your voting status."

"May I assume, then, that I'm also invited to your office at three o'clock tomorrow to hear the details of your proposal?"

"Of course. But there's another matter I'd like to discuss with you before then. Could you meet with me an hour earlier?"

"I'm afraid you've lost me, Mr. Soverel. Why would I do that?"

"There are matters you and I need to discuss, Ms. Malcolm. That's presumptuous, isn't it? I should say, there are things that *I* need to discuss with you. And I hope you'll do me the favor of listening."

Jane found this an interesting, if puzzling, gambit, but she had her own objective to pursue. "I have no problem with that, Mr. Soverel, but I do have a problem with tomorrow. I have to be in Texas to meet with a client's board. That can't be postponed. Are you willing to postpone the meeting with my family—and this one with me alone which you propose—until Monday?"

"That would mean changing the deadline I set for accepting my latest offer, Ms. Malcolm."

"That's correct, Mr. Soverel."

"What time is your Dallas meeting?"

Another surprise. He knew which client. Did he also know why the TSA board was meeting? "Nine o'clock."

"When might you be finished?"

"I hope to be back here right after lunch. Why?"

"You're taking the company jet, then?"

"That hasn't been arranged yet. What's that got—"

"Excuse me again, Ms. Malcolm. My point is this: If you're going to be done in Dallas that early, perhaps you *could* get here tomorrow afternoon. I think you'll find it worth your while."

"If I'm able to manage that, will you agree to postpone your deadline to the family till the day *after* tomorrow?"

"You're a very good negotiator, Ms. Malcolm. You want me to postpone that meeting against a promise that you'll *try* to get here tomorrow? Whether you make it here or not, the deadline is extended?"

"That's right."

There was silence on the other end. Jane did not make a sound. She even breathed away from the phone.

One beat. Two. Three. Then: "All right. We'll reschedule the family meeting for Monday, along with the deadline. Monday morning, ten o'clock sharp, in our conference room. I'll notify everyone directly. But I'm penciling you in for tomorrow at three. I'm on the sixtieth floor."

She didn't think he'd be found on the third or fourth floor.

"I'll call you from Dallas when my meeting is over and let you know if I can get to Chicago by then."

"I look forward to seeing you, Ms. Malcolm."

"I can't imagine why, Mr. Soverel."

She hung up, glanced around, noticed that the department was busier than when she'd arrived—and only then registered the rather flimsy suit still draped over her arm. She frowned, more in general than at the garment, and hung it back up. Whatever Tom Soverel was up to, she'd need something better to wear.

Damn. Where within twenty minutes of here do I find two terrific outfits, one really wow and one conservative enough to double for Texas?

45 MAPLE STREET
12:22 P.M.

"Anne, I think by all means you should go and listen to what the man has to say." Jane had just finished a tuna salad sandwich—and telling her stepmother about arranging a one-day postponement of the family meeting with Tom Soverel.

"All right," Anne said, pouring more tea into Jane's cup. "But that's not what I expected you to say."

They were having lunch in Anne's breakfast nook. Through the sliding glass doors, Jane could enjoy what her father liked to call his "aviary"—really only a grouping of plastic bird feeders. But it did make for fascinating watching, as the different varieties flitted in and out of view. Today, Anne had pointed out, they had a celebrity visitor, a male cardinal preening to display his full finery.

"Anne, how could I possibly advise you to do otherwise? I wouldn't want you to do anything to jeopardize your position."

"What position is that?"

"You may not realize it—though I suspect you do—but aside from the business, *you're* what's keeping this family together."

There was a fleeting sparkle in her stepmother's eyes. She had guessed correctly: Anne *was* aware of her influence and pleased that Jane had recognized it. Truth be told, she was slowly recognizing more and more about Anne Akers. Repeatedly Anne had a complicated role to play: first as the secret "other woman," surely not an easy position to be in for a woman with Anne's evident dignity; then as the public "other woman," when Joycelyn left Ocean Plains with such dramatic swiftness. That was followed by her marriage to Royal, which won her the unenviable dual roles of second wife and step-mother—which, Jane guessed, her father had probably seen as more of a sinecure than a challenge. Yet by some miracle she had found the inner resources to play all those roles in turn. She had over time soldered obvious links of affection to all the family members—to Aunt Louise and Grace, to both her step-sons. Even Jane had softened toward Anne once she gave herself the chance. Only Joycelyn seemed immune. But that, of course, said more about their particular situation than about either woman. The most remarkable thing about Anne's conquests was that they seemed so effortless.

"Well, I've tried," Anne said—as if to correct Jane's very thought.

"And succeeded. It seems to me that you've raised being nice to an art form. Why *are* you so nice, by the way?"

Anne laughed. "Why, that's been my job, what I always did." Over her teacup she studied Jane, considering whether to go on.

She did. "There was a saying my mother liked so much she worked it on a sampler: 'Wherever you go in this world, do what you can.' Maybe she worked it somewhere inside me as well. Or maybe the sampler caught my eye so many times that it sank in. In any event, looking back, I suppose I can see a pattern developing. It began with my sisters, growing up. It's how I was with George, my first husband. Working for Royal, it was definitely true. And when we married, that certainly didn't change. By now, I wouldn't know any other way to live."

"I'm not surprised. Which is why I could never put you in the position of damaging your relationship with Roy. He obviously wants all of us in Chicago as keenly as Soverel does. I suspect they've been working together on the deal all along. Doesn't matter. I'm going to be there, too, and I'm going to make a counterpresentation—why I think the company should remain intact. At that point, depending on your conscience, you can choose to vote with Roy or against him. And if you vote with him—please hear me, Anne—I will absolutely not hold it against you. Do you understand?"

Anne nodded. "And I believe you. Thank you."

"You're welcome."

"Now, may I offer *you* a small bit of advice, in more or less the same vein?"

"Of course—but first, I need some woman-to-woman advice."

"I'm not the expert you might expect in a woman who's had two husbands."

"I didn't mean relationship advice." Jane laughed. "Though God knows I probably could use it. Anne, you always look impeccable. I desperately need not one but two seriously presentable outfits for tomorrow and Monday—actually, one has to be at least on the cusp of smashing. I went to Akers this morning, but . . . " She held up her hands in a gesture of defeat.

"Is that all? As it happens, I *can* help. Molly Givens used to live in Boston, and she's got quite a sense of style. When her nest emptied, she started a small business right in her house. She doesn't actually have stock—it's not a store. Every season, she gets a selection of samples—dresses, suits, some dressy things—from a few manufacturers she deals with, and she has a kind of open house for her regular customers—where we can order what we want from the samples. Nearly all her business is special orders. But looking at you, I bet you'd fit into a sample size, and since spring is nearly over, I'm quite sure she'd be willing to sell you a couple. I'll call her right now. When would you like to go over there?"

Jane glanced at her watch. "Is she nearby?"

"Two blocks away."

"Do you think she might let us come over right now?"

Anne wasted no time in finding out. "She said now is fine. And she said the answer is yes if you promise not to tell anyone."

"Of course I'll promise."

"No need. I already did it for you."

When they left Molly Givens's house forty minutes later, Jane patted the plastic bag she carried over one arm and said, "If all your advice is that good, I don't want to miss a word of it. You were going to tell me something?"

"That's right, I was. Just this. If things don't go the way you want them to tomorrow—I'm sorry, I mean Monday, don't I? But if that happens, you mustn't hold it against yourself. Do you know what I mean? Don't punish yourself over it. What you've already accomplished with the company has been a miracle, and everybody knows it, even Roy. Especially Roy, though he won't tell you. But I see your nature, Jane. It's so much like Royal's. You strap on the blinders and drive yourself like a dray horse. I used to try to coax Royal to take a few days off—at least the odd day. But he never would. Even on our yearly family vacation, he'd manage to fit in some account he could call on, or a rival store he could case. He said that was what it took to build an empire. In a sense, his employees were his real family, not us. We always took a backseat. I'd just like to see you balance things a little more."

"Believe me," Jane said, "I *do* want to take a break after all this. And have a life outside the office. As for family, well—"

"I don't mean just getting married and having babies—"

"Oh, yes, you do!" Jane said, laughing. "But, you know, you've already given me back my own family, a lot of it, anyway. Beyond that, well, we'll just have to see, won't we?"

AKERS HQ
1:25 P.M.

As Jane pulled her car into the Akers entrance, the guard stood

outside the gatehouse, talking to a big man in a brown corduroy coat. The man's back was turned to her, but something
about the tilt of his head seemed familiar.

She nosed up to the gate, lowered her window, and inserted
her Akers ID card into the reader, then glanced sideways.

"Chevy!" she cried.

Chevy Johnson turned. "Ah, hell, Captain! You caught me!
Here I was trying to talk my way in past this diligent gentleman
so I could surprise you."

"You have, you definitely have." She opened her passenger
door. "Hop in."

Chevy complied, sliding in beside her and filling up the
front seat. In her head, Jane counted days. Only ten or eleven
had passed since their dinner atop the Sheraton Waikiki and
that crazy wonderful plan for an around-the-islands cruise. Yet
feeling that carefree seemed ages ago.

He leaned toward her casually. Jane turned her cheek toward
the kiss but followed up with a generous smile. "I hope you're
not here for your rain check on that cruise. I'm not allowed any
fun. Or are you flying into O'Hare now?"

"Neither. I'm here looking for work. A week ago, Orient Air
laid off about a third of their pilots and ground crews."

"Chevy, I'm so sorry! *And* I have to apologize for not returning your calls. I meant to, I've just been swamped. A couple of
times I got as far as reaching for the phone—honest!—but then
someone would page me or the phone would ring. And by the
time that emergency had been taken care of, I'd forget." She
made a wry face.

"Hey, it's okay. I read the papers."

"I do, too—even if it's at two in the morning. And I do
remember reading something about OA making 'restructuring
moves.'"

"That's the phrase," Chevy said. "Now that 'downsize' seems
to be a dirty word, companies 'restructure' instead. Whatever

you call it, me and a lot of other damn good pilots are out on the bricks. I was hoping maybe you could recommend me to your friends down in Dallas, get me hired as a freight dog."

Jane bit her lip. This wasn't the moment to tell him.

Chevy talked right past her silence, probably propelled by embarrassment. "When I couldn't get hold of you, I hitched a flight and came in person. Desperate men do desperate things." He grinned, but the dark circles under his eyes were no joke.

"Well, I'm glad you're here," Jane said firmly. She pulled into her parking space, and they climbed out together. "Welcome to the wonderful world of Akers."

Chevy glanced around. "Been a few changes since I landed here last. That building over there, for instance, with the big dish on top? That was the fighter-wing hangar."

"That's our data-processing center. I forgot you'd be familiar with Siler Field. Come on in, and we'll talk."

As they walked along the corridor of offices, Jane was aware of attracting more than the usual passing glance. Chevy was hard to ignore, with his wide wingspan, long, lazy gait, and Marlboro Man looks.

"Busy place," he said. "Hey, I know that guy."

"What guy?" she asked.

"Don't turn around. The crew-cut guy with the funny eyes who just threw you that cornball salute. Grooms, am I right?"

"Yes, Robert Grooms," she said, shunting Chevy into her office and closing the door. She cleared a chair for him. "You must have met Grooms in Vietnam, right? He told me he was with some marine outfit stationed at Da Nang."

"I don't know about that. When I knew Grooms—actually, I just saw him around a lot and heard things, didn't really *know* the guy—he used to be chief errand boy for the CIA station chief in Saigon. Grooms would ride around on this big, full-dress Harley. Word was, anything you wanted, Grooms could get it for you—at a price. What's he doing here?"

"My brother hired him to be chief of security."

"Oh, man, I don't think Grooms is the guy you want in that job, Jane, I really don't. You check him out?"

"Roy's his boss. All I know about Grooms and Vietnam is that he knew my grandfather—"

"Maybe he's just saying that."

"Trust me, Chevy, I can tell. I don't say they were friends, but he definitely knew Granddad. And I must admit I've enjoyed talking with Mr. Grooms. Besides, right now I don't have any inclination to get him removed from his job. I'll go further. He's made some very good suggestions about how we could improve our security around here. Besides, butting in like that on Roy's turf would take more tact than I have right now—not to mention time. Now, since I *can* only spare a few minutes, how about if we talk about you instead of Mr. Grooms?"

"Sorry, didn't mean to step on any toes."

She raised both legs slightly. "Toes are fine."

He smiled. "And I do realize you're busy, Jane. I hear you've been working your usual miracles around here."

"That remains to be seen." Time to tell him. "But, to get to your problem, unfortunately right now my stock isn't worth that much with TranSonic Airex. They may deep-six my whole China idea, and my firm with it."

"I thought it was a done deal."

"Well, it's come undone. I'm going to Dallas tomorrow morning to talk to the board and try to stick it back together."

"You are?"

"Yes. That's where you come in, friend. I obviously need to get there in a hurry, and it seems our company pilots—we've got five here full-time—are all busy tomorrow, flying regional managers around to check out how our stores are coping, even making some emergency stock deliveries of small but critical items like pharmaceuticals."

"If you want to get to Dallas in a hurry-up, I'm afraid you're gonna need something faster than—" Chevy stopped short.

"Faster than the old Beechcraft my father flew, right? That's okay, Chevy, your point is well taken this time. My father was death on corporate fat. That's why we have all those second-hand puddle jumpers you've no doubt read about. But he did finally okay an old Citation Two for commuting to our new stores in Arizona and the Northwest. Can you fly it?"

"Are you kidding? I'm checked out on every bizjet you can name, from the oldest Lears to the newest Gulfstreams. As for Citations, used to fly them for a living."

"Good. I've got to get my brother's okay to use it, then arrange to reimburse the company, since the trip has nothing to do with Akers. I need to be in Dallas by eight. What time do we need to leave?"

"Did you just hire me, Cap?"

"I'm sorry, Chevy. There I go again, my brain in overdrive. Yes, I'm offering you a job—I don't know for how long. My tenure here may come to an abrupt end day after tomorrow. It will definitely be over a few days after that. But maybe you can stay on. All I can offer for sure is a few days' work."

"Sounds good to me," he said. "To answer your question, figure the mileage to Dallas roughly at nine hundred. The Citation Two has a high-speed cruise around four hundred—that's miles, not knots. Which means—"

"We're about a hundred miles south of Chicago, don't forget."

"I didn't. Figure two hours each way. TranSonic Airex is downtown?"

"Yeah, on Commerce."

"Love Field is the closest airport. A morning cab ride from there during rush hour . . . better leave yourself twenty-five, thirty minutes to be sure. Means we take off at five-thirty. You still game?"

"Absolutely. How about you?"

"I'll be there in good time for preflight. Somehow, Cap, you always seem to be where the action is."

"Don't I, though?" She suddenly remembered his situation. "Money," she said.

"How about whatever you pay your pilots, prorated by the day?"

"I'll find out what that number is and see if we can do it. If not, I'll get as close to it as I can. Okay?"

"Deal," Chevy said.

As they shook hands across the table, Wade Crain walked in. "And here comes your guide now. Not his real job, you understand. By the way. Wade, what *is* your title?"

"Frankly, I've lost track." He extended his hand to the taller man, now on his feet. "Wade Crain."

"This is Chevy Johnson, Wade, an ex-air force major, ex-commercial pilot for Orient Air. Chevy and I met in the Gulf. When I called you from the car to tell you about Dallas, I didn't know yet"—she smiled at Chevy—"but it's Chevy who'll be flying me to Dallas tomorrow in the Citation. Got to check that with Roy, but I think it'll be fine. Fact is, I hope we can use him on some other company flights. Could I ask you to take him over to the hangar and introduce him around?"

Wade nodded, and she turned back to Chevy. "Wade's also a pilot, like a lot of our executives. And he's not shy about answering questions. Try him with any you have."

"Welcome aboard, Mr. Johnson," Wade said, with, Jane thought, less than his customary friendliness. Maybe he resented being handed yet another assignment when he was probably as busy as she was.

After the two men had exited her office, another possible reason for Wade's coolness occurred to her. If it was Wade who'd been watching her night before last, he might think she'd sent for an old friend to keep an eye on her.

Only he wasn't watching you.
Right. So he's a little jealous. That okay with you, Dad?
If it's okay with you. I've always liked the boy.

AKERS HQ
1:48 P.M.

"So how'd you two meet up in the war?" Wade asked as he and Chevy started walking in the direction of the airfield.

"Jane never told you about her Bronze Star?"

"Her father did, plenty, but Jane never mentions it. She helped evacuate a whole busload of injured people from our barracks in Dhahran, right, the one that got hit by a Scud?"

"Right. I just happened to be billeted next door and ended up driving the rescue bus. It was quite a night. And Jane's—well, I guess you know by now—she's quite a woman."

"Yes, she is," Wade said. "You two keep in touch all this time?"

Chevy cast an inquisitive glance at Wade but said, "I was married during most of those years, or I would have, believe me. Ran into Jane purely by accident in Hawaii couple weeks ago, Wade, that's all."

"None of *my* business."

"You sure about that?"

"Right here's our truck maintenance yard," Wade said. The entrance to the long line of service bays was currently blocked by bobtail trucks waiting in line for the diesel pumps.

Chevy was only too glad to drop Jane as a subject. "Back when all this was air force," he said, "that was the motor pool. Let's see, that the fleet manager's office?" He pointed to a small office window beside the garage area. Through vertical Levolor blinds a bank of glowing blue monitors could be seen.

"Ed's actual title is transportation manager," Wade said. "Pretty nifty setup he's got in there. A lot of our trucks now have GPS receivers—global positioning satellites—just like airplanes do, so Ed can basically track those shipments inside of thirty seconds and locate them on a computerized map to within a hundred meters."

"I'd like to take a look at that."

"I'm sure Ed'd be glad to show you if he's got a minute. Look, since you're familiar with Siler, we can skip the tour. Why don't I just introduce you to Marty Daniels, who runs our Aviation Department, and he can check you out on the Citation?"

"You fly it?"

"No way. I stick to our crop duster fleet. A hundred fifty knots is plenty fast for me."

"I like speed." Chevy grinned.

Wade gave him a level look. "I'll bet you do. There's Marty now."

JANE'S OFFICE, AKERS HQ 2:02 P.M.

"How bad is it?" Jane asked.

"I'm worried about Louise," Eliezer Grey said on the other end of the phone. Elie had volunteered to keep in touch with the family partners on Jane's behalf. He had actually been visiting with Louise Akers that morning, he explained, when Roy Jr. had phoned her. Elie had seen the fear spreading across her features as she'd listened.

"Honest to God, Janie, when she hung up that phone she was absolutely convinced that if you didn't sell immediately, you'd all be holding worthless stock in a worthless company. Wasn't anything I could say to change her mind."

"I wasn't there, of course, but my guess is she's not so much scared of losing her security as she is of even the slightest temptation not to be totally loyal to Roy. She's very devoted to him. I think she feels she has to stay in his corner no matter what. But maybe she'll listen to me. I already told you about

Anne. Mother is hopeless. I haven't talked to Alex yet. I think we've still got a shot, Elie. You heard about Soverel's limo?"

"Yeah, the nerve of that two-bit hustler! Gonna round up all the Akers family members like so many head of beef cattle. You got that stopped, right?"

"I decided it wasn't worth making a fuss over." Jane had already filled in Elie about her trip to Dallas. "I didn't tell you something else, and I need to. Soverel's asked me to meet with him in Chicago before the family meeting."

A beat. "You going to do it?"

"I don't even know if I can make it. But I *am* curious. I told him I'd try—bartered that for his delaying the deadline a day."

"You watch out, child. The man'll take you up to a high place and show you all the kingdoms of the world—"

"Oh, Elie, come on! Tom Soverel's office may be high up, but he's not Satan."

"I was speaking metaphorically." He smiled, then stopped smiling. "But the danger is real, Janie. He's after the whole she-bang, and now he sees you as the key to getting it."

"I'll remember to hold on tight to my immortal soul. If I go. That sounded flip. You do know I take my soul seriously?"

"I know it, gal. But if I've got Soverel pegged right, you may need to—seriously—hang on to it. Do you know *that*?"

Jane was depressed when they hung up. She went to the window, bent the venetian blinds to peer out at the sprawling complex. A gust of wind tumbled a weed across the cracked concrete. She had read with pride how Akers trucks had rolled onto the abandoned Siler Air Force Base, bringing it back to life as a thriving corporate headquarters. Now she could envision the whole place deserted again, her father's dream vanished with it. The town wouldn't last very long, either, not after the 4-Mor trucks emptied out the complex.

At least she wouldn't have to see it; she'd be back in San Francisco.

If she performed a little magic in Dallas.

**WAR ROOM,
AKERS HQ
2:28 P.M.**

"Don't just give me one more problem to add to the list, Wade. How do *you* suggest we cut down on wall-to-wall inventorying? If we're just plain wearing down our workforce—"

"I'm afraid we are, especially at the DCs, even though they've put on a lot of new people. But we can streamline."

"I know, I know," Jane said. "Another project in Khalif's in-basket. What about those good-news reports you were mentioning?"

"Sure." Wade grabbed a fax. "As of two o'clock today, Plainfield was up to seventy percent of capacity."

"Fantastic! I'll get onto Khalif about getting us more hand-held scanning terminals to read the bar codes. This will give us greater accuracy and productivity."

"Fantastic!" Wade smiled.

At the door he stopped and, without turning around, said, "Impressive guy, your pilot friend."

"Yes, he is."

Wade kept on walking.

**AKERS HIGH-RISE GROCERY WAREHOUSE,
PLAINFIELD, ILLINOIS
2:34 P.M.**

They gathered to celebrate in the computer room above the warehouse floor, touching Styrofoam cups of sparkling cider. In the cavernous space below, visible through the long window, pallets were rolling on all the conveyers, to and from the ten-story steel racks, past the outbound inspection station, down the sloping track to the selection warehouse next door. Over there, in every dock door, Akers trailers were backed up, getting stuffed with groceries for stores in all the Lake Michigan perimeter states and some as far as Iowa and Ohio.

"Job damn well done, troops!" saluted Hub Smith, the distribution manager.

"Amen to that, in spades," said Jack, the assistant manager.

After four days of excruciating paralysis—Monday through Thursday—while they retrieved and rescanned every pallet in the racks in order to rebuild the inventory database in their automated storage retrieval system—they were back on-line and moving food out the door. And doing it at nearly three-quarters of standard throughput.

"I don't know how long we're gonna have to keep doing it this way, Anita," Jack told the woman who managed the ASRS, "but we're all real proud of you."

"Thanks, but that Saudi guy in Ocean Plains deserves most of the credit," she said. "He's the one who really solved our order-entry bottleneck with that stripped-down fax-and-scan input. Otherwise we'd have a whole roomful of data-entry folks, half of 'em rekeying all the store orders and the other half checking their input and making corrections. The really encouraging thing is, according to Khalif, with some small modifications to our IDCS forms, we can do even better."

There were several weak links in the chain, of course. A store manager might check the wrong box on the order form. More errors could be generated during faxing, scanning, or the software conversion from graphic map into data file. But the end result was a whole lot better, all agreed, than getting corrupted downloads from the Ocean Plains mainframe. Plainfield might still end up shipping cornflakes when detergent was wanted, but if that happened, the culprit might be, say, the misplaced point of a No. 2 pencil. The saboteur was out of the loop.

A warning siren screamed. The celebrants all rushed to the window that overlooked the floor below. Manny, the section manager, spotted the problem first. A poorly stacked incoming pallet had somehow gotten past the straightener, then tipped its load, causing a blockage and an automatic conveyer shutdown.

There were some audible sighs as a warehouseman cleared the track, then routed the offending pallet to the nearest stacker crane for a restack and rescan.

Anita took another slug of cider. "So, Jack, how about springing for some actual champagne?"

"Get us up to eighty percent of capacity, and you've got yourself a deal."

TRANSPORTATION MANAGER'S OFFICE, AKERS HQ 3:40 P.M.

"When I first started in fleet management, Chevy," Ed Basinski said, finishing up a quick tour of his small, high-tech office, "if we wanted to get hold of a driver, we waited for him to finish his apple pie in his favorite truck stop and call in. Now I can page him, call his truck phone, even flash him a computer message. He's got a display unit mounted on his dash—comes, believe it or not, with a sealed, coffeeproof keypad.

"Except I probably don't *need* to talk to the guy. I sure don't need for him to tell me his engine's running hot and that I should maybe schedule some preventive maintenance. With electronic sensors and automatic data collection, I can sit right here and see a whole lot more of what's going on in his vehicle—engine *and* trailer—than he can. I'm talking engine performance, fuel mileage, fuel pressure, oil pressure, tire pressure, oil and water temperature. You name it, Chevy, we can get a real-time readout, send him a message, and tell him what to do. Someday we won't even bother the guy, just let 'em go on listening to his radio while we correct the condition from here."

"Ed, you make it sound like Mission Control down in Houston, turning the astronauts into glorified monkeys."

"You think you're kidding, right? Suppose a wheel bearing is running warm; we get an alert. Not now, but day after tomorrow, say, we'll be able to activate air-suspension valves in that truck and take weight off the axle. It's coming. But AVL is already one hell of a tool, not to mention one hell of a toy."

Basinski jerked a thumb toward a line of large-screen computer monitors, each of half a dozen displaying a digitized map of a different region of the country. "It's PC based, running our own modified version of AVL software."

"What's AVL?" Chevy wanted to know.

"Automatic vehicle location. We don't have all our trucks wired up yet, but on those we do, I can tell when a left rear taillight burns out, when a reefer unit—a refrigerator trailer—needs repair, when to schedule engine maintenance. Do your preventive maintenance at the right time, and you can make diesels last damn near forever."

"Okay," Chevy said, "but with a few thousand trucks on the road, how can you monitor all that incoming data?"

"We can't. We swallow it whole, crunch it, spit it out in reports or queries on demand. Critical items, though, they do get brought to our attention real fast. Thank God they didn't pull the plug on *my* system."

"So all your vehicle location is done by GPS?" Chevy said, referring to the global positioning system, the satellite-based radio navigation developed and operated by the Defense Department.

"Let me show you," Basinski said. "Because you're a pilot, you know how GPS works, but I bet you've never seen anything like this."

He swiveled his chair toward the nearest display monitor. "Most of our trucks have a built-in GPS receiver and a microprocessor linked to a nationwide cell-phone network. I'm going to call Unit 1605, see what's cookin'." The fleet manager punched a sequence on his keyboard, and a map of the western

U.S. appeared on the monitor. "Aha, there he is, South Dakota, traveling west on I-90. See, this window here gives us a time stamp with readouts of current speed, direction, and exact location."

"By exact, Ed, you really mean give or take a hundred meters, right? I'm afraid that's all the DOD will allow us civilians."

"I mean *exact*—down to *two* meters. We do our own differential correction, pretty close to military accuracy. Now let's get up close and personal with 1605." Basinski hit a key, and the map display zoomed in, showing a tiny truck icon and a larger dialogue box, with vehicle and driver ID, velocity, heading, latitude, longitude. "He's thirty-two miles east of Rapid City. Now watch this." The dialogue box changed to show vehicle statistics.

"I wish I had this in every damned truck. You know why? About a week ago, we had a driver ditch his rig in western Pennsylvania—a reefer unit loaded with food. It took us most of a day to track it down. The governor of PA got called in on that one. *If* that same truck had been wired up, we'd have pinpointed it immediately, and we'd have saved its load.

"Another time—this is about a year ago—we had us a brand-new driver pick up a load of goods in Memphis, bound for Nashville on I-40. Only the damn fool headed west instead of east. Guess how we found out? The driver got himself killed in a five-car pileup outside of Little Rock. You can't believe what can happen out there on the road."

"I don't doubt it," Chevy said. "I'll have to tell you some pilot fuckup stories sometime when you're not planning to fly anywhere. But what about those little handheld GPS units over there on the shelf? They look like the one I keep on my boat. What do you use 'em for?"

"They're more than GPS units. We call 'em MEBs—mobile electronics boxes. Like I said, some of our trucks aren't fully wired. In an emergency I can hand those out. Each box has a cell-phone we can activate to send us a GPS position. Got a couple double-A batteries, last about twelve hours. What kind of boat you got, Chevy? Power or sail?"

"Bit of both. Fifty-foot motor-sailer, ketch rigged."

"You keep that on Lake *Michigan*?"

"More like Lahaina."

"Maui! Now you're talking. Me and the missus head out there every winter. We gonna look you up one of these days, you watch out."

"Do it. Better yet, come in the spring, and I'll take you both whale watching in the Auau Channel. So what are those little units beside the MEBs—or whatever you call 'em?"

"External antennas for the GPS receiver. High performance, low profile, got a magnet base. You just slap 'em on the vehicle roof. Some of the MEBs are also magnetized, so they can stick on anything metal. We can call the unit, activate it, patch it into our dispatching software. Better than LoJack, if you want to track down a car. Works nationwide."

"Hey, I oughta buy me one of those."

"Tell you what, Chev. Why don't you just borrow one?"

**AKERS PARKING LOT,
AKERS HQ
4:02 P.M.**

Chevy strolled through the lot, just behind a line of cars nosed against a concrete-block wall that flanked the general offices. This was executive row, obviously: not too many fancy imports, mostly pricey Detroit models, with a name stencil sprayed on the wall in front of each car. Jane's rental Pontiac and Wade's big black Blazer were two exceptions to the luxury lineup. Another came almost at the end of the long line, a full-size blue cargo van. The slot was marked "ROBERT GROOMS."

Chevy took a moment to admire the vehicle, a windowless Dodge Ram V-8 of recent vintage. It was in immaculate condition and expensively accessorized. Aluminum wheels and running board gleamed in the afternoon sun. The high roof—at least equal to Chevy's six four—sported luggage rack, whip antenna, and a bar with four quartz halogen daylighters. Roof access was via a ladder mounted on the left rear door.

Probably got a Jacuzzi inside. Chevy glanced around the deserted lot. A video surveillance camera atop the general offices looked to be scanning the other way. He reached into the canvas bag Ed Basinski had given him, fingering the little metal disk—an inch and a half in diameter, maybe a quarter-inch thick—that housed a wireless GPS patch antenna.

Go!

Chevy stepped onto the van's passenger-side running board. He could see over the top now. He slapped the matte black disk down in the small space between the light bar and the base of Grooms's whip antenna, felt the magnet grab metal. As he stepped down, Chevy kept lowering into a full crouch, then stuck the magnetized MEB—with GPS receiver and cell-phone—as far as he could under the car, feeling it bond.

He raised his head just enough to check the position of the surveillance scanner. Looked safe.

He stood and walked away, humming the U.S. Air Force hymn.

CAFETERIA, AKERS HQ
6:15 P.M.

Wade Crain sat hunched at a far corner table, dividing his attention between a corned-beef-and-cabbage dinner and the latest sales velocity reports generated by Khalif al-Marzouki's PC workstations. The printouts were fairly rudimentary, compared to the mainframe computer's instant information arrays—such as fifty-two-week rolling histories of every Akers SKU, available by store, district, or region. But Wade found these reports useful nonetheless. Khalif had done fine work under horrendous pressure. Wade turned a page, then realized somebody was standing over him with a tray.

"Mind some company?" Jane said.

"Please. Don't usually get celebrities in here."

"You can thank my devoted stepmother for that. She's been smuggling in home-cooked meals for me, care of Alex. Tonight my sweet tooth got the better of me." Jane set her coffee and dish of chocolate ice cream on the table, then sat down.

She took a spoonful of the ice cream, let it melt slowly.

"I'd love to see your expression if that was some gourmet brand of ice cream."

"I only eat that in the privacy of my home." Jane laughed. "The truth is, this sudden craving for chocolate wasn't the only reason I wanted to join you. It's personal."

His mind made an absurd leap of hope. A ridiculous leap back to a long-ago fantasy. Wade had often sat in the school cafeteria, watching Jane Akers slide her tray through the line, talking and waving to her friends. In his fevered imagination, but never in real life, she would wend her way through those clamoring throngs till she arrived at his table, then ask to sit down.

He told himself he wasn't a teenager anymore, and neither was she.

But since when was there a statute of limitations on dreams? "Yes?" he said.

"It's something I heard while I was talking to Anne a few minutes ago. About you and my father. She says he put you through college, and I was wondering, your coming to work at Akers, was that part of the deal?"

He took a deep breath, let it out. Let the dream go.

"If that's too personal . . . " Jane waited.

"Of course not. And what Anne told you is absolutely true. And he hinted he thought my working at Akers afterward would be a fair exchange, but the only absolute condition was that

I get A's." He grinned. "I did pay him back, by the way."

"How come you never mentioned any of this?"

"You've never mentioned your father—except in meetings. I got the idea you don't talk about him much."

Jane ate a little ice cream. "I'm working on changing that. So if you don't mind, tell me a little about you and my dad."

Wade told her, starting with the hook slide into third base that had torn up his knee and ended his college baseball scholarship and moving quickly to Royal's offer to bankroll a marketing degree at Northern Illinois University in De Kalb.

"Believe it or not, he even came to my graduation. Now, I don't say I wouldn't have made *something* of myself without your dad's help, but God knows what. As a high school senior, I really hadn't a clue where my talents lay, beyond turning a double play and hitting behind the runner. Your father saw something else in me, I guess."

"I'm glad. For both of you. His plans for his own kids didn't work out that well. But in your case, his vision panned out. He got a terrific manager."

"It did turn out okay from his point of view, but that doesn't change the fact that he took a heckuva gamble. He had no way of knowing what I'd do once I graduated."

"I think he did." Jane's smile went all the way up to her eyes.

And made Wade feel about as good as he could remember feeling. Sunrise in the Sierras, something he'd especially savored on his recent backpack trip, had offered no grander benediction.

"Your ice cream's melting," he said.

"Taste's just as delicious this way." She ate some. "Good things keep. Don't you think?"

Maybe. Just maybe, he thought.

He glanced at his watch. "I *think* I better get back to work."

He stood, picked up his tray. "But, yes, I definitely think some good things have quite a long shelf life." He grinned at what was left of her ice cream. "Or, in this case, dish life. See you."

"See you."

The Citation Two launched itself into the prairie night, banking and pointing its sleek nose south-southwest toward Dallas. Back in the jet's eight-passenger lounge, Jane was already talking on her sky-phone with Wade, going over further favorable reports from regional distribution centers.

"We're definitely turning it around," he was saying. "Only nobody but us seems to know it."

"Our in-house PR people may not be up to that big a job," Jane said. "Maybe we should hire one of those big New York PR firms to get the word out. Especially to investment bankers and industry analysts."

"I like that idea! Whenever we had a new stock issue, Royal used to fly a bunch of Wall Street types down here for a dog-and-pony show. We could organize another junket like that."

"Get Alex to take 'em all over to Medinah Country Club for a round of golf." Jane laughed. Stopped. "I'm serious."

"Damn right. Round up some celebrities for a pro-am."

A beat. Jane gave him the thinking room.

"Of course, you know the trouble with all these ideas—" he said after less than a minute—"we just don't have time. Not with Tom Soverel putting a gun to our head. Jane, have you decided about tomorrow?"

"If I'm going to see him? I haven't decided. If there's time, maybe. I'll call you when I get in tonight."

She switched off the phone and stared out at her windowed square of night and her reflection in the Plexiglas, grateful to Don Landsman for suggesting she fly tonight instead of predawn. Better yet, he'd arranged a company reason for the flight, reducing the amount she'd need TranSonic Airex to reimburse. As a result of Don's quick maneuvering, Travis Blackwood, Akers's regional manager for East Texas, was now sitting across the cabin from her, alerting his district managers over the sky-phone that he'd be calling on them a day earlier than originally planned. Chevy, meanwhile, was at the controls up front, with Drew Chastain, company check pilot, in the right-hand seat, overseeing Chevy's every move.

Don's way was better all around, Jane thought. Instead of a journey at an ungodly hour tomorrow, culminating in a rush-hour taxi ride into downtown Dallas, she'd spend the night at the Adolphus Hotel in the financial district, then walk a block to TranSonic Airex headquarters.

And if she used these two in-flight hours to review her South China Air proposal, during that short walk she'd feel clear-headed after a good night's sleep—maybe the best idea of all, considering what lay in front of her tomorrow. Mark Gibian had faxed her the last complete printout, along with notes she used during a follow-up presentation. As she glanced over them and the list of points Dave Ellsmere wanted her to hit even harder this time, she felt spikes of frustration over having to make a sale again that she'd already closed.

TRAVELING SOUTH ON THE STEMMONS FREEWAY, DALLAS
10:23 P.M.

Veiled by the limousine's smoked windows, the shiny skyscrapers of downtown Dallas made only a modest display against the night sky. In any case, Jane wasn't paying attention; her mind was still hiking among the interwoven intricacies of Asian cargo routes—until the cell-phone went off.

Slumped beside her in the backseat, Travis Blackwood slapped at his suit pocket, then smiled sleepily at Jane as he realized it was her call and resubmerged in his commuter doze.

Was it Dave Ellsmere being a Nervous Nellie, she wondered, checking to see if she'd made it? Or Don Landsman, with news so bad it couldn't wait till morning?

But the genial male voice belonged to neither. "Good evening. Is this Jane Akers?"

It was too late at night to start correcting someone about her name. Besides, she was feeling more and more like an Akers every day. "Right," she said. "Who's this?"

"Ah, good! Because the president wants to speak with you. He'll be on the line in just a moment."

"The president?" Beside her, Travis Blackwood jerked upright, eyes wide.

"Yes, *that* president." A chuckle. "If you'll just hang on a moment."

Somehow Jane knew this wasn't a put-on. And that it had to be something bad—she hadn't just won the Super Bowl. Korea again. Had to be—after yesterday's out-of-the-blue call from Stickley What's-his-name at State. But what could be urgent enough to involve the president? And why would he want to talk to *her*?

"Good evening, Ms. Akers." The famous folksy voice sounded tired, but it was definitely him. Jane imagined the big campaign-poster face, the fierce squint that turned so easily into a twinkling grin. Was he keeping late hours in the Oval Office, with his necktie loosened, or was he upstairs, pacing in PJs and robe?

Tired or not, he bounded ahead. "I understand that Treasury and State have been talking pretty much nonstop to Mr. Landsman and to your chief financial officer, but I wanted to touch base with you personally, Ms. Akers. This whole shebang has blown up so quickly that—"

"Excuse me, Mr. President. But *what* shebang?"

"Where have you been the last few hours, Ms. Akers?"

"In the air. Right now I'm in a limo just taking an off ramp into downtown Dallas. Mr. President, I apologize for being out of the loop. I should have been contacted immediately about any crisis. If this concerns Hyon Group, the last I heard—when I talked with your undersecretary for East Asian affairs—we had a handle on things."

The man sighed—big, the way he did everything. "A whole lot has hit the fan in the last couple hours, I'm afraid. You can get the details from your people. Basically, though, looks like your bankers in Seoul didn't have the funds to pay Hyon today for some big shipment. Word of that hit the streets, where there were already trade union protests going on about Hyon shut-downs in front of their headquarters and flag burnings in front of our embassy. The protests ignited into a full-scale riot. CNN's got video now—ground-floor offices in flames, armored personnel carriers in the street—"

"Oh, my God! What a horrendous mistake! Mr. President, of course we have funds! We have cash problems, but we're not insolvent! This should never have happened!"

"Well, it did—and it isn't done yet, not by a long shot. *For whatever reason*, your bankers over there definitely held up on a thirteen-million-dollar payment to Hyon today. Enough zeros to break the camel's back, I guess. Not only labor unions are involved now. As I understand it, Hyon's suppliers were already in a panic, convinced the company was about bankrupt because of its close ties to Akers and your well-publicized problems. That, plus this latest rumor about your virus infecting all of Hyon's computers and assembly plants—"

"Mr. President, there's no truth to any of that. And there was only *one* Hyon plant shut down."

"Nope. All of 'em got shut down as of noon today, make that tomorrow, Korean time—ten o'clock eastern. Company's just about belly-up. It's stopped ordering from suppliers throughout Korea and Southeast Asia, all of whom will have to cut their production drastically, throwing thousands out of work. Ms. Akers, we can't have that happen! So it's not a question of thirteen million now. I'm told it'll take around two hundred million to get Hyon back on its feet."

"Mr. President, Akers can't pay that, it's not our indebtedness."

"Not asking you to, though it looks like you folks gave 'em a nice push downhill. The thing is, *somebody's* got to ante up the funds fast. If Hyon collapses, that doesn't threaten just the entire economy of South Korea, but their political structure—which, I don't have to tell you, is awfully damned precarious right now. Hell, the president of Korea told me his government would gladly put up the money in a flash, only he can't—he'd end up in prison on one of those bribery-for-loans corruption charges that took down some of his predecessors. No, *we* have to solve this thing tonight." He took a long, loud breath. "And I think we have."

The limousine had entered the downtown business grid, which was deserted at this late hour, but Jane was seeing other streets, half a world away, aswarm in afternoon disaster. She listened closely as the president explained that an emergency agreement was being hammered out over the phone by congressional leaders, cabinet secretaries, and the chairmen of congressional banking oversight committees. With their support firmed up, the president was going to invoke executive authority to pledge the necessary loan guarantees from an executive-controlled fund, thereby circumventing the need for legislative approval.

Since part of those funds would be discharging Akers's obligations to Hyon, the president explained, the government was expecting an equivalent chunk of Akers's convertible preferred—"the same deal we gave Chrysler, only on a lot smaller scale."

"You're calling *me* to okay that exchange?" Jane asked.

"No, Ms. Akers, the treasury secretary is working all that out with Mr. Landsman and your board. That's not why I called you. I wanted *you* to know that we're not doing this just to prop up South Korea. We're also doing it because *you* folks matter to America, just like Hyon does to Korea. We need Akers not only to survive, but to prosper. Akers means a lot to *my* hometown, I can tell you that, and to towns all across the country. And I'm told that with your father gone, *you're* the key player out there in Ocean Plains, the one who's gonna make it come out right."

"Well, that's very flattering, Mr. President, but—"

"I'm not quite through. I want to remind you that the taxpayers made money on that Chrysler deal, Ms. Akers. And I want your pledge that you folks will work night and day to make sure this deal turns out the same way. 'Cause if it doesn't, I'm gonna be taking the political heat, and it's gonna be *intense.*"

Only one response seemed appropriate: "You've got my pledge, Mr. President. We'll do whatever it takes."

"That's all I wanted to hear."

As he hung up, the limo was gliding into the Adolphus. Jane let out a vast breath. Tomorrow's showdown with TranSonic's board of directors had suddenly shrunk to real-life size.

"Well?" asked Travis Blackwood.

She shook her head, punching the speed-dial for Don Landsman's number. "Sorry, Travis, I don't have time to recap."

DAY 7

SUNDAY, MAY 16

TRANSONIC AIREX CONFERENCE ROOM,
DOWNTOWN DALLAS
10:14 A.M.

Jane cooled her heels for almost forty minutes in a Texas baron-style waiting room before being called in to make her presentation to the TranSonic Airex board. Ten of the twelve directors she recognized. Board chairman Max Garvelmann introduced her to the two who'd been added since her last Dallas visit.

And she was on.

But as she began talking, she found it hard to retain her focus. Her mind kept wanting to go to other, even more urgent locations—like the Korean peninsula, the White House, and the War Room back at Akers. She was high on coffee, her feet not quite grounded. She'd been up half the night in her Adolphus suite, phone in hand, talking with Don Landsman, with Elie and Wade, then with Isaiah Davis, Akers's CFO, her brother Roy, Madeleine Archer, and several other board members.

Don had apologized profusely for letting her be blindsided by the White House call. The Korean crisis had erupted only moments after she'd taken off for Dallas. Don had found himself fielding one emergency call after another. After a frustrating shouting match with their Korean bankers, he'd been dragged over the coals himself by a battery of Washington's power elite. Two hours of telephonic bludgeoning had culminated in a withering crossfire of harshly worded ultimatums from the secretaries of state and treasury. All of East Asian security, it had been made clear to Don, stood in imminent jeopardy, and all because of Akers's incomprehensible series of blunders. "Believe it or not, some of these guys were giving *me* a lecture on logistics," Don had said incredulously. Telling him how it wouldn't just be Hyon's just-in-time first-tier suppliers—the makers of power supplies and processor boards and disk drives and monitors—that would be devastated by a total Hyon shutdown. That the chain reaction would cripple the suppliers of raw materials to the first-tier suppliers—producers of plastics and copper and glass and semiconductors.

"I couldn't argue," Don said. "But the fact is, Singh had already investigated our EDI links with Hyon, and has dialed into their mainframes, and he assures me that despite all the corrupt data we've fed them, what they have over there is not a virus plague but a virus panic, exactly like those that periodically spread like wildfire across the Internet. Unfortunately, the effect can be just as disastrous as an actual virus, triggering all kinds of unnecessary shutdowns and wasteful operations by frightened users, the way a rash of phone calls from the IRA can shut down London, bomb or no bomb."

Akers was indeed hard-pressed financially, he'd agreed, having thrown so much money at so many problems. He'd acknowledged that their credit rating had been lowered twice, but he'd made clear that they weren't broke. Their Korean bankers had just overreacted, joining the panic. But even if the $13 million had been released promptly, Don told her, in his opinion it would have been too late for Hyon. The company was already in a financial free fall. An emergency U.S. bailout seemed the only way to stave off the disaster.

"Jane, let go of it for a while," he'd scolded her during their last call, around two A.M. "With these new loan guarantees, Hyon will be able to reopen their plants and get the workers off the streets. What *you've* got to do is get some sleep—okay?—so you can hit a home run tomorrow."

Well, sleep or no, she was here taking her swing, though the stakes definitely didn't seem as high as when she'd agreed to come to Dallas for this meeting.

Which wasn't fair—to TranSonic, to Dave Ellsmere, or to herself. *If you don't care passionately about the Macao deal, girl, neither will the directors. Wake up!*

She made a conscious effort to invest more of herself in

each point, but her usual knack for gauging her impact had deserted her.

Maybe that was a good thing, she decided, because it was making her try even harder than she otherwise might have to bring new emphasis to phrases she'd used before. Because they were the most powerful in her repertoire, it would be a mistake to substitute others, but she could say them somewhat differently. She told herself that an actor had to keep his performances alive without rewriting his lines. And she was definitely on stage here.

There was one thing she could do that an actor couldn't: she could maintain eye contact with her audience. The price would be dispensing with her audiovisuals. She risked it. With her eyes on them, the directors couldn't very well *not* pay attention.

If part of her remained watchful and detached, another part finally let go of all the issues not on *this* table and, freed up, became deeply involved in what she was talking about. She truly believed that South China Air was the right way for them to go.

More quickly than she'd anticipated, it was time to sum up.

"You already know the reasons for expanding into the Asian air cargo market," she said. "And most of you here can remember why, after looking at a great number of proposed solutions for achieving that expansion, you finally settled on Malcolm and Associates. Our concept was a middle course, between the all-out integrator route—like FedEx or UPS—requiring massive investments in infrastructure, building a complete door-to-door air and surface delivery system, and, on the other hand, just getting your toe in the door as a small player, hoping for a niche presence by partnering with an existing Asian package delivery company."

Jane paused. Their faces said she had described the polar options fairly. "I understand this lesser option—with reduced risk and reduced rewards—is being reconsidered by some of you. I hope you will also seriously reconsider our plan.

"The fact is, the joint operating agreement signed with South China Air not only retains most of the advantages of the partnership route—less investment, fewer downside risks—but actually provides many of the benefits of a leading-edge expansion. It will introduce TranSonic Airex throughout Asia as a name and a presence to be reckoned with. As I mentioned earlier, it will also net you more air routes into China than even FedEx has, under current laws.

"Of course this plan isn't risk-free. No enterprise is. TranSonic Airex would still be Lone Star Delivery if Max Garvelmann hadn't glimpsed a farther horizon than most of the people around him."

At the head of the long table, Garvelmann tipped his head slightly in acknowledgment.

"Friends."

I hope.

Maybe, maybe not. But you're not here to make friends, Janie. You're here to make them follow your advice—follow you.

I've given them the facts, Dad. What else can I do?

Why, just play a rousing tune, piper.

Her father was beside her even when it wasn't Akers business!

"That horizon is still moving," Jane told them, her voice infused with new hope. "It's farther out every day. I hope you'll all agree with me—and with Dave Ellsmere, and with most of your top management—that a partnership with South China Air is the best way—not tomorrow, but right now—for TranSonic Airex to head toward that bright horizon lighting the way to the future. Let's move toward it confidently—and make it ours."

She stopped. If she had them, there was no need to say anything more. If she didn't, there was no point.

Dave Ellsmere stood and asked if there were any questions. There were two, and she responded factually. There being no more, she thanked them for hearing her out, and then Dave Ellsmere was escorting her from the room.

"Terrific job, Jane," he said as the heavy boardroom door

caught behind them. "Can you stay around for the verdict? Hopefully have a victory lunch with me?"

"Dave, you know I'd love to. Unfortunately, I've really *got* to get back. You will call me as soon as you know . . . the good news?"

ABOARD THE CITATION II
11:25 A.M.

They had been airborne for only ten minutes, headed back to Siler Field, when Jane got Dave Ellsmere's call.

"I'm a proud Texan, ma'am, so I do believe that my threatening to walk out the door if they backed out of the China Air relationship had *some* effect on the board," he said. "But I've got a hunch it was your speech, especially that powerful ending, that put us over the top."

"I'm delighted, Dave. And, I have to admit, a little relieved."

Ellsmere chuckled. "Just between you and me, that makes two of us. Nice to know I get to keep my job. You, incidentally, will get a rather handsome bonus."

Jane felt a lot better about things in general as she went forward. The check pilot got up and relinquished the co-pilot seat to Jane. "Just for a minute, now," he said.

"A minute's all I need, thanks."

"Congratulations," Chevy said when she'd given him the good news, having decided to omit, for the moment, the Korean debacle. "So how about that job with TranSonic Airex?"

Jane laughed. "Sorry, I haven't laid you off yet. In fact, I've got another flight for you today."

"You mean to Chicago?" Jane had mentioned the possibility yesterday. "The Citation will make quick work of that."

"We won't be using the Citation. My old Saudi colleague, Khalif al-Marzouki, needs the jet to visit our west import hub tomorrow and figure a way to get our garment-sorting operation running at full speed again, without downloads from the mainframe."

"So, what, do I get to pick any plane in the barn?"

"I think we'll take the little Tiger Wade usually flies. He's going along, so he can check you out."

"Sounds real cozy," Chevy said.

"What does that mean?"

Chevy glanced at her, eyes masked behind amber-tinted Serengetis. "It means, Jane, that you're quite a woman. In more wars than one."

More wars than one. In the jet's cramped bathroom, Jane played back Chevy's compliment as she maneuvered out of the conservative navy suit she'd worn for the Dallas meeting and into the teal number she'd chosen for her meeting with Soverel, which perfectly combined tailored lines with high-style detailing. *Thanks again, Anne.*

Jane touched up her makeup carefully, then inspected the final result as best she could in the small mirror over the sink. For a moment a face smudged with soot looked back at her. That soldier's desert camouflage blouse had a tear in the shoulder.

Another war, another uniform.

The image faded and Jane saw herself as she looked now. With a subtly made-up face, in a uniform appropriate for the battle with Soverel.

Still, under her chic suit her heartbeat was unsteady. A planner by nature as well as by training, she hated the fact that she was going into combat without a strategy.

EN ROUTE TO CHICAGO
1:51 P.M.

Jane had called ahead from the air so Wade could get started on his preflight with the AmGen Tiger. As a result, she and Chevy spent less than ten minutes on the ground, transferring to the four-seater. It was snug quarters after the roomy Citation; Chevy slipped behind the controls, Wade to his right, Jane behind Wade.

"The Tiger does a hundred forty knots," Wade said, swinging around to Jane as she pulled on her headset, "which makes it about a thirty-five-minute flight to Meigs Field, right on the lakeshore beside Soldier Field. From there, it's about a three-mile cab ride to Soverel's office. We'll be in plenty of time for your three o'clock with him."

"And don't you go wandering off when we get there," she said, pitching her voice above the Tiger's warm-up roar.

"No, ma'am. I'll be right there waiting for you."

The banter was definitely a little strained. Before they'd climbed into the plane, Wade had beckoned her aside. "You look so great, you definitely need a bodyguard, but you don't need two pilots. I've got a lot to do here—especially now, helping to deal with the Hyon bailout—but if you need me for job one, I can also fly you."

"I know you can, Wade. But I want Chevy to stay with the plane and have it ready for a quick flight back. Just for the record, I appreciate the compliment, but I don't need a bodyguard, I can take care of myself. But if this meeting turns out to be . . . Wade, I want you there because I may want your input. Okay? And I may want you both to do the whole thing again tomorrow. That okay?"

He had just looked at her for a moment, then nodded and helped her aboard. Now she leaned forward, her head and shoulders between the front seats, and watched Wade give pointers on the Tiger's nonsteerable nosewheel.

"Till you get the hang of it," he was saying, "taxiing can be a little tricky."

Apparently Chevy got the hang of it quickly, for a few minutes later they were charging down the runway, then lifting smoothly above the spreading farmscape.

They'd be cruising at five thousand feet, Wade explained to

her, with good visibility the whole way. Their course was north-northeast, following Interstate 57 to the southern edge of Lake Michigan, just west of Gary. At that point, he said, Indiana was only about a dozen miles from the Loop. Jane nodded, watching Chevy switch on the Tiger's autopilot, then punch in the three-letter code, CGX, for Meigs Field.

"I like the way she handles," he told Wade. "Almost like an aerobatic plane."

Jane watched the dark brown, cloud-shadowed acreage below gradually give way to town-house developments, industrial parks, and shopping centers. There were several Akers outlets along the interstate, she knew. In her mind's eye she saw them with new facades and signage—"*4-Mor Shopping, Grand Opening!*"

If Tom Soverel got his way.

Yesterday, in Dallas, after the late night bombshell of events in Korea, Jane had regained a clear sense of mission, of an opposing force to be engaged and overcome. She wished she had that kind of clarity today.

What seemed to her all too clear was that by having accepted Soverel's transportation to Chicago later today, the family already appeared to be in Soverel's pocket. It was almost as though he already had their signed proxies. What kind of free ride did he have in mind to offer her? When she'd called to confirm her arrival, he had sounded pleased but noncommittal.

Suddenly Jane felt a surge of energy. She'd never accepted a free ride from anybody. She'd hardly let Soverel's be the first.

Don't you get too damn confident, mister. Who'll own Akers is still up for grabs—and I'm going to do my damnedest to see that your grab isn't the deciding one.

As Chevy and Wade discussed climb rates and cruise power, Jane phoned Don Landsman for a status check. Morale in the War Room continued positive, he told her, especially with the news that calm had returned to the streets of Seoul—though, of course, Akers employees were unaware of the impending show-down with Soverel and its possible repercussions. At least Don *hoped* they were.

"That makes two of us," Jane said. "Now, if we can only see to it there are *no* repercussions, they won't ever need to find out. What can you tell me about Mr. Singh?"

"I'm afraid he has no real leads. He said we might want to bring in the FBI. It's obvious the man is extremely frustrated."

"But he's still keeping at it—he hasn't given up?"

"Oh, he's still trying. I think part of his mood comes from the fact that he usually solves problems pretty damn fast."

"Not good," Jane said. "Dare I ask how our stock's doing?"

"Still dropping. The first stories about the president's bailout of Hyon, and our involvement, are moving on the wires, and they'll probably lead the evening news. Some congressmen are predictably outraged. Definitely not the kind of PR campaign we need right now."

"What diabolical timing! Despite the disaster with Hyon, we really *do* have good operational news, dammit."

She hung up more frustrated than encouraged. "How much farther?" she inquired over the single-engine drone.

"We're in the neighborhood," Wade said, gesturing ahead. "The Sears Tower is already sticking its nose up over there, about eleven o'clock. That's lake haze dead ahead, beyond all the smokestack smudge from the Gary steel mills. We'll be on the ground in fifteen minutes. I'm about to call Meigs flight ser-vices now and order us a taxi."

"Great," Jane said, counting on the plane's noise to disguise her absence of enthusiasm that they were nearly there. And if Wade didn't share her dread, well, that was understandable. Whatever happened, she thought, he would be all right. Tom Soverel would want to retain Akers's top regional managers, at least for a while, to keep the stores running during the transition. If not, plenty of other merchandisers would be eager to hire a youthful, intelligent hard charger like Wade.

Except that Wade belonged at Akers. Her father saw that long ago. And Jane intended to keep him there. She was less sure if she wanted him in her life, but there was plenty of time to find that out. If she and Akers won today. *And you, Dad.*

She gazed ahead at the vast blue of sky and water, as Lake Michigan unfurled to the shimmering horizon. Off to the right was the industrial sprawl along the Indiana lakeshore. Thousands of feet above them a streak of silver passed over.

"Wade, what's that?"

"Commercial jet en route to O'Hare," Chevy said.

"All right, Major Johnson," Wade said. "Time for you to talk to the *tower* and start landing this contraption."

Behind them, Jane smiled. *Men will be boys.*

They swung northwest, following the lakeshore toward the rapidly rising Chicago skyline. Chevy got a northerly down-wind approach from Meigs, which took them directly over the single-runway airport on its landfill peninsula, with McCormick Place and Soldier Field on their left across the yacht basin.

"Nifty little field," Chevy said. "Glad they saved it."

"Yeah," Wade agreed. "It was touch and go there a few years back."

They flew on past Grant Park and the sun-burnished sky-scrapers of Michigan Avenue. At the Hancock Building they turned onto their base leg, banking out over the shining lake and coming full around, with the Loop now at their right wingtip. Their final approach was over the half-mile-long Navy Pier straight in to runway 18, touching down gently and rolling along past the tethered private piston planes and corporate jets that frequented Meigs.

Chevy turned around to Jane. "Another city, another victory. Go get 'em, Cap'n."

CHICAGO
2:35 P.M.

Jane and Wade hopped into a waiting Yellow Cab, leaving Chevy to arrange a tie-down and top-off. She gave the address of Soverel's headquarters on North Michigan. "It's just above the Hancock Building."

"Whichever way it goes, today and tomorrow," Wade said as they sped north, "I want to say that I've been proud to work with you. And I know your dad would be proud, too."

"Thanks, but don't sound so solemn, Wade. We're going to win."

"You're right, I didn't say that right."

"Never mind that, what you said felt good. Your opinion, and my dad's—I know my dad's one of the reasons I'm doing all this. It makes me feel closer to him."

"None of my business, Jane, but I never really understood what kept you apart all those years."

"Actually, I'd like to talk to you about it . . . another time."

Michigan Avenue's Magnificent Mile was in full springtime flower, with milling sidewalks, flying flags—and crawling traffic.

Jane turned to Wade. "We're not going to get there any faster by my staring at the back of the driver's head. Why don't I tell you now? What happened between Dad and me was, I blew it. Elie says it was Dad's fault, too, that he should have tried harder, kept trying. I sure wish he had. But lots of it was my fault. I want to think he would applaud my efforts to save the company." She paused, deciding whether to tell Wade.

"There've even been a few times when I could actually feel him cheering me on." She glanced at Wade, but he wasn't laughing. "A couple times he set me straight, too."

She supposed that the only reason he wasn't laughing was because she was, for the time being, the boss. "I've probably really lost it, huh?"

"Don't think so. Knowing how he loved the company, wouldn't surprise me if your dad is keeping an eye on the business even now."

"Thanks. I have this strong feeling I could use some of his advice today. Still, even if I do everything right today, that won't make up for my not having picked up a phone six months ago, or six years ago, to tell him I loved him. But what else can I do?"

"You could do *that*," Wade said quietly. "If you think Royal's . . . well . . . around, why not tell him now that you love him."

Wade's hand touched hers momentarily. She saw it but hardly felt it, so strong was the flash flood of emotions his words unloosed in her. Grief, remorse, and hope washed over Jane, knocking the breath out of her.

Just then the cab pulled to an abrupt stop in front of the Soverel Building. Jane blinked hard, returning to the present danger, and scrambled out. She was still feeling somewhat limp as she caught a storefront reflection of a close-cropped brunette in a chic teal blue suit.

She strode toward the building.

SOVEREL BUILDING
2:58 P.M.

Wade caught up with her in time to open the lobby door.

The lobby was Sunday deserted, except for a uniformed guard who had Jane's name, ushered them into an elevator, and pressed "60."

As the elevator began its silent ascent, Jane turned to Wade. "Small favor?"

"Anything."

"Save any more emotional bombs you have for me until after this crisis is over, okay?"

"Sorry, Jane. I didn't mean to—"

"I know. And I'm fine."

She took deep breaths during the rapid rise to Soverel headquarters. The elevator door opened. *Here we go, Dad.* She preceded Wade onto polished terra cotta beneath a barrel-vaulted, neoclassical atrium.

"Ah, Ms. Malcolm," said a stunning blonde seated at a desk on a raised reception island. Jane recognized the English accent from her phone calls. "Mr. Soverel is expecting you. But I believe he was under the impression you were coming alone." The young woman tilted her perfect head ever so slightly in Wade's direction.

"Mr. Soverel's impression was correct. Mr. Crain has been kind enough to come along and wait for me."

"Of course. We'll make Mr. Crain comfortable."

TOM SOVEREL'S OFFICE
3:02 P.M.

"I'm honored and delighted," said Tom Soverel, loping across his considerable office toward Jane. Floor-to-ceiling corner windows provided panoramic views of both lakeshore and Loop. They shook hands in the center of the room.

"You managed to arouse my curiosity, Mr. Soverel. I suspect you're very good at that."

"Rest assured I'll do my best to justify the extraordinary demand I've made on your busy schedule." Soverel beckoned her toward a blond Biedermeier couch along one of the window walls.

Jane, not quite ready to sit, strolled over to the window and motioned with her head toward the Drake. "I understand you're going to be putting my family up there tonight."

"Yes, it's handy." His grin betrayed a front-toothed gap and split his homely, Lincoln-like features into beguiling asymmetry. The bastard did have charm.

"But the Drake's also special," Soverel added, "because the management still remembers how to spell 'service.'"

Jane nodded. "I always stay there when I'm in Chicago. So did my father. Service is something our family values very highly, I assure you."

Jane walked slowly back to the sofa and sat. But she didn't

relax into it. "Before you begin," she said, "let me state the obvious, for the record. My intention is to persuade my family not to sell you our company."

"A sale has not been discussed, Ms. Malcolm." Soverel splashed some Evian into two glasses. "May I call you Jane?"

Jane shrugged assent. "It won't make any difference. And let's not be technical. You want them to sign over their proxies, which, with the proxies you've already rounded up and your current Akers holdings, would give you control of, what, close to sixty-five percent of the voting shares? A takeover is a takeover, Tom."

Soverel spread his big palms. "Okay, I'll admit that. And, *for the record*"—he smiled that 4-H winner's smile again— "I consider your effort to stop me an act of filial gallantry."

He seated himself in one of the chairs flanking the sofa. "I'm not sure your father would applaud you, however. Sentiment is fine, but business is business, and my offer is really the only viable financial option left—from your family's standpoint. The risk is all on my side. In fact, I confess to serious second thoughts about being able to turn your operation around."

"We've already turned it around."

"No," he said. "You've made some damned good progress, from what my people tell me, but you're operating on a kind of continuing crisis basis, aren't you, with cost controls out the window? How long can you keep that up? And now, with this Korean fiasco . . . " Soverel gave his head a forlorn shake.

Damn those bankers! And double damn that saboteur!

"Even in the best-case scenario," Soverel was saying, "your stock isn't going to recover till Hyon pays Uncle Sam back those guaranteed loans with interest, *and* all your computers are back on-line. And at that point, I'm afraid, your cash flow will still be woefully inadequate to repay the kind of serious debt you'll have built up."

"Forget about Korea," she said. "That's a political issue now.

But if that's your current prognosis of our situation, I don't think you really know what we've accomplished—especially these last few days."

"Perhaps not. But I'm impressed by what I do know. I don't underestimate *your* abilities, Jane, or how well you've performed under severe circumstances. That's why I've invited you here." He leaned forward in his chair, as if to put a gift into her hands. "I'm seriously thinking of making my offer for the company contingent on you coming along."

"Excuse me?"

"Maybe I should be more direct. I gather you'd prefer that?"

"Always."

"All right, then. Jane, you think you can save the company. Fine. Come work for me and I'll give you the opportunity to do just that."

"I already have that opportunity, thank you—and despite anything you've heard to the contrary, I'm taking full advantage of it."

"Let's try it this way. You say your operation has turned around. All right, for the sake of discussion, let's say that's so. I doubt your stores can return to their former levels of profitability, and certainly not reach new levels, unless you're calling the shots."

He might be right—and the only company where she *was* going to be calling the shots in a few more days was her own, back in San Francisco.

"I'd like to be perfectly clear on this. You want to take over my family's company and then hire me to help you reap the benefits of a bigger and better Akers? I don't think Malcolm and Associates is available for that particular job."

"I guess I'm still not making myself plain enough. Jane, to put it perfectly bluntly, I'm not interested in Malcolm and Associates. I want to hire *you*—full-time, as head of logistics."

"You want *me* to help you turn all our stores into 4-Mors?"

"Frankly, that's a pretty small part of it. The job I'm offering you is vice president of logistics for Soverel Corporation, not just one division."

Jane took a sip of mineral water and tried to collect her thoughts. She knew 4-Mor Shopping was only one of many divisions under Soverel's global umbrella. The man was into ocean shipping, trucking, warehousing; his air freight company had moved aggressively into the Asian cargo market along with TranSonic Airex; and he was rumored to be partners in several broadcast satellite units.

"You have to say something," Soverel said. "It's only polite."

She managed to return his smile, but a thinner version. "I'll say this much. I'm surprised."

"Not altogether pleasantly."

"You obviously enjoy fast moves. Sudden turns. Speed. I bet you don't drive that stretch taxi of yours when you're out for fun."

Soverel laughed. "An astute observation. If I admit that weekends I drive a Lamborghini, will you hold it against me?"

"I never hold an honest admission against someone."

"Well, then, I'll also admit that I've had you in mind for this job for some time. A friend and colleague of mine worked with you in Saudi Arabia and recommended you highly for the job." Jane expected Soverel to mention one of the logistics officers; it turned out to be Khalif's uncle, Prince Rakan al-Fayez, who had helped underwrite much of their food service.

"You demanded frankness and I've obliged you. May I expect the same from you? How does my offer strike you?"

"First I'd like to hear a little more of what you have in mind. And why you think I'd be suited for the job."

Soverel got to his feet, began pacing in front of his Biedermeier desk. "Initially, there would be a lot of Akers work for you—during the changeover—getting them up to full speed. But I really see you as a global troubleshooter. Somebody who can look at all our businesses, and businesses both here and abroad we might like to acquire, in terms of integrated logistics and supply-chain management—see how we can streamline operations. I need someone who can sit down not only with our own executives, but with our trading partners as well, to figure out ways we can drive down total costs by working together—beyond electronic data interchange or vendor-managed inventory—by intimately studying each other's operations. By tearing down the proprietary walls, where it makes mutual sense. I'm convinced that a team coalition approach is the next big step in logistics, and I want Soverel Corporation to be on the cutting edge. I've been looking a while now for the person who can take us there. And, Jane, I'm convinced you're the one."

Jane was still puzzled. There had to be a trap here, at the very least a flaw in his proposal—aside from the minor fact that fifteen hundred Akers stores wouldn't exist as Akerses anymore. As the balance scales teetered in her mind, Soverel showed more of his talent for laying out genuine temptations. "If I may boast a bit, I think you'd have a hard time finding a bigger challenge, Jane, than the one I want to throw your way. And I don't mean to disparage any of your work—or your excellent clients."

"Would I work here?" she asked.

"*Out* of here. You'd be traveling quite a bit. That wouldn't be a change for you, would it?"

"No, it's actually something I enjoy."

"I have to say, right now you don't look as if you're enjoying yourself at all."

"A serious offer merits *serious* consideration."

"Well, I'm glad to hear you're considering my offer seriously."

"You're getting ahead of me again. I haven't heard your whole offer."

"All right, let's talk money. Frankly, I didn't broach compensation before now because I suspect that wouldn't be the decisive factor with you. I do know about your father's trust fund—don't frown like that, I make it my business to find out everything I can about the mind that created a business I'm planning to own. And the way your father handled this particular matter happens to have impressed me a lot. But think, Jane. If the price of Akers stock keeps dropping, you and your family all stand to lose a great deal."

"I'm aware of that. I also realize your per-share offer is generous."

More than generous, considering the last trading price she'd gotten from Don Landsman. Her newfound fortune was diminishing day by day. If she joined Soverel, she and her family would recoup nearly all they had lost. And, in the process, she'd be gaining a dream job in logistics, with a scope she could never attain as a consultant.

But Malcolm and Associates is mine.

I know how you feel. Akers was mine.

Oh, Daddy.

"So," Jane said coolly. "What terms are you offering?"

Soverel perched on the corner of his desk. "For starters, a three-year employment contract." He quoted a salary easily three times her current annual earnings, with performance incentives that could bring it substantially higher. Next came an option agreement to buy ten thousand shares of Soverel common at half the current rate. Her starting title would be vice president of corporate logistics. "How's that? Worth flying in from Dallas on short notice?"

"Interesting," Jane said. "Both the challenge and the compensation."

"Well, that's my offer. It's on the table and will remain there until the meeting tomorrow morning."

"Ah, another ultimatum."

"I'm sorry I can't wait any longer, Jane, but I have to make immediate plans for the acquisition and operation of all the Akers properties. So, yes, I'd very much like to announce your favorable decision at that time."

"I'm not at all sure I can make such a big decision overnight. There are things involved besides the appealing challenge and the satisfactory compensation."

"Now you're surprising *me*. What have I left out?"

"The human factor."

"You still don't think your father would approve?"

"I will have to think about that. But the person I had in mind just now is quite alive. I don't see working side by side with my brother."

"Why do you think you would be?"

"My brother doesn't have an offer from you?"

"I will be keeping many Akers people on. Your brother wasn't going to be among them."

"Does Roy know about this?"

"No one knows. It's no one else's business, frankly. Until you say yes."

Jane couldn't help but picture her brother's reaction if she did accept Soverel's offer. The satisfaction of seeing his jaw drop would be priceless. Well, almost priceless.

She stood up. "Frankly—and you've been frank with me, so I must return the courtesy—this has not at all been what I expected. If you intended to astonish and dazzle, you've certainly succeeded. If you're surprised that I haven't leapt to accept your generous offer, well, your research about me wasn't as thorough as you think. But I would be a fool not to consider your offer seriously. And I'll be here tomorrow either way."

"If there's anything you need to know before then, you can reach me at any of these numbers." He handed her a card.

He escorted her out a different way, bypassing the ornate reception area and ushering her into a different elevator, this one faceted with marble and mirrors, and pushed the Down button; Jane noticed that there were only two buttons; the other was marked Up. She smiled at the logistical sense of that as she was whisked downward to Debussy's *Clair de Lune*. Her thoughts, carefully channeled till now, began suddenly freewheeling. Suppose, just for the heck of it, she worked her three years for Soverel, then exercised her stock options, then went out on her own again . . .

The door sighed open on a Chinese-carpeted alcove. Had Tom sent her down to a different level? But no, just beyond the rapidly approaching guard, Jane could see the main lobby floor.

"I have a cab waiting, Ms. Malcolm. Where will you be going?"

Going? Where *was* her head? She'd forgotten all about Wade!

"Back upstairs, please."

3:39 P.M.

Jane returned to the sixtieth floor via the main elevator, obviously startling Wade, who was standing by the reception desk, cup in hand, chatting with the camera-ready blonde.

"How'd you slip by me?" he asked.

"Never mind that," Jane said. "We're done here." She stepped back into the elevator and held open the door.

"Okay, sure." Wade started toward her, then wheeled back, setting down his cup on the receptionist's desk. "Thanks, Diane."

"Ta-ta now, Wade."

"Ta-ta, Wade," Jane mimicked as the elevator closed.

"Cute. I got tired of reading *Business Week*, and she got me some coffee. What happened where something *was* happening?"

"Nothing yet."

She looked up. They were passing the fifty-second floor. Fifty-one. Fifty. She hit the button for forty-eight. When the door opened, she pushed the button to keep it open and looked at Wade. She had his attention.

"Something did happen. Soverel offered me a job."

"I thought that might be what he had in mind." He hit the Close button. "So what did you say?"

"That I'd think it over. That sounds rotten, doesn't it? After everything that's happened? I should've given him a flat no. Why didn't I?"

Wade didn't say anything. God only knew what he thought

of her now.

Moments later the door opened onto the lobby, and Jane hurried ahead out of the building.

Wade caught up with her. "Knowing Soverel's reputation, it must have been one heck of an offer," he said. "It wouldn't have made sense not to *listen*. More than likely that's what I would have done."

She threw him a skeptical glance. "Not if my dad was around, you wouldn't."

"Well, he's not. And I wouldn't do it if *you* were going to be staying on, either." Wade paused. "So now what? Back to work? Or do we hike up the street and wait for your family at the Drake?"

"Wade, don't tell anyone."

"Of course not."

"Back to work," she said.

"Right." He flagged down a cab, and they moved toward it, through the afternoon tide of shoppers and sightseers.

"Meigs Field," Wade instructed the driver.

"The worst thing about what happened," Jane said as they sped away, "was that I rode up that elevator fired with purpose, and came down second-guessing myself. I owe it to the family to fight for Akers as passionately tomorrow as I would have an hour ago. But how?"

"You'll find the 'how' inside yourself," Wade said. "That's where you keep your passion, isn't it?"

She glared at him.

"Jane." He shook his head. "I meant for your work."

She nodded. "Sorry. Let's hope it's still down there some-where, and I can locate it before tomorrow's meeting."

There was no way to talk to her family this evening without telling them about Soverel's offer. What could she say? That Tom Soverel didn't seem so bad after all? That she might *possibly* be going to work for him? That she *might* even end up running the Akers empire—only, of course, it wouldn't be called that anymore?

No, she had to figure out which side *she* was on before telling them anything. Right now she was in no position to try to persuade anyone of anything.

As they swung off Michigan toward the lake, her cell-phone trilled. Jane flipped it open. It was Jessica Zhuo, director of Akers's International Department.

"Jane, I'm hesitant to drop another problem on you, but—"

"Don't be, Jessica. Looks like we're going to need a whole task force to deal with Hyon's problems."

"It's not Hyon this time. But I do want you to know—about our bank in Seoul holding up that critical payment? I had absolutely no warning. That was completely—"

"Yesterday's news, Jessica. Let's hear your *new* problem."

"You're right. We just had a call from Rotterdam, the Office of Municipal Port Management. They suspect contamination in one of our shipments of frozen meat. If you have a moment—"

"Go *ahead*, Jessica." Jane felt her reservoir of patience dwindling.

"Sorry."

"No, *I'm* sorry. Go on, now."

"This is pretty convoluted, but here goes. Apparently, the first sign of a possible contamination problem occurred early this morning in Rotterdam. A Turkish-immigrant dockworker checked into the emergency room of Dijkzigt Hospital, complaining of chills, abdominal cramps, and diarrhea. The diagnosis was *clostridium perfringens*— they tell me that's a spore-forming bacteria found in meat and poultry—"

"How contagious is it?" Jane asked.

"It's not. And the symptoms usually last only twenty-four hours."

"Whew!"

"But I'm told the diarrhea can be *very* severe—life threatening to old people or children, and harmful to pregnant women."

"Oh, God! Go ahead."

"Finally, and only after repeated questioning by the emergency room doctors and public health officials—and, you better believe it, considerable coaxing from the Turkish consul general—the dockworker admitted that, the day before, he had swiped a box of frozen meat from a refrigerated ocean container on the Waalhaven docks, where he worked—"

"Swiped? If he *stole* it, he *stole* it."

"He swore it was the first time he'd ever . . . stolen anything, naturally. Anyway, right after he ate some of that meat, that's when he got sick. Unfortunately, here's where we come in. Authorities at the Waalhaven frozen-food container terminal determined the meat was from an Akers shipment. The municipal port authorities immediately notified Akers's continental shipping agents and us at International.

"Now, something could have gone wrong with the refrigerated container," Jessica said. "It's *supposed* to keep the meat frozen at zero degrees Fahrenheit, but—"

"Don't they check those reefer units in transit?" Jane asked.

"Yes. Periodically—whatever that means. Maybe the unit malfunctioned just long enough for the temperature inside to rise. I'm told it can hit a hundred degrees in twenty-four hours. The cooling unit could have kicked in again and refrozen the meat, so everything would appear normal on arrival and inspection at Rotterdam."

"If that scenario is right," Jane said, "just one container may be contaminated. But, not knowing which one, we've got to trace the entire shipment. What have you done so far?"

"The shipment originated from our regional distribution center in Maywood, New Jersey. Let's see . . . " Jane could hear papers being shuffled. "Ten containers of beef, shipped out of New York destined for Rotterdam Europoort and onward to our RDC in Madrid, to be consolidated with other products and then delivered to our outlets in Spain, as well as to other retailers."

"From what you say, the containers probably were unloaded in Rotterdam yesterday. Which means—"

"Yes, I'm sorry, Jane, the shipment did clear port yesterday, ten twenty-foot full-container loads, all mounted on truck chassis and hauled off the pier. Based on our transit-time schedules, those container trucks should have arrived at our Madrid RDC around noon today their time—they're seven hours ahead of us. I've got Madrid on the other line now, trying to confirm that—"

"Jessica, listen up. Tell them no reefer trucks leave that facility until we pinpoint that entire shipment. We need to know exactly where every bit of that meat is. We've got to throw a net over those damned meat boxes before they get parceled out into other loads."

"I already told them that."

"Good. Then go ahead and notify the Ag Department. As much as I'd like to keep this under our hats and out of the news, we can't worry about PR with lives at stake. The USDA can mobilize meat inspection people in Spain."

"Will do. I'll pull out all the stops, Jane."

"Does Don know about this?"

"He will the instant he walks out of a meeting. His cell-phone and pager are off."

"You're doing all the right things, Jessica. But keep us both posted—the instant you get anything new."

Jane hung up, filled with nightmarish visions of a killer truck stalking the Spanish hinterland.

Before she could fill Wade in on this latest crisis, or even fold her phone away, it summoned her again.

Elie Grey was on the line this time, his voice reedy with

agitation. "Jane, you in Chicago?"

"Uh-huh."

"Get back here as quick as you can."

"What's wrong, Elie?"

"Gotta be *here* that I tell you. I don't care what you're doing, this is more important."

"Okay, Uncle Elie. Actually, we're on our way back now."

"Who's 'we'?"

"Wade and I."

"You got that new pilot you hired with you, or is Wade flying the Tiger?"

"Chevy Johnson's waiting for us at Meigs Field."

"Tell him he can land it here. We got us a flat tractor road alongside our school farm. My nephew sets his Cessna down there whenever he flies up from Little Rock, no problem."

"It's *that* urgent, Uncle Elie?"

"Jane, just get here!"

He clicked off, and Jane leaned over the front seat. "Could you please try to go faster? We've got an emergency." *Make that several.*

"Almost there, lady."

"Please, could you just swing out and pass those cars?"

"All right, all right, hang on." The driver changed lanes and sped ahead.

"What is it?" Wade said.

"I don't know." Jane shook her head. "Elie says we have to come back right now. He even wants us to land the plane on his farm. But that's all he'd say." She could tell Wade about the tainted-meat crisis once they were airborne. And maybe call Jessica back, have her contact their liaison with the Spanish embassy in Washington to involve local authorities and relieve tensions.

"Shit! I hope nothing's happened to one of the kids."

Wade's tone startled Jane out of her own thoughts. She

turned to look at him. "I didn't think of that. And it would be terrible, but I don't think he'd be asking us to rush back if that's what had happened."

"You're probably right." His frown became shallower. "But what could it be, then?"

Jane shook her head. She had no idea, except that it had to be something very serious. Elie didn't know that her own future was suddenly hanging in the balance, but he knew that the future of Akers Corporation would be decided here in Chicago tomorrow. What could be more urgent back at his farm?

DATA PROCESSING CENTER, AKERS HQ
4:11 P.M.

Rikki Singh watched Ron Hagstrum as the assistant packed up most of their equipment. The diagnostic phase was done. All that remained was the ongoing comparison of virus code to programming samples gathered from Akers and project team staff. Any match could finger their predator. But time was rapidly running out. In fact, according to the rumors running amok in headquarters hallways, it might already have expired. A takeover might be in the works.

It was therefore with real regret that Singh now left his office, en route to a status meeting with Don Landsman and Chip Bragan. He didn't mind that his consultancy here would end; he had a standby list of clients eager for his services. But he couldn't abide such an unsatisfactory exit. To be sure, his reputation would survive this failure; but he would be diminished in his own eyes. And then there was Jane. Over the past week he had noted her efficiency, the respect she showed the expertise of others, her fortitude, and the moral energy he discerned in her. It would be terrible to let her down.

Even as he was reinforcing his determination, a number of Singh's faculties kept on working. Abruptly, now, he halted.

Something had caught his eye.

But it was his mind that blinked.

He had almost reflexively stopped alongside the window of the main computer room. And now his eyes were on Mimi Takuda. The truth, which he had acknowledged to himself only two or three times in rare unguarded moments as gray turned to lavender in the early morning sky, was that whenever the opportunity presented itself to observe the assistant system administrator, he tarried longer than was absolutely necessary. She seemed to combine quietness with vivacity in a way he found particularly appealing. Just now, she was screwing something into a modem rack. He admired how busy she was keeping without a computer system to administer.

But this time it was not Mimi *in toto* that arrested Singh's eye. Rather, it was a tiny tattoo on her exposed shoulder.

As he entered the machine room, Mimi glanced around, smiling and brushing the bangs away from alert black eyes.

"Hi there, Rikki."

While he preferred to be addressed as "Mr. Singh," he had made an exception for Mimi, thereby affording him reciprocal familiarity.

"Hello, Mimi. What are you doing there?"

"I know you said we didn't really need security lockboxes on the dial-in ports, but Lenny thought it couldn't hurt. These guys"—she slapped the faceplate of a black box she'd just screwed to the rack—"are supposed to authenticate all remote systems."

"That's excellent, Mimi." Singh sought a conversational transition but came up short. "Um, excuse me for asking, but I don't recall seeing you with a tattoo before."

"Does it . . . offend you?"

"Of course not, Mimi. It's very discreet, isn't it?" He smiled. "They'd have to stencil my skin with white ink to make an impression."

"Actually, I don't usually go sleeveless, what with it being so cold in here. That's probably why you didn't notice."

He ventured closer. "It's a crescent moon, isn't it?"

"No, it's the Cheshire cat—just his smile, actually—from *Alice in Wonderland.* You know, the Disney movie?"

"I know the book, actually, with the Tenniel illustrations. Yes, of course, the Cheshire cat."

"Anything else you'd like to know about me?"

"I apologize, Mimi. How rude you must think me."

"I was kidding. And the idea of you being rude, that's impossible to imagine."

"Thank you, Mimi, how nice of you to say so. Now I'd better leave you to your duties." Singh made a smiling retreat, with a certain quickening of the pulse. Could it be that Mimi was flirting with him? Even that stirring possibility must not distract him from his chief purpose—and the truly significant thing he had just discovered.

For Singh had seen that leering Cheshire smile before—in a computer virus. It had been a tiny scrap of compiled code he'd stumbled across in a backup tape cartridge. He couldn't quite recall the name of the executable file—it was something nasty, like "bomblet" or "boomboom." He'd taken the precaution of copying it over to his own laptop before running it—bracing for disastrous consequences. The program had self-destructed, leaving only an evanescent on-screen dot pattern of a dentate crescent—exactly like the Cheshire cat's vanishing smile.

Back in his office, Singh phoned Don Landsman's secretary and excused himself from the afternoon status meeting. Turning next to the nearest cardboard carton, he began to exhume stacks of printouts.

"What's up?" asked Ron Hagstrum, in the midst of packing another box.

"Possibly a great deal. Could you run back to the computer office and ask Mimi or Lenny for some drive-one backup disks?"

"Sure. What for?"

"Tell them I want to check some of their library routines."

"Do you?"

"I do. I intend to analyze as many admin programs as I can—even rudimentary shell and C stuff for cleanups and user menus and so forth."

"I don't get it."

"There may be nothing to get, Ron. If there is, you'll be the second to know. I'm in a bit of a hurry, actually," Singh added, looking around for his decompiler.

Part of him truly hoped his comparison of virus strains to Mimi's own programming would uncover no signs of common authorship. After all, a smiling tattoo and a smiling screen flash might be nothing more than a silly coincidence.

But Singh's gut told him it was more. He had the distinct smell of fresh blood in his nostrils.

The chase was finally on.

He polished his bifocals, then mouse-clicked a favorite CD selection on his laptop, Erik Satie's *Gymnopédies*. Singh found melodious monotony conducive to extended concentration.

And this could take all night.

THE GREY FARM
4:45 P.M.

The parking lot of the Ocean Plains Akers store glittered with cars as their small plane flew over. Good, Jane thought. The locals were coming back to shop in force. Of course, there was unlikely to be an adult who didn't grasp the stakes: if the company folded, the town would soon follow.

The Grey farm bordered the southern edge of the town's commercial zone. In the Tiger's front seat, Wade pointed out the landmarks to Chevy—the apple orchard, house and outbuildings, silo and corncrib, the little demonstration farm with its nostalgic red barn and tidy vegetable patch. Running between was the long, empty tractor road. Chevy began jockeying for a straight-in approach.

"As much as I've flown this puppy," Wade said, "I'm glad it's you putting her down here, not me."

Chevy nodded. "I've landed on a lot rougher terrain than this, and in lots bigger planes. Let's see now, flaps up, airspeed down, is that about it?"

As they flashed over the treetops beside the main house, Jane saw something she'd missed before. Wade jerked his head around; he'd seen it, too. Three vehicles were drawn up in front of the farmhouse. One was white and unmarked. The other two were green and white, with light bars, whip antennas, and big decals.

"County sheriff," Wade said. "Not good."

Jane felt her stomach tighten. Calm down, she instructed herself. It *could* be good news. Maybe they'd caught the saboteur. Out here? Besides, Elie's agitated voice, fresh in her memory, told her otherwise.

Then they were down and jolting along the farm road. Jane relaxed her death grip on the back of Wade's seat.

"Nice work, Chevy," Wade said as they slowed gradually.

"You two want to run ahead while I spin the gyros down and chock the wheels?"

Not stopping even to answer him, Jane released her seat belt and moved toward the door. Wade, moving equally quickly, helped her step up onto the wing, then down to the ground. They turned together in the direction of the house.

A man was standing at the end of the tractor road. It turned out to be the sheriff. Red Bevins was a big, amiable sort, his face as ruddy as his nickname.

"Where you folks fly in from?"

"Chicago," Jane said. "We got an urgent call from Mr. Grey. What's happened?"

Bevins shook his big head. "Afraid I've got some real bad news for you both."

Jane shook her head. She already knew.

"Yeah. His grandson found him—Down's kid? Saw that old John Deere going round and round in the field over there. Ran out, saw his grandpa slumped over, locking the wheel. Massive heart attack, is Coroner Holling's best guess. Probably went real fast, ma'am. Mr. Grey wouldn't have felt a thing."

Blindly Jane followed the sheriff inside. The living room was exactly as she had last seen it. The afternoon paper lay on the table beside Elie's favorite chair. His presence seemed palpable.

"Funny, him calling you like that," the sheriff was saying. "And you being in such a hurry to get here that you landed here instead of the airport."

"He *told* us to come straight here. Besides, he's not . . . wasn't . . . the sort of man to say it was urgent if it wasn't."

"But *what* was wrong?"

She shook her head. "He said he could only tell us in person."

"You're telling me you have no idea why Mr. Grey asked you to rush back here?"

"I'll tell you this, Sheriff. If it's humanly possible, I'm going to find out why."

Sheriff Bevins studied Jane's face for a long moment. Then he dropped into a rocker, gesturing them toward the davenport. "Let's go over what he told you one more time."

Jane did her best, even mentioning that the company had been having problems.

"That's not exactly news to me, ma'am. Our whole family's got Akers stock. We were actually thinking about early retirement before it started heading south. Lots of folks been hurt just as bad, and hoping you can patch things up."

Jane nodded bleakly. If they only knew about the Soverel deal, she thought, they'd be taking out loans to buy more Akers shares.

She heard a car start up and pull away, crunching gravel. Then Chevy walked into the room.

"The coroner just left," he said. "Who died?"

Jane burst into tears.

A few minutes later, outside in hopes that some fresh air would make it easier to breathe again, Jane saw two cigarette butts where the coroner's sedan had been parked. Elie was finicky about his property. He wouldn't like that. She stooped and picked them up.

"Please don't touch anything, ma'am," the sheriff said. "What you told me about that phone call . . . I don't know. But we need to take every precaution, just in case . . . This may be a potential crime scene."

"Sorry, Sheriff." She put them back, then spotted another filter tip in a wide tire track nearby. The track definitely didn't match the county vehicles. It wasn't a tractor tread, either. She pointed it out to the sheriff.

"Don't know what coulda made that, ma'am. Too wide for Big Ed's pickup. He might recognize it, though."

Big Ed Cherry, Elie had told her, was the man who worked the Grey farm on a contract basis.

"Will you ask him?"

Bevins scratched at his notebook. "Okay, ma'am, I just made a note to myself to do that."

"How about right now?"

5:22 P.M.

Jane went back inside to look up Cherry in the book. His wife answered, gave Jane the number for his tractor phone. Jane dialed again and got him. There was baseball play-by-play in the background. "I'm Jane Malcolm, a close friend of Mr. Grey's."

"I know who you are," the gravelly voice said. "You're Royal's girl."

"Yes, that's right." She took a breath, then asked hesitantly if he'd heard the bad news about Elie.

He hadn't, and she told him. He didn't say a word, but mixed in with the drone of the baseball game, she heard the sound of choked-back weeping. Jane felt wretched, wishing she could leave the poor man to his grief.

"I'm so sorry, Mr. Cherry," she said. "But I have to ask you about something. Please don't take this question wrong. But we found some odd truck tires here, and we're trying to place them. Sheriff Bevins said you're over here a lot and might have seen something."

"I *was* there—not an hour ago, Jane. Eating at me something fierce. Elie was out in the field on his old tractor. I even seen him turning circles, dammit. I figured he's playing games again with his kids, lettin' 'em steer, you know? Maybe if I'd thought harder, I coulda saved him. Damn, damn, damn."

Big Ed blew his nose, then regrouped. "But what you asked, did I see some kind of big vehicle? Answer is I did, right around then, or maybe a few minutes earlier. It was just pulling off the property, a big wrecker from Prairie-Town Recycling—"

"A wrecker?"

"Big flatbed wrecking truck, had the remains of your father's plane strapped down on the back. I waved at the driver—don't know him by name, but I seen him before at the Red Knoll swap meet. I figured Elie was finally tired of all that scrap. And I was glad myself to see it go, 'cause I need that shed space for seed and chemicals. You don't think there's some connection between them carting the plane away and Elie's dying?"

"It's too soon to have much of an opinion about any of this, Mr. Cherry. But I appreciate your help. And, again, I'm real sorry to break such sad news."

"I told Cherry I didn't have any opinion yet, but that's not altogether true. The timing is too tight." She'd just finished filling in the three men on what she'd learned. "Elie's call to us, Ed Cherry arriving, seeing that wrecking truck, then Elie dead on his tractor. We have to talk to that driver."

"Maybe one of the deputies could run us down to that

wrecking yard?" Wade suggested. "We could also find out who authorized hauling out your dad's plane."

"Right," Jane said. "And stop them from crushing the wreckage. Can you go, Wade? Chevy has to take the plane back, and I want to check something here."

Sheriff Bevins okayed their idea and detailed a deputy to take Wade.

"Now that that's being taken care of, I'd really like to talk to Elie's grandson. You did say he found the body?"

"Go on over to the farm, if you like. But Miz Cantrell wouldn't let me talk to 'em. Them kids are pretty broken up."

GREY DEMONSTRATION FARM
5:56 P.M.

"I can't let you do that, Ms. Malcolm," said Kerry Cantrell, the special-ed teacher on duty, a very imposing presence with her gray-streaked ponytail, aquiline nose, and strong arms folded across a neatly pressed ranch shirt.

"As I told the sheriff, these children are already very upset by what happened to Mr. Grey. The way this place is run, each one of them has gotten to know him—I'd be surprised if there's one who didn't love him. Now, I've finally gotten them quieted down a little—by talking to them, letting them talk, having them feed the sheep. But they're too deeply frightened to share all their anxiety even with me yet."

"What about Mr. Crain?" Jane asked. "Wade Crain? I understand he's spent quite a bit of time out here with the youngsters. Do you think *he* could ask them a few questions?"

"Wade's a good man, Ms. Malcolm, but the answer is still no. Not today, anyway. The county's sending a grief counselor down early tomorrow. We'll see after that."

"Fair enough. Let me ask this. How's Zach taking it?"

"Badly. But that's to be expected, of course. His grandfather dead—and he found him? You have to understand Zach's response whenever something goes wrong—either he's done something, or thinks he has, or someone's done something to him—no matter, what Zach does is hide somewhere until he can bear to deal with it. *This?*" She shook her head. "I tried soothing him, but I had to help the other children, too, and when I turned back to Zach, he was gone. Terry—he's my aide—is out looking in some of the boy's favorite hiding places now."

"He's run away?"

"It's not that bad—Zach doesn't run off. He could even be watching us right now. He'll turn up. But my guess is, unless we find him, it'll take a while. He's been scared really bad."

Jane crossed back toward the main house, thinking that the number of unanswered questions seemed to keep going up. But she was weighed down by grief as much as by troublesome questions. She'd only just gotten to know Elie grown-up to grown-up, discovered why her father had valued his friendship more than any other man's, hoped when the Akers crisis was over, one way or the other, to get to know him better. In the meantime, she had in the space of a single week come to rely on his perspective, his judgment, his bone-deep honesty—on *him*.

And what about his youngsters, how *they* relied on him? Dear God, what would happen to the school now?

Jane swiped at her eyes.

It wouldn't be the same without Elie. But it would go on.

That's my girl.

Dad? God, but I could use a hug right about now.

AKERS HQ
6:17 P.M.

Jane found Don Landsman in his office, his face bleak. "I wish to hell I hadn't let you talk me out of retirement," he said. "Then maybe Elie wouldn't have gotten involved in this mess. The strain must've been too much for his old heart."

"Don, we don't know that." She ran through a quick list of things about Elie's death that didn't gibe.

"What we do know is that Elie wouldn't want us to give up. The man went to a lot of trouble to bring you and me together as the nucleus of our little team, and he was proud of how hard we were going at it."

"Yeah, I know." He offered a wry grin. "We've done a heck of a job, mostly thanks to you. But I'll be honest, Jane. With Elie gone, a lot of the fight's gone out of me. I hardly care to find out what happened today with Soverel." He paused. "What did happen?"

"See?" She smiled. "We have to go on caring. As for the Soverel meeting, nothing's happened yet. Won't till tomorrow." She looked away, then back at him. "Unless you count my being offered a job."

Don nodded thoughtfully. "I'm not as surprised as you might expect. Happens Wade and I talked about that possibility. Another time, I'd be right curious about how he presented his offer. Anyway, sure not a bad career move for you, Jane."

"If you don't include the fact it would mean throwing in the towel here. After fighting back with all we've got and being in position to win."

"I didn't think you'd forget to take that into account. Still, there comes a time to look out for yourself. It did for me, remember? So what did you tell him?"

"I haven't given him an answer. He's expecting one before tomorrow's meeting starts. I didn't say I would, for sure, but I did think that might give me time to think it over. Then Elie died, and it's hard to think about anything else. So the truth is, I don't know what I'm going to do."

She slumped into a chair, caught up in her private thoughts. Don attended to some papers on his desk.

Three, four minutes later, Jane sat up straight. "What about Mr. Singh? Has he found anything?"

"Not the last time I checked," Landsman said gently. "He had his turban buried in computer printouts, snoring. Hagstrum begged me not to wake him. Guess our genius came up short. We don't have any rabbits left in the hat, Jane."

"Don't be too sure. The rabbit may *be* in the hat. Trouble is, the damn critter's disguised to look like something else."

"Jane, you've lost me."

"There was something very big that Elie wanted us to know."

"Maybe so. We'll never find out now."

"Don't say that! Don't even think it!"

Don stared at her.

"I'm sorry," she said. "I had no right to yell at you. But, don't you see, I have to find out."

OUTSIDE TOWN
6:32 P.M.

Desperate to get some perspective, and knowing she had to be alone to do it, Jane had left a message for Don that she'd be back in half an hour. She stopped at the condo long enough to change into comfortable pants and sneakers. Then she'd driven out of town, just far enough so there wasn't a building in sight. She parked, got out—minus her cell-phone and pager for once in the day—and walked. Began to jog. But she wasn't moving fast enough to shake loose the problems vying for a place in her head. So she began to run. Full out.

She ran faster and faster, until she couldn't breathe anymore, and then she stopped. Her whole body trembling, she looked around at the surrounding plain, flat as lake water.

How was it possible, in the midst of such flatness, to find herself standing on the edge of a precipice? A precipice she had climbed, inch by inch, problem added onto problem, tragedy by tragedy, for a solid week. Now she stood at the narrow top, swaying in the sharp wind of apprehension. Which way to go?

She could accept Soverel's offer, seize that extraordinary

professional opportunity, create an exciting, productive, profitable existence for herself—and spend the rest of her life trying to forget how she got there. Or she could climb slowly down the precipice, very very carefully so as not to slip. Jane shivered, because if she stumbled and went into a free fall, she knew that her body wouldn't be the only thing shattered.

Elie was wrong about one thing. Tom Soverel wasn't Satan, he was just a canny businessman. The temptation she had defeated originated within her. Which didn't make her evil, any more than Soverel was evil. Being tempted was part of the human condition. Saying no to it was what counted.

Jane shivered again, but this time not from fear. She was trembling all right, but in anticipation. And in awe. Because she knew now that—somehow—her father and Elie and God would be guiding her every step. She *would* make it down from the precipice safely.

She suddenly understood something else as well. Win or lose tomorrow, she *belonged* down on the plain. The air was too thin up here. Jane took a shallow breath and began the climb down.

JANE'S OFFICE
7:15 P.M.

She checked her messages. One from Jessica Zhuo in International. Three from Mark Gibian in the last hour. One from Wade's cell-phone fifteen minutes ago. As much as she wanted to call him right back, Jessica had to come first.

"Good news and bad, Jane. Our distribution center in Madrid has quarantined most of that meat shipment sitting in reefer units in their lot—"

"*Most?* What does that mean?" Heartsick as she was over Elie, Jane knew that this news held its own urgency—and right now it demanded her full attention.

"One truck left their site several hours ago."

"Dammit!"

"Amen," Jessica said. "But the RDC has a printout of the delivery schedule—locations and times—and they're contacting every place along the route."

"What about just contacting the trucker?"

"It's some independent guy, no phone, no nothing."

"Then we should have everybody out driving that same route, looking for him—our people, Spanish police—"

"Apparently they've been doing just that—only they can't find the guy. I'm staying right here in the office tonight, until they get him. I'll keep you posted."

"I appreciate that, Jessica. Thanks." She hung up and immediately dialed Wade, hoping he would still be at the wrecking yard.

He was, but his news wasn't good, either. "Got here too late, boss. They'd already compacted your dad's wreckage. I saw the cube. You want it?"

"I don't know. Are you still at the wrecking yard?"

"I'm on my way back to Akers. Deputy's driving me."

"Did you find out who authorized it, or released it to them, or whatever?"

"They showed me a bill of sale. Supposed to be Elie's signature. Could be, hard to say. The dispatcher'd gone home."

"What about the driver?"

"Gone, too. Took off in his own car. I called his trailer, no answer."

Jane cursed. The driver could tell them who'd signed Elie's name. They'd track him down eventually. But, dammit, she wanted to know now! "I'll be in my office."

And that was that, Jane thought. The temporary end of the line—except for her offer from Tom Soverel, ticking toward its own deadline.

She called her San Francisco office.

"Mark, what's happened? TSA do a flip-flop again?"

"No, the magic you worked there has stuck. But we've got

another label screwup by Healthy Hills. Not actually Healthy Hills this time, but their East Coast subsidiary, Champlain Preserves in Burlington, Vermont. Seems they shipped half a containerload of their whole line of jellies and relishes to France. Turns out France requires a special label that includes the production date. Champlain's Art Department produced nifty new labels, but carelessly used an old production date. The shipment got rejected in France. Champlain is asking Joe Dawes of Healthy Hills what to do, and Dawes is asking me, and now I'm asking you. Junk it, or salvage it?"

"Salvage," Jane said. "HH can still squeeze something out of the fiasco." She checked her Rolodex and gave Mark the names of two East Coast salvage outfits to call. "Get bids from both on the cost of returning the cases to the U.S.—where are they now?"

"Sitting in a warehouse in Le Havre."

"And the cost of delivering them to the salvage company. And Mark, after the dust settles, let's schedule a meeting with HH—Champlain included—to see if we can install some system to keep this from happening so often. By my count, this makes three expensive mistakes that could—should—have been avoided. A couple years ago, didn't they park a whole bunch of cases of their Healthy Trails snacks in a warehouse long enough for the peanuts to turn rancid?"

"Cruelty to health nuts?"

"I gather you're back on Snickers. Anyway, trust me. If it keeps happening, we're not earning our money."

"Okay, I'll pick a date a couple weeks out so you're sure to be available. How's it going, by the way? You sound . . . beat?"

"I am beat. Not beaten, though. Still, today's been real rough, I won't lie to you. I lost Elie today." She realized the name didn't mean anything to Mark. "Eliezer Grey. My dad's best friend—"

"I'm sorry. I wish I could—"

"—and tomorrow I may lose my dad's company. Other than that, hey, no complaints."

"—cheer you up."

But except when it came to his own capabilities, Mark was a realist. "I can't do much from here, I guess, except let you go. But Jane?"

"Hmn?"

"I just wanted to say, if *anyone* can save Akers, she's in the right place right now."

"You really think that?"

"I know it."

"Thanks, Mark. Bye."

Jane hung up. He did mean it. But he was there, and from where she sat, she was about out of time—and hope. If this wasn't rock bottom, she hoped never to encounter the real thing.

There was a knock on her door.

"It's open," Jane called out.

A thin man, with hollowed cheeks and sorrowful eyes, opened the door and leaned into the doorway. He had a vendor's badge pinned to his short-sleeved white shirt. "Ms. Malcolm? Do you have a moment? It's real important. Concerns Mr. Grey."

"Are you aware that Mr. Grey—"

The man nodded somberly. "I just found out, Ms. Malcolm. Makes my reason for seeing you that much more important."

Jane felt an adrenal surge. The man's face was familiar, but from where?

"Please come in, Mr . . . ",

"Harlow. Lynford Harlow." He handed her a business card. "Private Investigator" gave her another jolt. The next line triggered her memory: "Investigator—FAA—NTSB—Retired."

That was where she'd seen Harlow. Inside Elie's barn, when he'd tilted up his welding mask! He'd been the one working over the wreckage of her father's plane. Could he be the man who'd called the wrecking yard? The one able to clear up the mystery of Elie's final moments? Who knew what Elie had wanted to tell them?

"Mind if I close the door?" he asked.

"Go ahead. Then, please, take a seat."

Jane made herself fold her hands on her desk.

Harlow sat stiffly, obviously unused to sit-downs. For a split second, Jane envied him that. "You remember me from the other day?" he asked.

"Eliezer Grey hired you. You'd worked with the NTSB man investigating my father's plane crash."

"I had what you might call a dissenting opinion, which they weren't interested in for their preliminary finding. Elie was. He pulled some strings to get the wreckage trucked to his shed, asked me to go over it again, with a fine-tooth comb."

"And you did?" She glanced down; her knuckles were white as bone.

"I cut that plane apart with an oxyacetylene cutter—wings, tail, fuselage. Even the engine."

"But you weren't there when Elie died?"

"No. Like I said, I just heard about it."

"I don't understand. If you don't have any information about what happened . . . "

"I didn't say that. There's a lot more involved here than you're aware of, Ms. Malcolm. You don't know me, so I had to think twice before I came here like this. But Mr. Grey was a special kind of man, a gentleman. I had to come."

Jane nodded. "He sure was. What is it you know, Mr. Harlow?"

The man put his shoulders back. "I'm sorry to have to tell you, but I believe we're in the midst of something real scary here."

"Are you going to tell me what it is?"

"Early this morning I uncovered forensic evidence that your father's plane had been sabotaged. A connecting rod on the landing gear had been filed through."

Jane gasped.

"I'm afraid, Ms. Malcolm, your father's death wasn't any

accident. He was murdered."

7:28 P.M.

As shocking as Mr. Harlow's news was, Jane believed it on impact. The more experience she gained as a logistician, the more she valued common sense, and—of course—it made sense that there was deliberate linkage in the dreadful chain of events—computer sabotage, her father's plane crash, the Akers empire unraveling, now Elie's sudden death.

Murder motives? There was no need to search far, not with millions of dollars and vast power at stake. It required only one twisted, angry soul to target the tall man atop the corporate pyramid. Jane swallowed bile. No matter how much *sense* it made, it still sickened her to think that anyone would want to kill her father.

Which suddenly conjured in her mind the names of Akers executives who were recent casualties of corporate infighting. None of the names swirling about in her brain lit up in neon. Besides, the list must be considerably longer than she knew, Jane thought, and any of them could have vengeful designs.

Of course, the person responsible for her father's murder, and Elie's, could be outside the company, some unglued business rival, even some antibusiness wacko. There was no denying that she still felt a leftover sense of helplessness. But the fury inside her was stronger than her apprehension.

And brought her to a new level of resolve. Whoever was responsible for the death of her father and Elie, she suddenly knew that she would see to it that the culprit—or culprits—was caught and brought down.

"You have no idea who could have done it?" she asked Harlow.

He shook his head. "I think Elie had an idea. That's why I suspect his heart attack was no accident. That definitely needs to be looked out for by the coroner. They *are* doing an autopsy?"

"One more thing to find out." Jane took Sheriff Bevins's business card out of her wallet, reached for the phone, then halted her hand in midair. "God, I forgot! The wreckage of Dad's plane was hauled away—it's been crushed. Your evidence is gone!"

"You know this for sure?"

"I sent someone to the wrecking yard. He saw the cube."

"Definitely not good—but not the end of the world. I've still got evidence I can take to court. I took a lot of pictures."

Harlow opened the manila envelope lying on his lap. "I spent a good part of the day at your local Akers's one-hour photo lab. That's why I wasn't at the farm, or maybe I could have saved Mr. Grey's life. I had 'em make different kinds of prints, then these enlargements, which I just got."

He handed her several eleven-by-fourteen black-and-white matte prints. Jane hung up the phone.

Harlow pointed to a black, cylindrical shape with a jagged end. "That's the connecting rod to the landing gear of your father's Beech Bonanza. The ruler beside it is just to show scale. These were all shot with a 105 mm f/2.8 macro lens, lit with four strobe pops at full power, so the definition's pretty good. You can see right there, that's no clean break. No way that rod just snapped off. Those serrations are file marks."

He looked from the picture to her. "You're looking at sabotage."

"Why didn't you or the other NTSB investigator spot this before?" Jane asked.

"We should have, obviously. But the landing gear got pretty well mangled in the crash—it took me a while to find this. Plus which, the other guy had it all figured that the gear collapsed because your father inadvertently retracted it during landing, or just hadn't slotted the handle into full down landing position. Anyway, that's water under. What needs to be done now is blow this photo up some more, which I'll do when I get back to my own darkroom. Then the details will be even clearer."

Jane reached again for the phone.

"Whoever you're calling," Harlow said, "for God's sake, don't spell it out."

"Don't worry, Mr. Harlow. I understand the importance of this find of yours."

EN ROUTE TO SHERIFF'S SUBSTATION, OCEAN PLAINS
7:43 P.M.

Don Landsman steered his Lincoln Town Car through deepening twilight. Jane sat in front beside him, Harlow in back. All three were now burdened by Harlow's revelation, all anxious to share that burden with the authorities in the person of Sheriff Bevins, who had agreed to meet them at the local substation.

"Don," Jane asked, "who profited by your *not* becoming CEO?"

"Hold on. First, your dad never actually got around to naming me. He only *told* me he was going to do it, and I guess he hinted the same thing to other people. Second, that's a helluva stretch, Jane, saying somebody killed him just to keep me from running the company."

"Just let's not rule it out, okay? I'm asking you to speculate. Obviously Roy profited, since he got the board to give him immediate control. He brought in Wally Conover, which sent you packing. Who else?"

"Well, you could include Flynn Emerson. He was another guy some folks figured to become CEO. Only Flynn walked out and joined Soverel the instant rumors started to circulate about Royal favoring me. That was weeks before Royal's plane crashed."

"But he could still be bitter, right?"

"Hell, yes, he was bitter. He told me so himself. That doesn't mean he'd go out and commit murder, or hire someone to do it."

"Or sabotage our computer system, to set up a slam-dunk takeover by Soverel?"

"Anything's possible." Landsman sighed. "Hell, *I* could have done that—not that I'd know how to sabotage a computer or an airplane. But I had a grievance. Jane, a lot of people fit that profile."

"And one of them did it."

SHERIFF'S SUBSTATION
OCEAN PLAINS
8:02 P.M.

Sheriff Bevins listened to what they had to say, particularly to Lynford Harlow, whom Bevins had met during the NTSB investigation of Royal's plane crash. He examined Harlow's photographs, handed them back.

"So the NTSB hasn't actually filed their *official* report yet, is that what you're telling me?"

"Right," Harlow said. "Their ruling of pilot error was only preliminary."

"Well, you got a lot of interesting material here. Maybe enough to get the NTSB to reconsider their verdict." Bevins paused. "That's assuming your reputation is good enough, Mr. Harlow, so nobody will suspect you of maybe doctoring these photographs or tampering with that crash evidence."

"I think you'll find my credibility intact," Harlow said evenly.

"Didn't mean . . . Looking at all sides is part of my job." He turned to Jane. "Now. What you hinted at, Miss Malcolm, about the computer problems at Akers, the FBI Computer Crimes Division ought to be interested in that. But I don't see how anything you've got here justifies my ordering an autopsy on Eliezer Grey's body. Family hasn't requested one."

"Speaking of family, what about Elie's grandson?" Jane asked. "Do you know where he is yet?"

"The Down's kid? When he turns up, we'll get his statement. And whatever you turn up on this sabotage angle, you keep us posted, hear? But unless you come up with more evidence, I'm gonna stick with Coroner Hollings's findings. We got a seventy-year-old man died of a heart attack."

They had left Lynford Harlow at the substation, sitting at a deputy's desk to call the NTSB regional office in West Chicago and request a reopening of the crash investigation. Harlow had also promised to contact the wrecking yard first thing in the morning and talk to the truck driver and dispatcher.

Jane, sitting beside Don, didn't feel like talking. Problems seemed to be multiplying. She didn't know what to do next—let alone first. All the balls were in the air, but none was coming down—or would till tomorrow, it seemed. She wanted desperately to hear more from Wade, from Elie's grandson, from Rikki Singh, and from that truck driver who'd picked up the wreckage of her father's plane. In the meantime, lacking such vital information, she still had to make her decision how to handle refusing Soverel's offer, then arrange to get to Chicago in time for the ten A.M. showdown with him and her family.

Chevy was the obvious choice to fly her there, freeing Wade for whatever came up back here. If there was a killer out there—and Harlow's photos sure pointed that way—then she was now in position to pose a threat to that killer, and flying in an Akers plane right now would be foolhardy. At the very least, she had better tell Chevy to take extra time to check out the plane.

Don Landsman was also thinking about Jane's safety, and as the Lincoln neared the condo complex, he finally spoke up. "I don't think you should stay here tonight. Why don't you come out to the lake?"

Jane was on the point of saying yes when her cell-phone trilled. It was Wade.

"Where are you?" he said

Before she could answer, Don's car swung into the condominium parking area and its headlights swept over a black Chevy Blazer, then skewered Wade standing beside it, phone in hand.

"Looking right at you," she said.

OCEAN MEADOWS CONDOS
8:42 P.M.

"Heck, you're *both* welcome to spend the night at my house," Don said as he pulled his big car alongside Wade.

"Appreciate the offer, but I'll be fine here," Jane said.

"If you insist on staying here, I think you should stay at Wade's place. He's got room. More important, he's got a shotgun. I've been hunting with him. I'm being very serious, Jane."

"I know you are, Don. Thanks."

"Take my advice, then—so you can thank me in the morning."

A moment later he drove off.

Leaving her facing a frowning Wade.

"You look shook up," he said. "What's happened?"

"Don says you have a shotgun."

"Seriously? Yeah, my dad's old Winchester 12-gauge in the back of my closet. I've got a couple boxes of shells somewhere."

"What about liquor? Not beer or wine. Booze."

"Wild Turkey?"

"That'll do. Pour me two fingers straight up, and I'll fill you in."

He handed her the Wild Turkey, took a cold beer for himself, and sat down across from her. "Okay, talk," he said.

"Not until you bolt that door and get out your shotgun. Don't look at me as though I've suddenly gone paranoid. If I sound scared, I am. But not for no reason."

Without a word Wade bolted the door, got out his Winchester, and began to assemble it. "I can listen while I do this," he said.

"Right." Sipping bourbon, she began to pace the room and bring him up-to-date. Wade didn't interrupt. When the Winchester was loaded, he placed it in the corner, popped open his beer, and took a swallow.

She stopped walking, glanced toward the windows, set down her glass, hugged herself.

"There aren't any windows in the dining room," Wade said. "Would you be more comfortable in there?"

Jane nodded and followed him.

"Have you eaten?" he asked.

"No. And I *must* be hungry."

He stepped into the kitchen and opened the refrigerator. "How about breakfast for dinner? That's about all I got here."

"Sounds good."

"Keep talking, Jane. Breakfast is the one meal I can prepare without needing to concentrate on a recipe."

She picked up the story at the part where they'd shown the photographs to Sheriff Bevins.

"I understand the sheriff saying what he did," Wade said from the kitchen. "But I don't buy the heart attack. Elie's heart stopping just when it did, after calling us, then that big hunk of evidence disappearing? Uh-uh. Someone's behind the scenes, yanking levers."

The smell of bacon frying wafted her way. Its homeyness was somehow comforting. Still, she had to ask. "What about us?"

Wade came to the kitchen door, mixing bowl in hand. He continued to fork whisk the eggs. "You mean, are *we* in danger? Too soon to say, but not too soon to be careful—you were right. Now, give me five minutes and I'll give you perfect scrambled eggs."

True to his word, not six minutes later Wade set two plates of bacon and creamy scrambled eggs on the table, followed almost immediately by toasted bagels with cream cheese and a pot of decaf.

Jane surprised herself by devouring everything in sight.

"I wasn't hungry so much as famished." She tried for a bigger smile than she had in her.

He refilled their coffee cups. "Jane, somebody's gotta catch this bastard quick. I'm betting Harlow gets a solid lead tomorrow morning at the wrecking yard."

Tomorrow! She'd barely survived *this* day. Its events, flashing neon, tumbled in memory—up till two A.M., in the aftermath of the president's call, jawboning the Hyon-Korean crisis; then the six A.M. wake-up call in Dallas; strategy breakfast with Don Ellsmere; her pitch to the TranSonic Air board; jetting back to Ocean Plains; jumping into the four-seater for the flight to Chicago; reacting on the fly to a public health menace, a containerload of contaminated meat somewhere south of Rotterdam; the meeting with Tom Soverel, surprise topping surprise; the flight straight to Elie's farm with its dreadful ending; and, finally, Lynford Harlow walking into her office with *more* dreadful news. No wonder it felt like a day and a bloody half.

And tomorrow would bring more highly charged events—none of which she was prepared to contemplate now. She felt her head nodding toward the table—and it wasn't even nine o'clock!

"You're about ready to clock out," Wade said. "I'll sleep on the den sofa and you take the bedroom. On second thought, there's a window over the big bed, though, and no windows in the den. What'll it be?"

"Sofa's fine for me," Jane said, rising unsteadily. "I'll clean up in here."

"No, you won't."

"Maybe I won't, then," she said as a dizzying tide of fatigue swept over her. "I know it's early, but if you don't mind, I think I'll go to sleep now. Maybe if I'm lucky, I'll sleep right through tomorrow. Guess I sound pretty sorry for myself, huh?"

"What you sound is pretty exhausted. You've had too many days with too little sleep. Come on."

She followed Wade into the den, watched as he removed the sofa bolsters.

"I'll just get some sheets and blankets from the closet to put over the cushions. Be right back."

"That'll be lovely."

He returned moments later with a double armload of bedding, to find Jane already curled up on the sofa, facing the back, knees drawn up. He crossed the room, knelt down, listened to slow, resonant breathing.

Wade slipped off her shoes and, very carefully, covered her with a blanket.

OCEAN PLAINS
9:46 P.M.

The day the mountain bike had been delivered to Rikki Singh's office by the local Akers store, it was obvious that it was decidedly inferior to the Italian model he kept on campus at Berkeley. The suspension especially didn't measure up, but the bike was sturdy—definitely serviceable. He'd used it every day since, around the Akers property, where if he hadn't been under such time pressure he might have walked instead.

Tonight he really needed it.

Singh pedaled out the gate toward the dark town.

A wind chased him across the railroad tracks on Main Street, and he turned up the collar on his windbreaker. There were few signs of life in the town center. He swung left on First, as directed, his headlamp needling the dark storefronts.

At the next corner, just where Lenny Cristofaro said it would be, Rikki could see the flicker of neon. A second later, over the snickering of his tires, came the bass thud of nightclub speakers.

"Mulekick Saloon," flashed the sign. Cars and pickups nosed diagonally against both sides of the street, with a big rig parked parallel on the next block.

It was, Lenny had said, a place Mimi Takuda liked to go—and had gone this night.

"But frankly, R.K., I don't think she's your type," Lenny had added, bouncing rubber bands off a wastebasket across his office.

"Frankly, I haven't given that any thought. I just need to talk to her about something."

"Sure you do."

If you only knew how I wish I didn't.

Beyond the swinging door, the Mulekick was skinny and deep, a long bar on one side, dance area in back. It used to be a four-lane bowling alley, Lenny had told him, now reopened weekends only as a country-western dance club. Blue-jeaned bottoms filled the barstools. Hats stayed on, male and female alike, some ranch straw, others fancy felt.

"Looky here," someone called out. All along the bar cowboy hats swung around—just as in western movies Singh had seen, when a strange gunslinger entered the saloon. Except in a western the stranger never wore a turban.

He tried to look genial as he walked on past, toward what sounded like a combined hoedown and cattle stampede. He plucked a cocktail napkin from an empty table, tore off strips, wadded them up, and inserted them as earplugs. The decibel reduction was significant. Singh smiled.

The small dance floor was packed with high-spirited couples, men in straight-leg jeans, women in swirling denim skirts, all shuffling and stomping their high-heeled boots. It was a carefully choreographed stampede, Rikki saw, the various bodily maneuvers requiring almost military precision. The more he observed, the more he admired.

And right there, in the midst of the wheeling and whirling

couples, was Mimi, all dressed up and yelling, "Yeehaw!" with the rest of them. Singh watched her ponytail lashing to and fro beneath a tufted black felt hat. She wore a cowhide-print dress, matching cowhide boots, and a look of utter delight. Towering beside her, and spinning her like a top, was a gangly blond guy with long knobby arms, dressed entirely in black—from T-shirt and vest to jeans and boots.

Suddenly Mimi glanced over at Singh, smiled, waved, then spun away. Her partner *kept* looking Singh's way—and glaring.

When the music stopped she hurried over, out of breath, face flushed. "Rikki, this is great! Do you like line dancing?"

"I came to see you, Mimi."

"Super! Wanna try?" She jerked a thumb at the dance floor. The tall guy stood, solitary, a little way off, drinking from a long-necked bottle, his eyes fastened on them.

Singh shook his head. "I certainly enjoyed watching you, however. You're most proficient."

"You're sweet, Rikki. All you need is a lesson. And they're having one next. The cotton-eyed Joe."

"I don't think your boyfriend would approve."

"Delbert's not my boyfriend. Besides, he's basically harmless. C'mon."

Singh found himself dragged onto the floor and inserted in the middle of the front row, facing the instructor, a potbellied man with tobacco-stained mustache and clip-on microphone. Spotting Singh front and center, the instructor did a funny double take, then with a broad wink said, "Howdy, Swami. Guess sometimes the twain does meet, eh? More power to ya. Gonna do a little cotton-eyed Joe tonight. Just watch my big feet. Let's have us some music . . . . "

Amazingly, after a few diligent minutes, Rikki found that not only could he follow and execute the basic steps, he was actually enjoying himself. Mimi made no attempt to hide her pleasure at his progress, shouting encouragement with every new step. Implausibly, a number of strangers also offered loud verbal support.

When the cotton-eyed Joe was completed, Singh moved to the side to relish his small triumph for a quiet moment before he took advantage of the fact that Delbert was nowhere to be seen—and tried to speak to Mimi. But her intentions were more physical. She began coaxing Singh back on the floor, wanting to show him the next dance herself. She tilted her head to the side and smiled at him. He gave in. The Texas two-step, alas, was considerably trickier than the cotton-eyed Joe, but Rikki trod on her toes only twice, and Mimi claimed her metal boot tips kept her from feeling it. Then came the cowboy waltz.

A half hour later, hot faced and sweat soaked, Singh was seated across from Mimi at a small table, sipping a Coke and wishing fervently that he *had* come here with only amorous intentions, because it was obvious that Mimi suspected as much—and was pleased. He hated to disappoint them both.

"Mimi," he plunged in, "you remember this afternoon when I asked about your tattoo, and you said it was the Cheshire cat's smile?"

She stopped sipping and gave him that tilted look of hers, minus the smile. "Of course I remember. Why?"

"I'd like to ask you something that will certainly seem none of my business. I am apologizing in advance for any impertinence, you see? Now here is the question: What made you decide to get that particular tattoo?"

She straightened her head. "I thought you came here because you wanted to see me, Rikki. And I thought you had as much fun dancing as I did. Now it's just like this afternoon in the computer room. All you want to talk about is my tattoo. Why are you being so weird?"

"I apologize if it seems that way. I am merely curious as to what might motivate a lovely girl to mark her body—I mean, *adorn* it—with this particular design."

She glared at him. "I don't get it."

Was Mimi really this angry over his slight impertinence, or was she hostile because she had a good idea where his inquiries were leading? In any case he felt wretched, knowing he'd just sacrificed any romantic chance he might ever have had with her. Worse yet, if she refused to answer, chances were she was the saboteur. How he had hoped, as he'd pedaled there, that Mimi would persuade him otherwise.

"I apologize again," Singh said, standing. "Now I must return to my office."

She didn't say a word to stop him. He had no choice but to leave and contact Jane. He turned, to find his passage blocked by a wide man bent over the next table, nuzzling an apparently willing girl. "Excuse me," Singh said.

"When you get there, why don't you ask Lenny your burning question?" Her words, sharpened by fury, lashed at his back.

Singh swiveled and, carefully monitoring his expression, asked, "Why Lenny?"

"Because he's the one who talked me into it, last year at the Compu-Con at McCormick Place." She was angry.

"Please, Mimi," he said. "Please tell me more."

She grunted. "Oh, all right. One night we went to this North Side pizza place with a whole group of people. We had a few beers—a few too many. There was this tattoo parlor next door, and afterward some guys were going to get one, even one of the girls, and they dared me to, too. I said, 'No way.' But Lenny made a big deal out of it, saying how movie stars were getting them, that it was very hip. It ended up we all did it."

"Lenny, too?"

"I just said that."

"I mean, did he get a Cheshire cat, like you?"

"It was his idea, wasn't it? Lenny even drew the design for the tattoo guy. His is actually a lot bigger than mine." She made a sarcastic face. "So, now you know the whole tragic story. I knew Lenny was hoping it would make us closer somehow—off duty, if you know what I mean. Which will never happen. Not that you care. Now, are you satisfied?"

"But I do—Mimi, please, let me explain why—"

Too late. She was already on her feet, squaring her cowgirl hat, her back turned, looking for the best route away from him. Which she then took, quickly vanishing into the crowd.

Lenny!

At the bar, ten feet away, a man in a red flannel workshirt tugged his dirty Stetson lower, smoothed his walrus mustache, and slid off his stool. He wore just bought Akers-brand jeans, so stiff they whistled as he walked. At first he followed Mimi, but she detoured through the door with the cowgirl cutout. The mustached man went on past, unobtrusively searching the crowd. It took him a few minutes to locate the blond, black-clad man called Delbert, but only a few seconds to mumble in his ear.

The bony head snapped up, eyes slitting as he scanned the room for a turban. Satisfied with this reaction, the mustached man headed for the exit.

Singh scolded himself for having constructed a fatuous syllogism: The saboteur had left a Cheshire cat clue in one of his programs; Mimi Takuda wrote software that closely matched the saboteur's linguistic template, had unlimited system access, and bore a Cheshire cat tattoo. Ergo, Mimi was the culprit.

Apparently, however, all those parameters also applied to Lenny Cristofaro—including a hidden tattoo smile. Singh was sure now that if he reexamined those incriminating programs, created by "root" or "sys" or "adm," he could determine that they'd been authored by Lenny, not Mimi.

He would definitely do that.

But he dared not wait for final confirmation. His every instinct told him Lenny was the saboteur. He had to call Jane now. Singh hurried toward the pay phone at the end of the bar.

Busy! A mountain-size man had the phone in his big fist. Singh waited, trying to concentrate on his breathing, as the man kept talking and talking and talking, his thick lips pressed against the mouthpiece. *Assert yourself.* Just then the man spat a glob of brown juice at a nearby cuspidor, missed it, and blotched the floor.

Gratefully Singh suddenly remembered that his flip-phone was zipped into his bicycle saddlebags, and he hurried outside.

His bike was gone! The cable lock had been snipped from the leg of the metal news rack. He'd taken precautions, even though he'd been told they were unnecessary in Ocean Plains. *Untrue!*

He ran out into the dark street. No sign of the bike or its thief. He felt a sharp tap on his shoulder and whirled around.

"Hey, Swami, I hear you called me a fairy. Wanna try it again?" Delbert smirked at the Indian's obvious fright.

"But I don't even know you—" Ocean Plains was a dangerous place!

"I'm Delbert Bowen, that's who the hell I am. Who the hell are you, Swami? And who the fuck told you to come messing around Mimi? Now she's bawling."

"She is? Oh dear, I'm truly sorry," Singh said. Perhaps this fierce man was a genuine friend to Mimi. He supposed that some of his own friends might look strange to others. "It was a misunderstanding. I was not, as perhaps you may think, trying to interject myself into your eve—"

Singh saw the punch coming and tried to duck, but not quickly enough. Delbert's bony fist crashed against his turban, just above the left ear. Rikki's last thought before he blacked out was, If I die now, Lenny will escape.

WADE'S CONDO
11:05 P.M.

The sound of an Akers semi downshifting on the street outside was enough to awaken Wade. After checking the clock, he decided to check on Jane as well. He hauled himself out of bed and tugged on jeans and T-shirt before padding out to the den.

It took him an anxious moment, in the darkness, to resolve the shapes and shadows. Jane lay facedown on the sofa now, blanket kicked off and right leg bent, revealing a slice of thigh above the raised hem of her skirt. Sleeping hard, like a child.

His heart went out to her in a rush and stayed there.

He wished there was more he could do to help her through this. There wasn't, at least not now. He knew that. Better get back to sleep himself so he'd be in shape for tomorrow's dung heap of surprises.

Still, he didn't move, rendered absurdly happy by the sight of a totally disheveled Jane.

Without warning, she turned over onto her back. Wade saw her eyes open and then widen in fear as she realized she was being watched.

"It's me," he said quickly.

She bolted upright. "Who's there?!"

"It's Wade. Everything's okay."

"Oh. Wade." She rubbed at her eyes. "I don't even remember going to sleep."

He switched on a lamp across the room. "You fell asleep before I could fix up the sofa for you. I just slipped in to make sure you were okay. I didn't mean to frighten you."

"I'm fine—except I feel as though I've been sleeping with a lot of other pretzels." She stretched tentatively. "Do you have some juice or soda?"

"Orange juice?"

"Fine. Bathroom through there?" She pointed.

"Yes. You'll find a jogging suit hanging on the door. It's clean, and the closest thing to pajamas I have."

She came back wearing the baggy two-piece garment, carrying her brand new teal suit in a crumpled-up bundle. "Thanks," she said, reaching for the glass in his hand.

"If you wait a moment, I'll fix up the sofa for you."

She stood by as he spread sheets out on the sofa cushions, then a blanket, tucked everything in, added a pillow.

"There," he said, straightening and turning.

Suddenly they were standing very close.

Don't make a move, he counseled himself. Don't do *anything*. He didn't even breathe.

Then he felt her hand touch his.

"Just hold me a moment?" Jane whispered.

His arms complied, drawing her into a circumspect embrace. "How's that?"

"Mmn." Jane really wanted only to be close. Didn't want Wade to go back to bed yet. Didn't want to be alone right now. She gave herself permission to relax.

Maybe she relaxed too much. Or maybe Wade lost his balance. Or maybe they both meant to topple together onto the couch.

She landed on top. *I could almost go to sleep like this. Just put my cheek on his chest.*

He raised up only enough to look at her inquiringly.

She was suddenly very aware of her nakedness under the jogging suit.

"Jane?"

"Mmn?"

"Could you turn your face just a wee bit? Not that way. Toward me."

She did.

How long had that kiss been waiting for them? Now that it arrived, it was certainly taking its—very sweet—time.

"What?" she asked after a minute. "You're mumbling."

"I was just saying your name."

"Oh."

He kissed her again. She contributed a generous half. Whatever nastiness lay in wait tomorrow, Jane thought, this was to be savored. She snuggled even more closely against him.

"Jane, could we—"

"Am I crushing you?"

"You're not, but maybe if we could just—"

They moved simultaneously, ending up side by side, faces still close. The movement pinned Wade's hand beneath her, actually beneath her right breast. He made no move to free it. Jane sighed, half in surrender, half in anticipation.

"We don't have to . . . " he said.

"Don't you think you've been gallant enough for one night? Besides, I want to." She raised her torso slightly and tugged at the bottom of the baggy sweatshirt, making room for Wade to slip his hand underneath.

"Better?" she inquired, as her breast compressed into his palm.

"Umn-hmn," he said.

To keep things equitable, she wormed her fingers under his T-shirt. If their hands were getting serious, their kisses continued to be playful. Which was good, Jane thought. All day she'd been in a frantic race. She didn't want to hurry this. Still, one thing she wanted without delay—Wade naked against her. Now. But her usual efficiency failed her. There was just not enough room to maneuver.

Which didn't prevent their mingled breathing from quickening. Then, as Wade shifted his weight, his elbow slipped off the sofa and he collapsed against her. He gasped an apology, Jane laughed aloud—and their teeth clicked painfully.

"This is silly," she said. "Take me to the biggest bed you've got—right now. I don't care how many windows are in the room."

Moments later, garments out of the way, they were clasping

each other full length, flesh to flesh.

Then they were atop the mattress. And the world.

Wade held Jane in the crook of his arm, still dazed by his staggering good luck. How many years had he wished Jane Akers was *his* girl? Sure, there'd been a decade or so there when he thought about her only when Royal mentioned her latest accomplishment. His own memories of the Jane who barely knew he was alive faded. Neither part of that seemed possible now that she was in his arms, where she revealed an earthiness that both surprised and delighted him, just as her undressed abundance had done. He understood that what they shared now was more than the chance reunion of near strangers. During long days of working side by side, they'd leaned on each other's strengths—and learned to rely on them. Still, his good fortune humbled him.

"Hey, you up there," she said.

"Sorry, just saying my prayers. Where were we?"

"About to start again," she said. "Doing this and that."

She showed him.

And he showed her. "We keep practicing, is that it, Ms. Malcolm?"

"Until we get it just right, Mr. Crain."

OUTSIDE OCEAN MEADOWS CONDOS
11:53 P.M.

His view was limited by the slant of the venetian blinds. Perched atop a nearby carport, hidden within the foliage of an overarching magnolia, he had a straight shot into the bedroom thirty yards away. But the down-slatted blinds sliced the erotic tableau into horizontal strips. That, and the fact that he was forced to watch through three-power, night-sight optics, rendered the reward insufficient for the risk.

But the acoustics more than made up for those deficiencies.

He'd arranged a simple laser surveillance system. The transmitter, bolted atop the tiny tripod beside him, directed its invisible infrared beam down at the first-floor bedroom window. Sound in the room caused the window to vibrate exactly like a microphone. From the glass, the laser could detect and collect modulated vibrations of a hundred-thousandth of an inch. Because the beam was slanted down, it bounced back lower yet, to a precisely aligned optical receiver he'd concealed below. The receiver converted the modulated beam into audio signals, which were filtered, amplified, and then transmitted to his headphones.

There was little conversation to overhear, but plenty of organic noises and percussive friction. He found the sounds of this obviously fully aroused woman an incredible turn-on. So stimulating was the ecstatic buildup in his ear, he became a silent partner in their ultimate satisfaction.

There was another satisfaction. His vantage point proved to him that these two were more of a team than they'd let on. How long had they been so . . . entwined with each other? How much more of a threat did they pose as a duo than individually?

Then there was the next question, a favorite of his. How best to eliminate that threat? This was not the moment, of course, despite the vulnerability of the melded targets who were having at it for a third time and the Glock 17 snuggled into its shoulder holster. A better moment lay ahead. *I'm dead certain.*

Still, laughing at his little joke, he took out the lightweight pistol and pointed it toward the enhanced images in the dark window, then cycled the slide to feed a round into the chamber. This was a pleasure few mortals had the courage to taste, holding death in one's hand, a finger twitch away.

But the game had to be brief. It was time to button everything up and call it a night.

DAY 8

MONDAY, MAY 17

WADE'S CONDO
6:49 A.M.

At some point they had both succumbed to deathlike slumber—from which, dark aeons later, Wade was cruelly exhumed by an insistent shrilling. He groped and eventually found his bedside phone, listened to its dial buzz, hung up. He groaned, tried to roll back to sleep, but lifted his eyelids just enough to get jolted awake.

Jane wasn't there.

Wade came surging out of sleep, adrenalin pumping, primed for danger—as she walked back into the room, cellphone in hand, without a stitch of clothing on. He took a deep breath of relief. Then a sharper one of desire. He held his hand out toward her. "Come back to bed," he said.

"Nice idea, but lousy timing, I'm afraid. We've got to get going. That was Lyn Harlow. He located that truck driver, got a description of the guy who signed the plane wreckage over to him."

"Yeah, okay. What time is it?"

"Almost seven. Harlow wants to meet us at my office as soon as possible. I told him fifteen minutes. Let's make it ten."

Five minutes later Jane walked swiftly along the chilly breezeway of the condo complex, Wade beside her stride for stride.

Her condo was two units over. Wade scanned ahead and side to side—carports and corridors, shrubs and flower beds. As they neared her front door, Jane had her key out. She tumbled the lock and stepped inside.

The phone was ringing. She picked it up.

"Jane Akers?"

Malcolm. But just now it hardly mattered. "Who is this?"

"I'm so glad I got you. Zach's had me up all night saying over and over that he has to tell you something—oh, I'm sorry, this is Bonnie Chambers. We've never met. I'm Zach's mom."

It took Jane a moment to refresh her brain cells with the particulars, provided several days earlier by Elie. Bonnie had divorced Thad Grey, Elie's youngest son, who, Elie admitted sadly, had been less than supportive of Zach. She'd remarried a cement contractor, Walt Chambers, lived in town, worked in a local office. According to Elie, she and Walt both spent a lot of hours volunteering at the farm school.

"Is Zach okay?" Jane asked. "I know Ms. Cantrell was searching for him."

"Yes, he was so frightened about . . . about what happened to Dad . . . " Her voice trailed off, came back after a moment. "He ran all the way home last night. I notified Mrs. Cantrell right away. But, like I said, he keeps saying he has to tell you something."

"Good. Tell me."

"But I don't know, Ms. Akers! He won't even tell *me* what it is, only you. But I'm afraid it has something to do with—his grandpa. I started calling you fifteen minutes ago. Ms. Malcolm, could you possibly come over here this morning?"

"Of course. Where do you live?"

The address was on Gates, a residential street that meandered behind the old high school.

"Be right there," she said.

She hurriedly told Wade of the stop they had to make and raced to her bedroom. After splashing some water on her face, she brushed her teeth and ran a comb through her hair. Jane smiled as she pulled on the navy skirt and was still smiling as she grabbed the jacket. Both garments could have profited from a pressing but were, unlike her more stylish teal suit, in wearable condition. *Amazing what a difference in your priorities twenty-four hours can make.*

"Call the office," she called out to him. "Have someone page Lynford Harlow. He needs to know we're going to be a little late."

She grabbed a quick glance in the mirror. No time for her customary dab of blush—but then her face didn't seem to need it this morning.

Ready or not, day, here I come.

OUTSIDE THE CONDOS
7:09 A.M.

As Jane was climbing into Wade's Blazer, her cell-phone went off again.

"Jane, it's Jessica—"

"Jessica, are you at work *already*?"

"I've been here all night, remember?"

Miraculously, thanks to Wade, Jane *had* actually forgotten about the meat crisis for *hours*. Now it all came flooding back.

"Right, right. So . . . did they find that reefer truck?"

"Yes, thank God, about an hour ago. I just got word. The Spanish police found the truck parked outside the driver's girl-friend's apartment in Almazán, a town about a hundred sixty kilometers northeast of Madrid."

"Thank God for *amor!*" Jane said, greatly relieved. And just a little embarrassed because she and that Spanish trucker had something in common: they'd both taken time out for a night of passion. "But why did it take them so long?"

Almazán's on a secondary road off the delivery route. The Spanish health authorities have seized all the meat boxes with UPC bar codes matching that shipment."

"Great! But what about any deliveries he might have made *en route* to his señorita?"

"Well, he must have been pretty eager to get to her, because he only made one drop-off, which the RDC tracked down yes-terday. It looks like we dodged a bullet, Jane. Or a microbe."

"Fantastic follow-through on your part, Jessica. Now go home and get some rest."

"Hey, with that new espresso machine in the cafeteria, I'm feeling no pain. Besides, I've got to close the book with a lot of folks in Washington and Madrid, then brief our PR people, in case we get some media queries. I'll probably even bug you a couple more times today."

"Okay, do that. And thanks again."

24 GATES STREET
7:16 A.M.

Bonnie Chambers kept Zach close beside her on the couch, a box of Kleenex in her lap. Mother and son shared red hair and freckles. Zach's oversize Chicago Bulls T-shirt hung almost to his red tennis shoes. Jane and Wade sat on a sectional sofa facing them. Zach had okayed Wade's presence, which surprised Bonnie somewhat but only confirmed for Jane that she'd made love with a good man that night.

"Zach, I loved your grandfather, too. A lot of people loved him. He was a very, very special person. But I bet nobody loved him as much as you did. So his death has to be especially hard for you. But your mom says you have something to tell me?"

Zach nodded vehemently. But each time the boy attempted to say what was on his mind, he was defeated by tears.

"Maybe this isn't such a good idea, Zach," Bonnie said, kneading her son's shoulders. "I can hardly talk myself."

She turned to Jane. "I'm taking the next few days off, of course—and Walt should be back any minute. We're taking Zach over to the farm to talk to the grief counselor there."

Jane nodded, then turned to the boy. "Do you want to wait till later to tell me, Zach? It's up to you. I know you had a terrible experience, finding your grandpa that way. It's so sad. He was such a wonderful man. Everybody loved him. I've been crying, too. So has Wade."

Zach shook his head fiercely.

"What does that mean, Zach? You want to wait?"

"No! I'm gonna tell you *right now!*"

"All right, then. You go ahead."

"You're wrong, Jane. Everyone didn't love my grampa.

He didn't!"

"Who didn't, Zach?"

"The man who killed Grampa. I saw him do it!"

OCEAN PLAINS CLINIC
8:24 A.M.

Rikki Singh had fallen out of his treehouse—really only a plank nailed flat across two boughs of the great deodar in his backyard in the foothills of the lower Himalayas. He landed on his head with sufficient force that his mother, hurrying out the instant she heard a boyish scream and snapping branches, had to shout many times to open his dark eyes.

"Rikki! My darling Rikki! Oh, my darling boy!"

Blinding light showered in. Singh cried out in pain and shut his eyes again, retreating back into the darkness.

"Rikki, it's me, Mimi!"

The swirling dream vanished along with his boyhood on the Punjab's northeastern frontier. Singh squinted, allowing a faint seepage of light. He saw, not his mother, but Mimi Takuda. First there were two of her, more than he could ever hope for. But the two lovely faces coalesced gradually into Mimi's single sweet face. Her hand grasped his tightly. He was lying back in bed. But where? He tried to see what he could without moving his head. Each pulse of blood struck like a battering ram as it thudded through his head. The room was a hospital room.

Oaths flew from his mouth.

"Man, I never heard Rikki talk like that before. Must be Hindustani. What if he got hit so hard, he injured the part of his brain that learned English?"

This new voice emanated from the opposite side of the bed, by a—terribly bright—window. Singh squinted into the dazzle, discerned a tall figure topped by a very familiar head from which sprouted a faint red goatee and a bushy, blond ponytail.

"I was cursing in Punjabi, Ron," Singh answered distinctly. "You've heard me do it before, believe me."

"Thank God!" said Mimi.

"That is not among our expressions, Mimi, but I appreciate the sentiment. What happened to me? Something hit me awfully hard."

"Some redneck maniac is what hit you, man," Ron said.

"Delbert Bowen," Mimi, with a professional preference for specificity, put in. "The big jackass I was dancing with last night. Who is right now just where he ought to be, in jail. Where I hope he stays. You have a concussion. I was afraid he'd killed you!" She squeezed his hand tightly, sending a tingle through Singh. "And I feel especially terrible that you got hurt fighting over me, especially when I remember some of the things I said to you last night. Can you forgive me?"

"I can think of nothing to forgive you for, Mimi." Which was unfortunately all too true. There were alarming blanks in his recall. Singh did get a quick flash of a wrecking ball fist hurtling out of the night, but he could not remember what happened next or much of what went before. But if Mimi wanted to credit him with physical bravery, a trait never before attributed to him, he would accept it gratefully. "As for this Delbert individual, he wore all black, I think, yes?"

"Yes."

"He's not your type, Mimi." Singh smiled, recalling Lenny Cristofaro using that same line on him. *Lenny!* Singh swore again in Punjabi and jerked upright—then screamed in pain.

When his temples ceased pounding, and his breathing slowed, Singh saw that a nurse now stood at his bedside. In one of her hands was a paper cup of water, in the other a miniature cupcake wrapper with a blue-and-yellow capsule inside.

"What is that?"

"Something for the pain, very mild. Doctor Ling doesn't want you going back to sleep. He'll be around to see you in a few minutes."

Singh stuck the capsule under his tongue, sipped water, thanked her politely. When she left, he took the capsule out of his mouth and secreted it under the pillow.

"That's not right," Mimi said. "You're in pain."

"I wouldn't lie to you, Mimi. I am—oh, my, yes!—in pain. But what if the pill makes me groggy? I cannot risk that right now. We must alert Jane at once—and the police."

"About what?"

"About Lenny, of course. He's the saboteur!"

"Lenny? Are you sure?" Mimi said. "Do you have any proof?"

"Indeed, yes. The smiling cat. It is his signature. I discovered it in one of his joke programs. Also, you see, his coding syntax, it is identical to the saboteur's."

"So that's why you kept asking about my tattoo? That means I owe you an apology about that, too."

"Yes, Mimi. No, Mimi—in that order. I made a far less understandable error than you. I mistakenly assumed this tattoo pointed only at you, though I did not wish to believe it. But now I know better, and we will put a stop to Lenny's destructive activities."

"Boss, did you keep a lot of that stuff on Elrod?" Ron Hagstrum asked, using the name Rikki had given his current notebook PC, a high-powered prototype he helped design.

"What 'stuff'? My investigative files?"

"Uh-huh."

"Of course, they are all there. Why?"

"Because I didn't see Elrod this morning. So I looked in your closet here, in case you had it with you."

"No, no, it's on my desk, back at Akers."

"Not anymore it isn't."

Singh swore once more, and somehow one did not have to understand Punjabi to get the drift.

"You have to stay calm."

"Please do not ask the impossible of me, Mimi. If Elrod is

missing, Lenny took it! He knew *where* I went last night, and *when* I went. You see what this means?"

"That he's got Pentium-plus response time?" Hagstrum suggested.

"This is no computer game, Ron," Singh said sternly. "It means he's never coming back! He must be on the run already—and we must catch him!"

Again Singh sat up, this time more carefully, his hands holding his head rigid. The gauze and tape that were wound around his head made for a very untidy turban indeed. The extent of the bandaging made him explore his face—tentatively. It felt . . . pulpy. Especially on the right side, from brow to cheek. With that much swelling, no doubt his face was also discolored. He glanced at Mimi, wishing she did not have to see him in such an unseemly state. But this was no time for vanity! He started to slide his legs from under the covers, then stopped. Except for the flimsy hospital gown, he was quite naked.

"I need my clothes, Mimi."

"No, you don't. Because you're not going anywhere. You're going to wait here for the doctor."

"You sound like my wife."

"You're married?" The disappointment in her eyes only made them more beautiful.

"Of course not, Mimi. But when I do have a wife, I certainly hope she'll sound just like you. Now, please, one of you fetch my clothes before I am obliged to make a public spectacle of myself."

**WAR ROOM,
AKERS HQ
8:33 A.M.**

With Chevy getting the plane ready, Jane and Wade had time only for a fast morning status check.

"If it weren't for the fact that it now looks as though the death of two very good men may be part and parcel of how our problems started, and the president blaming us for nearly bankrupting Hyon and jeopardizing the stability of all of East Asia, I'd have to give yesterday a rave," Don Landsman said flatly after being briefed by them on the latest developments concerning Elie's death.

"Khalif's been working all night to get our West Coast Fashion flow up to speed, thinks he's got it licked. They're still getting the occasional sabotaged containerload out there, but we're pouncing on them quicker, keeping them from getting any farther in the pipeline. On the international front, Nick King and George Parrish at the Gravesend DC tell me they've gone over every damned shipment currently in the European pipeline from our Far Eastern importers—and no more nasty surprises turned up, except for that malfunctioning reefer unit. Meanwhile, on the home front, our first-tier suppliers are meshing with our backup systems, filling orders one way or another. I suggest we hold off on our accelerated deliveries for the time being—except for emergencies—and make do with two-day turnaround on replenishment orders. Which means we have to beef up inventory levels at the RDCs, and staffing, but we knew that going in."

"But we're back!" Jane said. "That is what you're telling me?"

"Yes, ma'am. *We are back*." Landsman grinned.

"And what about Mr. Singh?" Jane asked. "Still at it?"

"You didn't hear?"

"Don't tell me he gave up and left?"

"Singh's in the clinic next door. He was beaten unconscious in that C-W bar last night in town."

"Oh, my God, Don! I can't believe it! Is he conscious now? I've got to go see him."

"Mimi only called a little while ago to tell me what happened. She and Ron Hagstrum were going over to check on him."

"Jane," Wade broke in, "we've got to get airborne if we're going to make the ten o'clock meeting with Soverel. You have to call Singh from the plane."

"Mr. Singh is here because I brought him here. I have to call him *now*." She got the hospital number and punched it up as they hurried out toward the Siler Field hangar. It took her a minute to get connected to a nurse, who sounded out of breath.

"I'm calling about Mr. R. K. Singh," Jane asked. "How is he doing?"

"He's not here."

"But I was definitely told he was a patient there."

"He is—was. He's disappeared. So have the two visitors with him. We've been searching everywhere—I'm afraid they kidnapped him. I just called the police. The man shouldn't be out of bed."

Jane told the nurse she would call back, then clicked off. What in God's name was going on?

Just ahead, Chevy was motioning them to hurry.

OCEAN PLAINS CLINIC
8:36 A.M.

The trio exited the back door of the clinic, which had taken over the bottom floor of the old Siler AFB Hospital. Only a chain-link fence separated the property from Akers HQ. Mimi led them to her VW Jetta, where she made one last appeal before opening the passenger door.

"This is crazy. Look at you! I should be arrested for helping you sneak out. You can't even see."

"Nonsense, I see twice as much as usual." He smiled, he hoped ingratiatingly.

"You think double vision is funny? I think Delbert scrambled your brains, that's what I think. Let the police catch Lenny."

"By all means," Singh said. "I merely wish to see Lenny's rooms before the police start tossing things about in their enthusiastic way. Besides, I may find Elrod."

He turned to Hagstrum. "Ron, please punch Jane Malcolm

on my speed-dial—she's four. I must inform her immediately about what we've learned."

"I don't have your cell-phone, boss. You must have left it in the hospital."

"So we're going back?" Mimi asked, her hand poised on the ignition key.

"No," Singh said. "Lenny must have swiped my cell-phone, too. We're going to proceed. The sooner we track him down, the better."

AKERS IMPORT DISTRIBUTION CENTER, CHINO, CALIFORNIA 8:03 A.M.

Chris Azarian, manager of the huge import hub, stood on the mezzanine deck overlooking the Fashion floor and thought about how the sortation process appeared to him on good days: like an intricate ballet, its mechanized choreography wonderful to watch. Garments on hangers were trolleyed in from trailers backed up to receiving doors, each garment handled once by human hands—in this case, by Norma Sandoval's now-capable hands—as it was pirouetted from trolley to sliding hook. From that point on, the dance was entirely automated; garments clicked along the overhead tracks, monitored by photoelectric sensors, sorted by order number, sent spiraling out along branching tracks, switched and shunted into lanes according to regional distribution center and, within that, to individual Akers stores.

That was the way the system was designed to work but had *not* been working for the last five days. Five days of sorting by hand—sluggish, high-reaching, back-wrenching labor, garment after garment, trolley after trolley, trailerload after trailerload. The result had been a lot of profanity, screwed-up orders, and delayed shipments.

"We moved nine thousand garments in the last hour," Chris Azarian announced. "That's damn close to top speed. I thank you, I know Norma down there thanks you, all of Akerdom thanks you." The manager turned and executed an elaborate bow to the elegantly dressed, tawny-skinned man standing beside him at the railing.

"What could I do?" Khalif said with a smile as smooth as the fabric of his blazer. "After they flew me out here, I *had* to produce." Then, more seriously: "Especially with you guys being off-line so long."

Khalif al-Marzouki had been parked down there on the warehouse floor at a PC workstation all night, writing and rewriting batch programs to import headquarters fax-and-scan orders *automatically*—and smoothly—into the hub's PC database, so that information flow could keep pace with garment flow through the automatic sorter.

"Well, we got a heck of a backlog to catch up on, Khalif. But, putting on a few extra shifts, I'm betting we can dig out pretty fast."

"Glad I could help, Chris."

Amazingly, Azarian thought, the young Saudi showed no fatigue, despite his night-long labors. His shiny black hair remained impeccably combed, his fawn-colored slacks retained their knife-edged crease.

"Look, Khalif. After this all-nighter, I think we've both earned a couple hours' relaxation. So if you like baseball, the California Angels are playing a day game today right down the road. I've got box seats behind home plate I hang on to for days when a baseball game is the only way to restore sanity. Heck, you've even got time for a nap, and we can still be there for the first pitch. Interested?"

"Thanks, Chris, but cricket's more my game."

"Can't help you there, my friend."

"But you have. By saying thanks so graciously, you make me feel very happy that Jane Malcolm remembered me from when she was helping *us* through a very rough time."

"Us?"

"My people." This time Khalif's smile was truly lavish. "I am Saudi. Your country stood by us, and we are taught from a very young age to be grateful for loyal allies. I am happy to say that to me Jane has become even more than an ally. She is my friend. So to help her now is a privilege. Perhaps that sounds . . . corny to you?"

Chris shook his head. "My father was in Vietnam. Got wounded pretty bad. He still keeps in touch with a guy from South Dakota who carried him for nearly a mile that day."

Khalif nodded and offered his hand. Silently Chris shook it.

"Now, I'd best be on my way. I'll take a little nap on the jet back to Ocean Plains. I've got at least a dozen more problems waiting for me. *Ciao.*"

With that, Khalif hefted his Gucci briefcase, filled mostly with computer diskettes, and headed for the parking lot exit.

THE SOVEREL CORPORATION CONFERENCE ROOM, CHICAGO
10:05 A.M.

The members of her family were already settled into beige leather chairs around a blond wood table of imposing length that looked like a Biedermeier conference table—if that style of decor had included conference tables. Jane wondered if the Biedermeier fixation were Tom Soverel's or his decorator's as her eyes traveled to the table's far end, where Soverel sat, back-lit by the azure radiance of Lake Michigan.

His greeting to Jane was a reserved, "Do join us, Ms. Malcolm."

Who could blame him? Only an hour earlier she'd called him to say, "Tom, I'm respectfully declining your offer."

"I'm sorry to hear it," Soverel had said without losing a beat.

And *that* fast, Jane had thought, he closed that particular door and moved on.

Nonetheless, she'd felt it appropriate to say just a little more. "It was definitely a generous offer, and in other circumstances I would have leapt at it."

"Yes. Well, I do appreciate your telling me before the meeting. You intend to be there, I take it?"

"Of course."

"Of course."

Now Roy, on Soverel's left, gave Jane a cursory nod, and a sharp-faced stranger in a chalk-striped suit who was seated across from Roy smiled politely. The man's features matched a photo of Flynn Emerson that Landsman had pointed out in last year's annual report. Jane wasn't surprised to see him. Soverel had told her Emerson would be attending today, also that he was slated for a key role in consolidating Akers's worldwide operations with those of 4-Mor Shopping.

Alex, swiveling his attention from a chat he was having with a Soverel executive whose gray silk suit did not entirely mask her pulchritudinous figure, tossed Jane a cheerful, "Good morning." Anne Akers sent her a small wave and large smile. Grace, Roy's wife, merely pursed her lips. But Royal's sister, Louise, glanced up from her needlepoint with a definite sparkle in her eyes.

"So, now we are all here," Tom Soverel said. "Thank you for coming, Ms. Malcolm. I believe you know everyone." There was a general chuckle.

"*We* haven't met," Emerson said in a baritone drawl, rising to his feet and supplying his name. "But I've heard all about you, Ms. Malcolm. I worked ten years for your dad."

"Of course. How do you do, Mr. Emerson." Jane came around to shake his hand, then detoured to the sideboard for black coffee and a brioche.

"Before we get started," Jane said, taking an open chair between Alex and Anne, "I want to apologize for being late, and then ask if everyone has heard the tragic news about Eliezer Grey."

There was a general murmur of assent.

"A deeply personal loss for all of us," Roy said. "Elie's passing is, in fact, what we were mostly talking about before you arrived. However, I think it would be very appropriate if we observed a moment of silence."

"Thank you, Roy," Jane said. In the silence that followed, she prayed for Elie. And them all.

14 CHURCH STREET
NEWKIRK, ILLINOIS
10:07 A.M.

Singh was frustrated. They'd left Ocean Plains at eight-forty en route to Newkirk, eighteen miles to the west, where Lenny lived. Lacking a mobile phone, Singh had asked Mimi to stop at the first pay phone along the way—which turned out to be at a McDonald's in Red Knoll, an old corn-canning center halfway to Newkirk. But he had been unable to reach Jane at any of her numbers.

The situation was not acceptable. In the midst of a crisis, Singh was out of the loop. Should they return to Akers and borrow someone's cell-phone, or continue toward Lenny's house? Shortly after Singh decided on the latter course, Mimi had realized her car was nearly out of gas, and they'd had to return to Red Knoll. Singh had used the time to pounce on the service station pay phone. Again, no luck.

By this time Singh was wishing keenly for that pain pill the nurse had given him.

With all the delays, it was almost ten o'clock when they'd finally pulled into Newkirk, which was really more a scatter of ramshackle houses than a town, Mimi had explained. There were a half dozen peeling white-clapboard houses on Newkirk's main street, several of which looked abandoned, including one, judging by a barely legible weatherbeaten sign, that used to be a church. Of course. New church—that was what Newkirk meant, didn't it, Singh thought sadly. The remaining houses showed various hints of habitation—butane tanks, a rusted metal swing set, blocked pickups, TV antennas. Things that outlasted hope.

"Lenny lives just around the corner," Mimi said, turning on the only cross-street. She pointed at a grassy lot. "Right there."

"Mimi, you get hit on the head, too?" Hagstrum asked. "There's no house there."

True enough, there was only a shingle roof, complete with chimney and vents, built about a foot off the ground.

"I thought I had the wrong address the first time I came here, too. But that's where Lenny lives, believe me. The owner never got around to finishing it. So he put a roof on the basement and rents it as is. What does Lenny care if he lives in a hole in the ground? So long as he's got cable and a T-1 Internet connection. But I don't see his car."

They parked. Mimi helped Singh get out of the car and down a flight of stone steps to the basement door.

"Locked," Hagstrum said.

Mimi squatted, lifted a geranium pot, held up a key. "Don't go jumping to conclusions, either of you," she said, but her eyes were on Singh. "I've been here to pick up stuff a few times, that's all. Wait'll you see what a dump it is inside."

Mimi had not exaggerated. Singh, still suffering dizziness, fought off a wave of nausea as, Hagstrum leading the way, he followed a few steps behind, with Mimi's hand against his back lest he lose his balance. Thus they picked their way single-file through a debris-littered kitchenette. Cockroaches skittered from the light in every direction, across old pizza boxes on the drainboard, out of crockery stacked in the sink, under the chipped enamel stove.

"Keep your eye out for a phone," Singh said.

The others preceded him into the partitioned living space—an equal disaster. Pizza boxes were there, too, stacked everywhere. Books were mostly on plank-and-brick shelves, but magazines were scattered across every surface. Most were computer related, but Singh also spotted *Soldier of Fortune*, *Juggs*, *Cruise World*, *American Cinematographer*, *Muscle Girls*, and *Scientific American*. The sleeping area was about the same.

Singh's missing notebook PC was nowhere to be found. In fact, what computer equipment he saw was mostly junk, a scattering of cannibalized parts. Empty footprints on worktables, empty sockets on power strips—all testified to missing components.

"I'm not really surprised there's no phone," Mimi said. "Lenny was inseparable from his cell-phone, and he obviously regarded this as something other than a *home*, so what would he need a home phone for? Anyway, you were sure right, Rikki." She pointed at the techie detritus. "None of his good stuff is here. Lenny's gone."

"Still, we may find something to our purpose," Singh said, sifting through an assortment of molded plastic and electronic miscellany. He recognized the skin and entrails of several disassembled cellular phones.

"Now *you* were right, Mimi," Singh said. "Lenny seems to have had a sizable collection of cell-phones. Hmm."

"You have that look," Hagstrum said. "What is it?"

"Just taking a quick inventory of components. Give me ten minutes, and I do believe I can make one complete phone from this collection. Ron, help me find some needle-nose pliers and a small Phillips screwdriver."

"I saw a toolbox somewhere," Hagstrum said. "Well, not exactly a toolbox, but a big box with lots of tools inside. There—by Mimi's foot."

While Ron searched through the box, Mimi said, "Why not just use a neighbor's phone?"

"Because there are a great *many* calls to make now, not merely to Jane. For instance, I must alert my friends in the FBI's

Computer Crimes Division about your *former* colleague—that he's definitely on the run. Probably left last night, I'd say. And I also need to monitor local police bands."

"How can you do that with a cell-phone?"

"You'd be surprised what the boss can do with a cell-phone," Hagstrum said, grinning with secondhand pride.

"Wait a minute!" Mimi said. "Lenny didn't take off last night."

"Tell me, please, how do you know?" Singh asked.

She lifted a newspaper from a sliding stack on a chair. "Today's *Trib* sports page. Lenny buys it at the 7-Eleven, I think, early in the morning. For the race results."

"He plays the horses?" Hagstrum asked.

"Kind of. He wrote this computer program for beating the odds by playing only long shots, nine-to-one or higher. He made like a hundred thousand dollars, he told me, playing on paper." She chuckled. "But the only time he ever went to the track and tried his system, he lost his paycheck and had to borrow lunch money from me."

"Excellent work, Mimi!" Singh said, inserting a circuit board into an earpiece and quickly securing it with screws. "So, Lenny may still be in the vicinity—or at least not far away. We must alert the authorities at once."

"Fine," Hagstrum said. "But any chance of finishing up that cell-phone in the car? I'd sure like to get out of this roach-pit before I barf."

MEIGS FIELD
10:10 A.M.

A stocky man wearing an Akers golf cap, aviator sunglasses, and safari jacket pushed open the glass door of Vollmer Aviation Services. Two men were behind the counter, one squatting and filing, the other staring at a computer screen. Both had their backs turned. The newcomer unslung a shoulder bag and bonged the counter bell.

The file clerk pivoted on bulging thighs, sprang to his feet. His skin was ebony, his smile dazzling, his physique impressive—made even more so by a tight T-shirt emblazoned MIGHTY MEIGS FIELD. "What can I do for you?" he asked.

The newcomer flashed a bucktoothed grin. "Name's Bannerman, Chuck Bannerman. I'm a freelance photographer, doing some work for Akers." He unsnapped a flap pocket on his jacket, fished out an Akers vendor's pass. "Their PR department asked me to hustle over here and take some shots of one of their planes. Supposed to be parked here today."

"*Aerial* shots?" asked the other man, swiveling from his PC screen and revealing a burn-scarred cheek.

"Naw. I don't even need to get inside. They just want it sitting on the tarmac, looking pretty against the skyline. Don't ask me why. I just do what I'm paid to do. I got the tail number somewhere."

"AG Tiger," the man said. "Only Akers we got at the moment. Came in this morning. Mind showing your ID to Reggie?"

The black man checked it, then a clipboard on the counter, thrust a Herculean arm toward the glass door. "It's parked down at the south ramp. One of our line guys will take you down in a minivan. The pilot's still hanging around here somewhere."

"Is he? Great. Much obliged." The photographer hefted his shoulder bag and left.

The only hard part so far, thought Robert Grooms as he climbed into the back of a Vollmer minivan driven by a blond-mustached youth, was putting up with the gum-pinching orthodontic appliance that created his protruding incisors. Of course, the pilot could pose a far more serious problem, but Grooms had come prepared for that.

After a half-mile sprint alongside runway eighteen, Grooms was dropped at the single-engine Tiger with the Akers tail logo.

"How long's this gonna take?" the driver asked.

"Fifteen, twenty minutes. Don't worry, I'll find my own way back."

The young man shrugged. "Gotta hang around till you're done, man. Security."

"Then I'll try to make it quick." Grooms smiled, then set immediately to work. He walked around the plane, clicking off frames with his Minolta Maxxum 600si. Then, for effect, he shot another roll crouching and moving, framing the plane against the distant Loop. He finished that roll as he ducked under the engine cowling. Grooms's view of the minivan—and the driver's view of Grooms—was effectively blocked by the fuselage and part of the wing. Grooms now brought something else out of his camera bag—a plastic case about the size of a cigarette pack, with four strong magnets and a strip of metal.

He removed the strip and used it to tighten a setscrew, which in turn sealed an interior pressure chamber.

The device was activated.

Now, where to put it so it would not be seen during preflight inspection?

Grooms settled on the nosewheel. The Tiger's gear, he knew, was permanently locked down. Gingerly he reached under the nosewheel fairing, placed the magnet side against the metal. There was no molecular tug. Shit! The damned housing must be fiberglass!

Then he felt a magnetic pull from the wheel itself. *Steel.* Grooms strained wrist and fingers upward, under the fairing, letting the device kiss and lock against the wheel hub.

It had taken only a few seconds, but a rivulet of sweat ran down his back. He duck-walked backward, out of the plane shadow, popped in a new roll of film, and shot a final roll, his heart pounding.

Now get the hell out.

He shouldered his gear and headed back toward the minivan.

"Mission accomplished," he said, hopping inside and sliding the door shut. "Quick and dirty."

"Hope you got some good ones." The minivan jerked and zoomed away toward the two-story terminal building. As they skirted the main parking area, the young driver asked, "Where's your car, my man?"

"Right here's fine," Grooms said, reaching for the door handle.

"Suit yourself." The kid braked, and Grooms was out.

He was only a few steps from his van when a tall, broad-shouldered figure stepped from behind it. It was Johnson, the pilot Jane had just picked up after he was laid off from his airlines job, a tidbit Grooms had collected from his voice-activated surveillance tape of her office. He decided to play dumb.

"Can I help you?"

"Yeah. What the hell are you doing here?"

"Doing?" Grooms slapped his camera bag. "The Akers PR department hired me to take some shots."

"You a photographer?"

"Obviously."

Johnson moved closer.

"The fuck you are, Grooms! You think you can hide behind buckteeth and dark glasses? You were a spook in 'Nam, and you're up to the same kind of shit here. You were fucking with our plane, weren't you, you son of a bitch?"

Grooms ducked the vicious left jab, then twisted up, swinging his shoulder bag as hard as he could. It caught the bigger man in the back, knocking him off balance—Grooms's intention precisely. Even as Johnson's knees hit the blacktop, Grooms was snatching the loaded syringe from his bush jacket and yanking off the tip protector. Then he pounced, sinking the needle into the underside of the pilot's thigh, thumbing the plunger. Johnson bellowed and rolled, slewing his legs around, spilling Grooms and dislodging the hypodermic.

Too late, asshole. I just killed you.

But the big pilot wouldn't die. He scrambled up, fury in his eyes. He bulled straight into Grooms, who went sprawling backward, dental piece flying. Johnson landed atop Grooms, flattening him. Grooms clawed at the pilot's eyes, but big forearms blocked the blows. Then powerful hands formed a vise around Grooms's neck.

Despite the muscular thickness of that neck, Grooms knew he was only seconds from blackout—and death. Unable to scream or breathe, he flailed feebly against his assailant.

Suddenly the throat-crushing pressure ceased. Oxygen and blood flooded Grooms's brain. He rolled sideways, cawing for air. Johnson's face was inches away, eyes wide, mouth agape, his shirt ripped open by his own hands. The pilot was choking out a gravelly gurgle. The familiar death rattle was music to Grooms's ears.

About time!

Grooms savored his close-up now, noting that the pilot's suntanned face had darkened further, sign of a massive coronary. Grooms didn't know for sure how many of those forty milli-equivalents had actually been injected into Johnson's bloodstream. No matter. Enough got inside him to produce cardiac arrest.

He reached out, found no carotid pulse, then no heartbeat under Johnson's torn-open shirt.

I think I see a note here on your chart, Johnson. "Do Not Resuscitate," right? I was sure you'd feel that way.

Satisfied—really well satisfied—Grooms crawled to his feet, caressing his neck, then scanned every direction. He heard a car moving slowly in another lane of the lot, but nothing moved in his lane.

Get out of here! Move!

He yanked open the rear of his cargo van, dragged, lifted, and levered Johnson's big frame inside, then slammed the double doors. He searched next for the syringe but couldn't spot it.

Probably had rolled under one of the cars to either side.

No time to look for it—forget it!

He jumped into the van, swiveling to see Johnson motionless back there, facedown among the litter of equipment. Grooms fired up the engine and switched into survivalist thinking—conjuring the dangers still ahead, not those safely behind. Only after he turned back onto Lakeshore Drive—tire squeal accompanied by the sound of the corpse sliding sideways as the van cornered—with Meigs now safely in his rearview mirror and the Loop ahead, did Grooms permit himself to relax a little.

He palmed his mobile phone, tapped a speed-dial number, got voicemail, and hung up. Next he dialed Meigs, recognized the lilting voice of the muscular counterman. Lowering his own voice by half a register, Grooms identified himself as Chevy Johnson, pilot of the Tiger.

"Did that photographer check in with you, Mr. Johnson?"

"Bannerman, right, no problem. He just packed up and drove off. But I have to leave, too. Something's come up, and I can't reach Jane Malcolm or Wade Crain, the Akers people I flew in. Can you give 'em that message—that I had to leave? They'll be okay. Wade can fly 'em back."

"Want to give a reason, Mr. Johnson?"

"Tell Jane I just had an old friend at United give me a buzz about a job there, and I gotta head for O'Hare right away. I can't turn this down—she'll understand. Thanks, my friend."

Grooms hung up and smiled. That should do it. She'd definitely be pissed off—but that wouldn't be bothering Johnson any. Wade would—of course, happily—fly her home. Partway.

Because when the little plane reached an elevation of 2,500 feet, give or take a yard, the altimeter-triggering device would detonate three ounces of Semtex inside the magnetic case. Grooms knew it had taken only eight ounces of the Czech-made plastic explosive to blow a 747 out of the air over

Lockerbie, Scotland. There would be no wreckage of this four-seater for the NTSB or anybody else to investigate.

And vanishing with the plane would be two meddlesome individuals who had suddenly shifted from being a minor annoyance to an intolerable threat.

You're going to have one final, shared, out-of-body experience, lovebirds.

Grooms drove past Buckingham Fountain, then swung left on Jackson, heading for Michigan Avenue. He had a final message to deliver, but plenty of time to get there before the meeting broke up.

THE SOVEREL CORPORATION CONFERENCE ROOM
10:12 A.M.

"Thank you, all." Roy ended the brief silence for Elie Grey. "Tom?"

"Excuse me," Jane cut in, "but before anything more is said, there's something I have to tell you all." She took a shallow breath, wished she had some other message for her family. "A critical situation has developed at Akers that demands a course of action that's entirely different from the one you've all come here to consider."

"You've got it just backward, Jane," Roy said. "The critical situation at Akers is precisely *why* we're all here today—to consider Tom's offer. I remind you, we're on the verge of completing a momentous agreement, one we've been discussing and refining for weeks—with, counting today, an additional eight days' delay due only to your impassioned request to attempt the impossible. Now, in fairness to Tom and to all of us, I think it's time to proceed."

"I appreciate all that, Roy," Soverel said. "But I, for one, am anxious to hear what your sister has to say. Especially about any new 'critical situation.' Please proceed, Ms. Malcolm."

"Thank you." She took a sip of coffee. "Let me preface my

remarks with something personal. I'm here today as a family partner because of Dad's legacy to me, a legacy I didn't know about and probably don't deserve. But I'm also here to defend Dad's larger legacy—his company. And that larger legacy, I think, is what we should all be keeping in mind here today. *Your* father's legacy, Roy, and *yours*, Alex. Your *husband's* legacy, Anne, and Louise, your *brother's*.

"Because of that, I want to announce that as of this morning, Akers Corporation continues a strong recovery from the crippling sabotage of its distribution systems and is operating at a level only a little below normal capacity. I'm extremely proud to have had a hand in that logistical turnaround, but I definitely didn't do it alone—a lot of members of the greater Akers family played a vital part, giving way above and beyond fair return for a paycheck. I'm proud to have worked with them—and I'm sure my father shares that pride."

"Hear, hear!" Anne said.

"And bravo, Janie!" Alex added.

"I, too, applaud all those efforts," Roy said, "especially my sister's. As for the hard results and the company's overall fitness, how I wish her words were true. The fact is, we're still crippled—operating on hastily drawn-up emergency plans, antiquated procedures that are draining profits as well as employee morale—and now we're involved in an embarrassing federal bailout of our main electronics supplier. I'm not saying this to weaken our bargaining position with Tom. He knows all this already—we all know it, and so, alas, does Wall Street."

"But we *don't all* know any such thing, Roy," Jane said. "This company is not crippled anymore. I'll say it again. Our logistics systems are operating pretty damn close to par. As for morale, well, my job has me talking to a lot of Akers people, and it seems to me their morale is a whole lot better than it was a week ago. I'm convinced that in a very short time, Wall Street will catch up to the fact of our recovery." She took a breath. "As for the rest of what's happening, of course restoring full computer operations remains a priority. I had hoped to be able to tell you this morning that we've located the saboteur and arrested that person, or persons. I can't deliver that good news yet. But I do believe we're very, very close."

"I'm afraid even *that* wouldn't affect the economics of this deal," Roy said. "Not now, not this late in the game."

"It would if people in this room were deeply implicated in that sabotage. People who stand to profit from—"

"That's by God enough!" Roy leapt to his feet. "I don't know about you, Tom, but I'm not going to sit here and listen to nasty and baseless innuendoes."

Jane looked sharply at her brother. "I didn't say it was *you*, Roy."

"I don't care *who* you're trying to smear."

"I do," Flynn Emerson said. "*I* stand to profit from this deal, both careerwise and from my Akers stock. But I object categorically to any implication that I might have been involved in what went wrong with the newly installed computer system. It was on course when I left. In fact, I even offered Roy my services gratis to help solve his system problems, but was turned down pretty—"

"Flynn, don't you see?" Roy cut him off. "Jane's just making wild accusations. *She's* the real saboteur here today, trying to wreck our business deal at the last moment with absurd charges. If we don't put a stop to it right now, it's going to be too damned late to salvage this company."

"Is that it, Ms. Malcolm?" Tom Soverel asked. "Is that what you came to say? You must admit, I've been a pretty good sport so far, but I do have my own agenda here, and other urgent business to conduct."

Jane hesitated, put off stride by Flynn's protest of innocence, but even more by his claim of having offered to help rescue Akers from the inventory disaster. She had persuaded herself—

Elie's thinking having naturally influenced hers—that Roy and Flynn were somehow together responsible for Akers's problems, both having been bypassed by Royal in favor of Don.

Now she scrambled to catch up. "Sorry, but I'm not quite finished. I said that I don't know who committed the *computer* sabotage. But *another* kind of sabotage has been uncovered." Silence descended on the table. Jane opened her folder and took out the photos taken and then enlarged by Lynford Harlow. She turned to Anne. "Anne, I have to warn you. This concerns my father and how he died."

"I'll stay," Anne said.

"So will I," said Louise, putting down her needlepoint.

"All right, then. These are pictures of a landing-gear rod on the wreckage of my father's plane—greatly magnified, as you can see by the ruler. They were taken by a man named Lynford Harlow. He used to work full-time as an investigator for the National Transportation Safety Board, and now works for them on a kind of freelance basis when they're shorthanded. Mr. Harlow was called in to work on the investigation of my father's plane crash and then continued his own private examination at Eliezer Grey's request after the wreckage was moved to a shed at Elie's farm. I'm going to pass around the pictures he took. You'll see clearly where the rod has been filed through. Mr. Harlow has officially asked the NTSB to reopen their investigation. According to him, there's only one possible finding— someone sabotaged the plane, causing the gear to pancake on landing."

Jane stopped. She hated to deliver so much pain. In the light from the window, Louise looked suddenly old. And Anne . . . well, there was in her eyes a new grief.

"Go on, Jane. Please," Anne said firmly.

"We do have to see this through, don't we?" Jane said softly, addressing only her father's . . . loving . . . wife.

She glanced once around the table and went on. "Often

such belly landings are not fatal, according to Mr. Harlow, but, tragically, this one was. And it was no accident. Which means my father was murdered."

Anne's face was ashen now. She'd known two minutes ago—Jane had seen it in her eyes—but the information had taken this long to attach to her heart. Grace Akers clutched Roy's hand.

Tom Soverel spoke first, his face a solemn mask. "I have one question. If this proof is as conclusive as you say, why haven't you gone to the police?"

"Janie," Alex interjected, "are you saying you know who did it? Or that it had to be one of us?"

"She obviously has no proof, Alex," Roy said, disengaging from his wife's grip. "Or, like Tom says, she'd be telling this tale to the police. I mean, look at this." He reached across the table, taking a photograph from Flynn Emerson. "Who the hell knows what this is? A magnified hunk of metal. So what? Which some friend of Elie's claims he found in a heap of plane wreckage. Come on! Let Dad rest in peace, for God's sake. Jane, hasn't Anne—haven't we all—suffered enough? Why are you being so cruel? What more can you possibly want?"

"The truth," Jane said. "I want the truth. And these pictures are part of that truth. I've got Harlow's affidavit here, which lists his credentials. Unfortunately, Roy, there's more. There was another murder. Yesterday."

"Who on earth!" Alex gasped.

"The victim was Elie Grey."

"That's just crazy!" Grace Akers was on her feet now. "My husband's right. We have to stop this. Elie died of a heart attack. The coroner said so. Elie's daughter-in-law, Bonnie, told me so herself."

Roy stood. He made a stab at calming his wife, then looked at Soverel. "Well, Tom, have you heard enough now? There's still time to conduct business if you'll put a stop to this nonsense. I suggest we adjourn for ten minutes, then reconvene without my sister."

"I second the motion," Flynn Emerson said. "I've heard enough wild accusations for one morning. If this woman has anything, let her go to the authorities."

Now Jane stood. "I think at least a few people are left in this room who would like to know who murdered Royal Akers's oldest and most loyal friend."

"Then you do know, Jane," Anne said. "Tell us."

"Did the same person kill my brother?" Louise had a new fire in her voice.

"We don't know that yet, Aunt Lou. But I have brought along an eyewitness to Elie's murder. He's waiting outside. With your permission"—she looked at Soverel—"I'd like to bring him in."

Soverel threw up his hands. "By all means."

McDONALD'S, RED KNOLL, ILLINOIS 10:36 A.M.

Mimi drove them east, back to Red Knoll, where they cruised the sleepy downtown grid, looking for Lenny's red Miata—to no avail. While they discussed what to do next, they headed for takeout McMuffins and coffee.

As they waited in the drive-through lane, Singh held his head in one hand, reconstituted cell-phone in the other. He'd already notified the county sheriff about Lenny Cristofaro and was now on a conference call with his friends Pete Petrossian of the FBI's Computer Crimes Division and Ernie Macomber, SAC—special agent in charge—of the FBI's Chicago field office. For the second time, Singh was detailing his logic for suspecting Lenny Cristofaro, then this morning's confirmation of that suspicion. Macomber was already running computer checks on Lenny, and Petrossian had ordered three of his men to fly down from Chicago to work with Sheriff Bevins. The Illinois State Police were issuing an all-points bulletin for

Lenny's Mazda Miata. "Which is probably dangerously over-loaded with computer hardware," Singh pointed out.

As he hung up, his concussion hangover, and his spirits, lifted a little. The chase was definitely on, and Singh felt himself part of it.

"Come on!" Mimi, behind the wheel, complained. Two cars ahead, some guy in a big, square-sided U-Haul truck was taking forever to order breakfast, holding up the whole line. Finally, when he got to the pickup window and the food appeared, she saw why. The bags just kept coming, five of them, plus a tray of drinks. *He must have a whole crew in there.* Curiosity kept Mimi's eyes on the U-Haul as it pulled ahead to the street and turned left—when you were supposed to turn right. She caught a single glimpse of the driver—a mustached guy in a cowboy hat and red shirt. Odd, he *was* alone. Why'd he buy all that food? And why was he still dressed that way this morning? Mimi had definitely seen that dude last night at the Mulekick in that same Stetson and red shirt.

The car ahead pulled out. It was Mimi's turn at the window. A hand reached out with her order.

"Lenny!" Mimi shrieked. "That was Lenny!"

Instead of taking the bags, she tromped on the gas. The Jetta leapt forward. The drive-through girl was yelling—what was she supposed to do with the damn order! Beside Mimi, Singh cried out as his head jerked back against the headrest.

"Sorry, Rikki, but he's getting away!"

Singh and Hagstrum were shouting questions.

"Are you sure it's Lenny?"

"How *can* you be sure?"

"Trust me!" She, too, swung left as she reached the street. The big U-Haul was at the end of the block, slipping through a yellow light. Mimi accelerated, slowed at the red, looked both ways, accelerated again. With any luck, she thought, a sheriff would spot her running the red and become part of the chase.

"Get that deputy back on the phone," Singh said. He tossed the cell-phone back to Ron. Placing his hands along his temples, he squinted through his fingers.

"I'm on it," Hagstrum said. "What is this street?"

"Knoll Road West," Mimi said. "Which turns into County Line Road outside the city. Looks like he's heading back to Newkirk."

Two blocks later, with a final gas station and a Dairy Queen, the town cut off and farmland resumed. Just as Mimi was closing the gap between them and the U-Haul, a tractor nosed out of a farm road, lumbering along at a good ten miles an hour. She was blocked from passing by a parade of oncoming traffic. She used the frustrating delay to tell Singh and Hagstrum about spotting Lenny in disguise at the Mulekick. By the time she could pull out, the U-Haul was a tiny gray rectangle in the distance.

"You don't have to catch him, Mimi," Hagstrum coached from the back. "Just keep him in sight. I passed the word to a deputy, and he's giving it to the state police."

"I'm just so mad!" she said, passing the tractor, then fighting the urge to maintain speed and overtake Lenny. "That little weasel, what he did to the system—to the whole company. While he had me working my rear end off to fix things. You know what else? Last night at the club, *somebody* told Delbert that Rikki called him a 'fairy.' That had to be Lenny, too. So he's to blame for Rikki's getting beat up."

"Plus," Singh said, "I believe he stole my bicycle and my cell-phone. What is that noise?"

They all heard it now, a full-throated, fast-rising snarl from behind. Hagstrum checked the rear window. "Yahoo! Looks like it's about to be payback time for Lenny Cristofaro."

The Jetta rocked in a double windblast as two white-and-yellow state police cruisers shot past them, closing swiftly on the U-Haul.

Jane wasted no time leaving the room.

"This has turned into a circus," Roy said, but his gaze, like everyone else's, was riveted on the door, waiting to see whom Jane would usher in.

It turned out to be not one person, but three—a redheaded, freckle-faced boy with shuffling gait; a woman with similar coloring and anxious features; and a large, sunburned man in an ill-fitting suit who filled the overscaled doorway.

Jane introduced Zachary Grey first, then Bonnie and Walt Chambers. "Zach lives at the school farm adjoining Elie's. He's a very courageous young man. Zach saw his grandfather attacked yesterday, saw the man who did it, and can describe him and what happened. He will be giving this testimony to Sheriff Bevins in Ocean Plains this afternoon. But Mr. and Mrs. Chambers agreed to let Zach come here and talk to you first. Go ahead, Zach."

The boy gulped, nodded, then gaped at the unnerving array of faces all riveted on him. His mother stood close behind him, hands on his stooped shoulders, whispering support.

Zach nodded jerkily several times, then finally said, "A man jumped onto Grampa's old tractor—and stabbed him in the leg! Then he ran away! And—and—and—" Zach was now blinking rapidly, his face beginning to crumple. "And Grampa died!"

His cry of outrage hung in the air as Bonnie enfolded her now sobbing son in her arms.

"Zach," Jane said, kneeling beside him quietly until he looked at her, "thank you for telling us. I know it's hard to talk about. Can I ask you to do one more thing?"

Zach blew his nose on the tissue his mother offered, then nodded. "You loved Grampa, too."

"Can you tell us what the man *looked like* who attacked your grandpa?"

But the question only brought on new tears. The boy was

lost in grief. His mother took him in her arms again. Walt Chambers shook his big head. "Me and Bonnie better take the boy home now. Sheriff can come see us if he wants to hear the rest of it."

"Of course," Jane said. "I'm sorry, I didn't mean to—"

"It's okay," Bonnie spoke up, steering Zach toward the door. "I take responsibility. Walt and I know it's real important, what you folks are deciding here."

"What we're *supposed* to be deciding," Roy muttered as Walt opened the door for his wife and stepson. "Jane, that was an extremely insensitive stunt. And the point was?"

Another gamble was what it was, Jane thought. And another setback, this one at Zach's expense. What was it Don Landsman had said yesterday? No rabbits left in the hat.

"That's him!"

Jane whirled at Zach's shout. He stood rooted in the doorway, his small freckled arm pointing toward the area outside.

Jane, closest to the door, was able to see past the couple and the boy. A man was standing out there—a bull-necked man with a crew cut.

"Grooms!" she yelled.

Jane hurried toward the door but collided with the Chamberses, both of whom had turned back toward the room. By the time she'd squeezed past them into the corridor, Grooms was vanishing around a corner.

She hurried in pursuit, calling ahead, "Wade! Stop Grooms!"

Reaching Soverel's reception atrium, she found Wade flat on his back, groaning, the blond receptionist kneeling beside him.

She rushed toward him. "What happened?"

Wade struggled to his knees, forehead on the terra-cotta floor, both hands at his crotch. Then he rolled onto his side.

"He was trying to stop some maniac. The guy kicked him, then ran down the stairs!"

"Wish you'd . . . given me . . ." Wade whispered between

groans, " . . . little more warning . . . . "

"And I wish I hadn't yelled to you to stop him. Crazy, asking you to tackle that murderous bastard. I'm so sorry."

"That was Robert Grooms?" Soverel said, arriving behind them, Roy two steps behind him. Turning he asked, "The security man you hired away from me?"

"Apparently," Roy said. "I didn't see him myself, Tom. Only heard what Jane said."

Jane glanced around. Tom Soverel was talking to the receptionist. Now he grabbed her phone, dialed. "What was he wearing?" he asked the reception area at large. "I'm alerting my security staff."

"Gray slacks and blue denim shirt," Jane said. "Crew cut, powerfully built, eyes of different colors—"

"*That* I remember."

"He's extremely dangerous," Jane said. "Probably armed. And someone should call the police. Zach will need protection."

Soverel passed along the information, then hung up. "They're calling the cops. I warned them not to take any chances. Told them Grooms knows this building pretty well. So, you're telling me *he's* the man the boy saw kill Elie Grey?"

"That's right," Jane said, helping Wade to his feet. "Zach had seen Grooms before—once at my father's funeral and another time sneaking around Elie's farm. He must have remembered the unusual eyes. The question is, what was he doing *here*?"

Soverel turned to Roy. "Well? The guy works for you now."

"Damned if *I* know," Roy said. Grace appeared beside him. Anne and Aunt Lou stood off to the side. "He's supposed to be back at Ocean Plains, doing his job. And I don't know why the hell he ran away, if that's what he did. Maybe I hired a screwball, Tom, but it was on your recommendation. You never know with these ex-CIA guys, right? On the other hand, how can you take the word of—" Roy glanced around just as the Chamberses and Zach appeared. "Just one person?"

"There's corroborating evidence," Jane said firmly.

"Oh? Well, then." He glanced at Grace. "I suggest we listen to it later. It's obvious we're not going to be concluding any business here today, and Grace is extremely upset and wants to go back to the Drake." He took his wife's hand. "Anne, I think you should come with us. Aunt Lou, too."

BASEMENT PARKING GARAGE, SOVEREL BUILDING 10:39 A.M.

Michel and Eliane Trunel of Fort-de-France, Martinique, were making a second circuit of the basement garage, having missed the exit ramp on the first pass. They had come down from the sixty-sixth-floor observation tower, where Eliane, a professional photographer, had been shooting the lakeside panorama. They were in the second week of a month-long motoring tour of the States, excited to be heading now into the American West.

"*Merde!*" Michel cried, slamming on the brakes, which sent his wife careening into the padded dash of their rental Toyota.

"Are you trying to kill me, Michel?"

"Look!" He pointed past her at the line of parked cars.

Eliane gasped. A few meters away a man's leg dangled below the slightly open rear doors of a blue cargo van. Now, as both watched, the doors were slowly forced wider apart, and the whole body tipped out like a rag doll, limbs aflop, sprawling lifelessly on the cement. A male corpse, face up, in leather jacket and jeans.

Then one arm lifted, fingers splayed.

Eliane screamed but wrenched open her door.

"Elle, don't!" Michel screamed. "This is Chi-ca-go, where all the gangsters live! Let's get out of here!"

But she was already outside, crouching beside the body, watching eyes definitely alive—and a mouth trying to work. His face was sweaty. Fingers plucked at her shirt.

"Don't worry," she told the eyes. "We will call an ambulance for you!"

The head tremored sideways, the eyes obviously terrified. Was the poor man asking her *not* to call an ambulance? But why? And where would she make such a call? Their car had no mobile phone; there must be a pay phone in the garage, but it wasn't in sight. She'd have to search for it or stay here while Michel drove around looking for it. And there was a more frightening consideration: Had someone *left* this poor man in the back of the van? If so, that person, or persons, might return at any second.

Michel, now beside her, echoed her fear. "Eliane, this is far too dangerous! We can't help this man now. *After* we leave, we can call 911."

Fingers slid around Eliane's wrist. Before she could react, the man used his grip to haul himself to a sitting position—with obvious and excruciating effort. He pointed at the Toyota, his eyes beseeching again. As she hesitated, he flopped forward and started crawling toward it.

"Michel, move the car closer!"

"Absolutely not! We can't take him! What if he dies?"

"If we do nothing, he *will* die! Michel, I beg you!"

Her husband blew out his breath in exasperation but scrambled up. Behind the seat of the rental Toyota again, he backed it up several car lengths, then brought it forward, angling close to the crawling figure. Eliane threw open the back door.

The injured man tried to haul himself up into the backseat, but could not.

"Michel, go in the other side and pull his arms. Not too hard, but steady."

She, meanwhile, lifted and shoved the long blue-jeaned legs. Somehow they got him folded inside and both doors shut.

A moment later they had located the ramp and charged up it, nearly running down a gray-haired woman walking a small gray dog.

"Forgive me, madame," Michel called out, lowering his window. "Could you direct us, please, to the closest hospital?"

"With a facility for emergencies!" Eliane added.

The woman stared at them critically a moment, bracing against the pull of her dog, then tipped her head in the direction of Lake Michigan. "You want Northwestern Memorial. It's on Huron and Fairbanks, about five blocks south and a couple blocks closer to the lakeshore. But the emergency entrance is actually on Superior. I know because I had to take—"

"*Merci!*" Eliane called back to the woman as Michel accelerated down the street. Then she twisted around to the man pretzeled into their rear seat. She reached out, squeezed his heavy hand, felt a slight return pressure. He looked gray faced and very frightened still, but there was also gratitude in his gaze. She smiled at him.

"You are very brave, my friend," she said. "I will say a prayer for you. A *psaume* of Da-vid, you must know it, '*Le Seigneur est mon Berger . . .*'"

KNOLL ROAD WEST BETWEEN RED KNOLL AND NEWKIRK 10:41 A.M.

Lenny Cristofaro drove the big, cab-over truck sedately, working on his third Egg McMuffin. He still had three hash-brown patties left, plus a large coffee. He didn't plan to stop for food again till Memphis, at which point he'd decide whether to take Interstate 55 south to New Orleans or I-30 to Dallas. He was feeling a little scared, a little euphoric. He'd maybe cut things too close. He'd been planning his getaway for weeks now, but only *seriously* since R. K. Singh showed up at the front gate. When the baddest gun in the West rides in, it's definitely time to get out of Dodge.

Of course, Lenny had known he'd have to leave eventually the first time Roy Jr. had called him in for a "private chat."

The only question for Lenny was, how much could he siphon from the Akers pipeline before he left? God knew the company owed him big time, for all the frigging years he'd clocked in this hick town.

Lenny remembered the way Roy had leaned across his big desk and dealt out those color shots . . . .

They were Nazi-costume kink, multigirl action, a few OB/GYN close-ups salted in, all laser-printed Net porn from some choice password sites. Lenny tried to stay cool, playing dumb and shaking his head when Roy asked, "See anything familiar here?"

"Definitely not my speed, Mr. Akers."

"What *is* your speed, Lenny?"

"Oh, you know. Supermodels, centerfolds—cool gals."

"Let me ask you this, then, Lenny. Where do you suppose a person could get material like this?" Roy tapped the nearest printout, depicting a whip-wielding woman, two men in chains.

"I wouldn't know, Mr. Akers."

"Well, I'll tell you where *I* got them, okay? I got them from a golfing buddy, who confiscated them from his eleven-year-old son, who bought them in Champaign near the campus from some 'weird old guy in black tennis shoes' who apparently hangs out at a video parlor there, outscoring the kids at any game and selling nasty pictures like these for five bucks each."

"You think that was *me*, Mr. Akers? No way."

"*Way*, Lenny, *way*. As you may imagine, my golfing buddy was extremely interested to find out more about this particular scumbag. After some paternal prodding, his son remembered seeing an Akers ID in this weirdo's wallet and that he drove a red Mazda Miata. Fortunately for you, Lenny, my friend came to me with this information rather than to the police."

He expected to be fired on the spot. But, surprisingly, after giving Lenny the scare of his life, Roy backed way off, suggesting that Lenny might be given a "second chance," as long as he stayed the hell away from those—or any—kids and any video parlor selling smut.

Amazingly, that seemed to be that.

Until a most unpleasant follow-up interview with the new security chief, Grooms, to whom Lenny's file had obviously been passed. But even that nasty conversation had ended all right, although not before Lenny had sweated a cupful. Grooms was this hard-nosed, supermacho militia type, who obviously got his kicks—but probably not all of them— watching another guy squirm like a pinned-down fly. Anyway, calling it a conversation was a laugh—almost a scream. This guy really had a way with him—Lenny would've bet a hundred this wasn't Grooms's virgin interrogation. When Grooms had concluded his sadistic version of Q&A, he'd issued marching orders to Lenny. It had taken Lenny only two seconds to decide that if Grooms guessed that Lenny *liked* the assignment, dangerous as it was, he'd find a way to take the pleasure out of it real quick. So he'd kept a real unhappy face on . . .

Even when Grooms got to the part of the interview Lenny had liked almost as much as the assignment itself. He'd never forget the tight-assed grin on Groom's face as he'd told him, "This is our first and last talk. If you do your job right, we won't need to talk again. If you don't, next time we won't be talking."

With those last three blasts from Grooms's bugle ringing in his ears, Lenny had set about bringing down Akers.

Now it was the whole beautiful plan that was being brought down—by Royal's long-lost daughter (who should've stayed lost) and her multicolored minions. But Lenny had escaped just in time, like blasting off in the last spaceship from a doomed planet. And, all in all, he was taking a pretty good haul with him, not counting his stash of Krugerrands. And leaving almost nothing behind.

Feeling proud of that, he smiled. When the Cheshire cat vanishes, only his smile lingers.

It was then Lenny heard the first siren.

SOVEREL CORPORATION CONFERENCE ROOM
10:44 A.M.

It was a reduced group around the big blond table—Wade, mostly recovered now, Jane, and Alex—sharing croissants and coffee. Mr. and Mrs. Chambers and little Zach were in the executive dining room one floor below, awaiting a police escort back to Ocean Plains. Tom Soverel, having listened attentively to a significant portion of Jane's "corroborating evidence," had excused himself to take care of "some business." *This* wasn't any of his, was the obvious implication.

"It was one of Zach's favorite games," Jane was saying. "Hiding in plain sight. He had climbed the ladder to the top of a storage bin, was watching his grandfather on the old John Deere tractor. Suddenly Grooms appeared. The boy saw him jump onto the back of the tractor and stab something into Elie's leg, then run away. Zachary was so terrified that he ran away, too—which he was terribly ashamed of later."

On the basis of that account, which they'd first heard early this morning, she explained, Wade had alerted the county coroner, S. J. Hollings, an old friend of the Grey family. An immediate autopsy had been ordered. Toxicological and tissue test results would not be available for several weeks, but they'd been phoned some preliminary findings just before this morning's meeting. Blood potassium levels in Elie's body were extremely high, easily sufficient to cause a fatal heart attack.

"A high dose of potassium chloride is known to cause heart attacks," Wade added, turning to Alex. "Jane and I are wondering if Grooms was bugging Elie's phone, overheard his urgent call to us yesterday, knew he was planning to tell us about the sabotage of Royal's plane, and decided it was time to silence Elie for good."

"Which makes knowing that Grooms is still on the loose real scary," Jane said.

"Definitely sounds like something from the CIA bag of tricks," Alex observed. "Maybe this Grooms guy really is just a nut case, like Roy said, and he pulled all this shit on his own?"

"Maybe," Jane said, sipping coffee. But the criminal pattern she was beginning to discern was far more intricate than that of a madman on a killing binge. She had been on the phone non-stop the last fifteen minutes, first briefing Sheriff Bevins, and next the head of the Chicago FBI office, who had called her at Singh's request. From Rikki himself, unfortunately, she had heard nothing.

Wade, meanwhile, had been giving status reports to Don Landsman and other company honchos.

Her cell-phone trilled once again. She picked up, nodded, then let out a whoop. "At last! Congratulations! Get back to me when you've got him!" She slapped her cell-phone shut. "Bingo! Lenny Cristofaro!"

"He's the saboteur?" Wade asked. "Really, Lenny's our guy?"

"Singh is dead certain."

"And they got him?" Alex asked.

"Not yet. The state highway cops are closing in. Lenny's in a U-Haul, probably heading for the nearest interstate. Singh will call back as soon as they pull him over and snap the cuffs on."

"The little bastard," Wade said. "Poisoning his own well."

"The thing is," Jane said, "just like with Grooms, I can't see Lenny doing all this on his own. He had to be taking orders from someone. Next question is, will he talk? Wade, you know him better than I do."

"If the prosecutors cut him a reasonable deal? Sure, he'll talk plenty."

Singh had been listening in on the police pursuit of Lenny. By punching a series of little-known commands into the keypad of his hastily assembled cell-phone, he had reprogrammed the device into a cellular telephone scanner, able to monitor police frequencies. So it was no surprise to him, or to Mimi and Ron, when they came upon two Illinois State Police cruisers parked alongside the road, with a sheriff's car on the other side. Lenny's U-Haul was lying on its side in an irrigation ditch, the left rear tire still turning.

They drove past and pulled over. Hagstrum jumped out and hurried back, while Mimi followed more slowly with Singh. Everything had happened so fast this morning, she thought. Could she be about to have her chance to tell Lenny what she thought of him? They saw the state trooper they'd been listening to, one booted foot on his car's doorsill, talking alternately to Hagstrum and into his police radio. Mimi heard him ask for an ambulance, then heavy tow trucks. "Got one of those U-Haul Super Movers sideways in a ditch."

"Is he badly hurt?" Mimi asked Hagstrum.

"I don't know. They're just bringing him out."

A moment later Mimi saw him, the fake walrus mustache hanging now from his upper lip, wrists cuffed behind him, being led out of the culvert by another trooper and a deputy. A second deputy hung back, one hand on his holster.

A moment ago Mimi had been ready to spit in Lenny's face, to curse him for what he'd done. But, as she saw him like this, her hatred, even her contempt, failed her. His hair was matted and stringy, his gaze myopic and bewildered. Cuffed, he was unable to wipe the rivulets of blood from his forehead. What she mostly felt was pity.

Then he glanced at her and shrugged, and rage at him broke

free in her again. It rose in her throat and stuck there as she watched him being led to the back door of the police car to wait for the ambulance. He started to get in, then looked back toward Singh. He smiled weakly.

"Hey, ragtop. I nearly outfoxed you, man."

But Singh was no longer interested in Lenny Cristofaro. His fingers were busy on the keypad, reprogramming his scanner back into a cell-phone, then punching Jane's number.

"Done," he told her.

THE SOVEREL CORPORATION CONFERENCE ROOM
11:07 A.M.

Jane concluded her phone conversation with Singh by asking if he was well enough to chair a three o'clock meeting in Ocean Plains on phasing in normal computer operations once the capability was restored.

"That would be my pleasure," he said, avoiding a direct answer to her question.

"You're sure now?" she pressed him.

"It is only the second time in many days I have been sure of anything," he said dryly, obviously unused to such a low score.

"Okay, then." She laughed. "Chip Bragan can help you draw up a list of who should be there," she said. "IT folks, hardware and software vendors, network specialists, consultants—"

"And don't forget Mimi," Singh interrupted. "She deserves some of the credit for catching Lenny, and you will definitely need her expertise. Of course, when this is done, I give you fair warning, I'm going to make her a job offer."

"Not before we're up and running smoothly, or *I'll* steal your bicycle."

Singh chuckled and promised to call back later.

She hung up. Beside her, Wade was folding away *his* cell-phone. "Don will schedule a series of management meetings with all department heads," he said. "Starting at four."

She told him about Singh's three o'clock technical powwow on start-up issues.

"Okay," Wade said. "If you want to sit in, we'd better get a move on."

Jane nodded and dialed the Drake, asking for Anne's room. There was no answer. She asked the desk to ring Grace and Roy's suite. No response there, either. Finally she requested the connecting room occupied by Louise Akers. Again, nothing.

"Something's very wrong," she said. "I'm going over there."

SHERIFF'S SUBSTATION
RED KNOLL
11:14 A.M.

The big U-Haul was towed to the parking lot behind the station. Sheriff Bevins unlatched the back door, holding the rope as it slid upward with a crash.

"Little prick was busy," he said. "You can take a quick look before we seal it."

Ron vaulted inside, while Singh and Mimi stood by the tailgate, peering in. Lenny's Miata occupied most of the interior. The remaining space was a welter of boxes and uncrated computer hardware, obviously strewn around in the wreck.

"Those look like Akers workstations," Singh observed.

"They sure are!" Mimi said. "Lenny must have smuggled some of that stuff out last night, faked out the gate guard somehow."

Hagstrum reappeared at the tailgate, grinning and wheeling Singh's mountain bike.

"I knew it!" Singh said.

"Plus Elrod," Hagstrum added, holding up the experimental notebook PC. "The creep had the sense to wrap it in blankets."

As Singh reached for it, the sheriff blocked him, shaking his head.

"You'll have to file a claim. It's all evidence."

Singh asked Hagstrum for his cell-phone. "Sheriff, perhaps

you won't mind discussing this matter with Mr. Petrossian of the FBI Computer Crimes Division? They're on their way here, to take charge of the Akers investigation. You see, most of the key evidence against Mr. Cristofaro is locked up on that hard disk. If you impound my computer, and it becomes lost or damaged, you may be responsible for this dangerous criminal going free."

The sheriff scratched his jaw, then gestured at Hagstrum. "Okay, give 'em his gizmo. But the bike stays."

SOVEREL BUILDING
11:26 A.M.

Jane shared the elevator to the parking garage with three men— Alex, Wade, and a white-haired, barrel-chested individual with a slit of a smile and steely eyes. He had a radio on his belt, a wired plug in his ear, and, Jane was sure, a gun under his suit jacket. Frank Kilfoyle was an ex-cop—and Soverel's security chief, whom Tom Soverel had offered as escort when Jane told him where she was going and why. "He knows people at the Drake," Soverel had said.

Jane liked the idea of having him along.

The hotel was only a block up North Michigan, at the top of the Magnificent Mile. But Kilfoyle insisted they drive. They waited near the garage elevator while he brought around the limo that had ferried the family north. Jane and Alex climbed into the rear seat. Wade sat facing them. For once in her life, Jane tried *not* to think three moves ahead. All the likely avenues augured bad endings for her family. Royal's death had been tragic; his murder was worse yet. The looming possibility that Roy might somehow be involved was too terrible to contemplate.

Still, the shadow of doubt hovered at the edge of her mind.

Kilfoyle went up the ramp, swung right, then made two quick lefts, taking them down a separate ramp into another subterranean parking garage. The trip took maybe a minute. Another minute saw them ascending the elevator to the Drake's marbled-floored lobby.

"Nine, correct?" Kilfoyle asked as they stepped out.

Jane nodded, repeating the suite number.

"Okay, have a seat." He indicated a sofa. "I'll be right back."

But then he spotted someone over her shoulder and called out, "Boggs!"

A silver-haired, gray-suited man heading their way, whose zigzag profile reminded Jane of Dick Tracy, glanced over, jerked a nod, but kept moving toward a just opening elevator. In his wake was a uniformed patrolman. "Sorry, Frank, can't talk."

Kilfoyle stepped into the elevator after them. "Going up to nine, by any chance?" Jane heard distinctly.

She also heard Boggs's response. "You know about it, Frank?" Then the door closed.

"I'm going up there," Alex said.

"Me too," Jane said, springing up.

"No." Wade took Jane's wrist. "Both of you, sit down. Let's all wait here like the man said. I know how you must feel, but there's a killer on the loose, and Kilfoyle looks to me like he's wearing a flak jacket. Are either of you?"

Jane pulled free of Wade's grasp. Alex sat back down and began nervously pounding his fist into his palm. Okay, maybe Wade was right, Jane thought. But, dammit, he didn't need to grab her like that. Her insides kept churning.

What was happening upstairs? She looked up in frustration at the ornately carved ceiling.

An eternity passed, the elevators disgorging and ingesting, the lobby trafficked meaninglessly. Every instant of that eternity, Jane was braced to hear a scream—or a shot—echoing down the elevator shaft.

An elevator settled again, wheezed open. Frank Kilfoyle beckoned them inside. He waited until they were all in, with the door closed. "Roy ducked out. The women are up there, all pretty shaken up—especially his wife. Boggs called for the hotel doctor."

"*Ducked out?*" Alex said. "What the hell does that mean? Where did he go?"

"He told them he was going out for ice. He didn't come back. They went looking all over the hotel. That must have been when you folks tried to reach them. They only called Boggs—he's the house dick—about five minutes ago."

Boggs opened the door to the suite's sitting room, then went back to the phone. Grace Akers was scrunched up on a wing chair, her hands covering her face, sobbing. Aunt Lou was kneeling beside her, stroking her, speaking softly.

Anne Akers stood close by, talking to the uniformed patrolman. She had little to tell Alex and Jane beyond what they'd already heard. "I'm sure Roy will be back any minute," she offered—less than convincingly. "He's obviously distraught, probably just out there wandering around, wanting to be alone."

"Let's hope you're right," Jane said, watching Frank Kilfoyle check out the adjoining room with the patrolman. Alex had joined Aunt Louise in her attempt to calm Grace.

Jane had not the slightest idea what to do next. Lenny had been caught, Grooms was on the loose, Roy missing. *Dad, tell me what I do now.*

There was a knock. Boggs opened the door to a svelte woman of an age where her blond hair definitely had help. "I'm Dr. Lindquist," Jane heard. The woman came into the room, toting a medical bag. Boggs pointed toward Grace.

The doctor leaned over Grace, one hand on the chair arm, her other hand feeling for Grace's pulse, and spoke quietly.

Jane walked over to Frank Kilfoyle. "I imagine you need to get back, Mr. Kilfoyle. But I want to thank you. And thank Mr. Soverel for me. I won't forget his kindness."

Kilfoyle shrugged. "I'm yours—your family's—for the rest of the day. Drive you back to Ocean Plains, if you like."

"Thanks, but that's way beyond the extra mile. I doubt Grace

Akers will be going anywhere tonight. She'll want to be right here, waiting for her husband—she'll probably be getting some light sedation. The other two women will want to stay with her. I suspect my brother will also want to hang around."

"That still leaves you two. Need a ride *anywhere*?"

Jane turned to Wade, who was closing his cell-phone again and checking his watch, obviously worried about making those afternoon meetings. "We could use a lift to Meigs Field, Mr. Kilfoyle, thanks. We have a plane to catch."

MEIGS FIELD
12:32 P.M.

The muscular black man behind the counter at Vollmer Aviation Services gave Wade and Jane a curious smile. "Kind of a confusing morning for you folks," he said.

"What do you mean, Reggie?" Wade asked.

"What do you mean, what do I mean? First, right after you leave, that photographer guy shows up, then your pilot, Johnson, calls in to say he's heading over to O'Hare to see about some other job in—"

"Whoa!" Jane said. "Chevy did *what*?"

"He's seeing about a job with United," he said. "Then another guy comes in and flashes his ID, turns out to be your president, and *he* wants the plane."

"Reg, I'm not tracking any of this," Wade said. "Can you back up to where we left here?"

The counterman retraced the strange sequence from the arrival of the photographer, Bannerman, to Chevy Johnson's apologetic phone call en route to O'Hare, leaving a message that Wade should fly the Tiger back to Ocean Plains, to the arrival of the company president, demanding the plane.

"Roy wanted the plane?" Wade frowned.

Jane wondered about that, too, of course, but at the moment she was unable to set aside fury at Chevy for ditching them in the middle of an assignment, no matter the reason. How dare he!

Reggie was checking a sign-in sheet. "Royal Akers Jr. That's the man."

"When was that, exactly?" Wade tried to read the upside-down time next to Roy's signature.

"Only about fifteen minutes ago." Reggie picked up the tower phone. "Gene, that Tiger 95X take off yet? . . . Uh-huh. Thanks." He glanced up. "He's on takeoff roll. You should see him any minute now on runway thirty-six."

"Jane, c'mon!"

The urgency in Wade's tone made her refocus quickly. She was right behind Wade, heading out of the office through the glass door in the east wall. Wade grunted in pain—any sudden movement needled his groin injury—then pointed at the lakeside runway.

"There!"

The sporty white four-seater came rushing down the concrete, gathering speed, then lifting lightly off the deck into the afternoon blue.

Jane ducked back into the office. "What's his flight plan?"

Reggie dialed the tower. "Gene, what's the Tiger's flight plan? . . . Okay, thanks." He shrugged. "He filed same as usual. VFR—visual flight rules—to Ocean Plains."

Jane ran back out.

"I heard," Wade said, squinting into the sky. "Trouble is, he's banking north, out over the lake. Doesn't look to me like he's heading home—no matter what he told them. What the hell does he think he's doing?"

Running away? Her first, strange thought.

And then an even stranger thought. *Did I ever tell you I love you?*

She asked, "How often does Roy fly?"

"Not often enough. He's rated for VFR only, single engines. That means he can fly in good weather—visually—but not by instruments. I don't remember him being checked out in the

Tiger, though. And it takes some getting used to."

Jane shielded her eyes, watching the small plane grow ever smaller on its wayward course, its white wings heliographing back the high sun. Her brother, her troubled brother. Was he running away? Roy wouldn't do that! He loved Grace, and he loved little Tyler.

She felt, more than at any time in her life, a link to him up there, not so much of blood as of a shared family crucible, the strife that had torn Royal and Joycelyn asunder. She and Alex had done their running long ago. Was Roy Jr. doing his today?

Jane watched the Tiger out of sight. She didn't know if her brother was involved in the crimes and catastrophes that had befallen her family and its enterprises. If he was, she might never know how deeply.

But he *knows.*

She blinked hard. Looking into the sun had made her eyes tear.

OVER LAKE MICHIGAN
12:31 P.M.

He had to kill himself.

The unthinkable option. But what other way out *was* there for him? He felt like a rat caught in a maze with only one exit.

Full of resolve, he had set off, his brisk strides taking him the full length of the walkway under Lakeshore Drive and out onto Oak Street Beach. He had stared for endless minutes at the sweeping lake horizon, then swiveled round to contemplate the luxury high-rises along the Gold Coast.

Plunge to his death from a high floor? Swim out too far? Roy shook his head vehemently, rejecting both possibilities. Yet die he must, and by some means swift and certain. He rummaged through his brain—fruitlessly, then took a long pull from the pint of Chivas Regal he'd bought in the Drake gift shop, his eyes on a small plane droning out over the lake.

And in that instant he saw another door out of the maze! He didn't have to die. He only had to disappear. Fly straight north, all the way to Canada.

The beauty of the solution caused him to laugh out loud. *Fooled you, old man!* His father would *expect* Roy to take the easy way out, kill himself. But Roy wasn't going to do what dear old Dad would expect. Not this time.

As if to nail down the perfection of his sudden inspiration, he remembered that a company plane was just sitting there waiting, two miles away at Meigs Field, fueled and ready for takeoff.

Now here he was, making good his escape, soaring right out of the rat maze. The blended Scotch was almost gone. Just as well. Roy didn't need liquid courage now; he needed his wits collected. More to the point, he needed to keep this aerobatic little plane nosing north-northeast, straight down the middle of the wide blue watery fairway, well below any commercial traffic patterns. He had the radio off, airspeed showing 130 knots, altimeter just under 2,000 feet.

In about two and a half hours, as he crossed Michigan's Upper Peninsula above the Straits of Mackinac, he'd have to make some swift decisions. Air Traffic Control on both sides of the border would be trying to contact him. The first Canadian airport of entry would be Sault Sainte Marie. But Roy had no intention of landing at *any* airfield and ending up in custody. He'd veer off over Whitefish Bay into northwestern Ontario, look for a deserted road to drop onto, even a lake to plow. A spot remote enough to allow him to hike into the woods. And vanish.

There, in hiding, he'd stage a symbolic suicide. He'd burn and bury every scrap of ID—credit cards, driver's license, wallet photos. He'd be left with just the cash in his wallet—a little over $600. He'd camp out, grow a beard, lose weight. Acquire a new identity—using either fake documents or someone else's.

Brand-new name, brand-new self.

As he contemplated his impending metamorphosis, he experienced a euphoria greater than any he'd ever known. In his new life he would stride along interesting streets, encounter exciting strangers, greet with open arms any adventure coming his way.

But there was an even greater reward awaiting him.

He'd no longer be an Akers!

How about that, Dad? For once, your wishes and mine match up!

Jane had done it long ago, coming back a war hero, beating him to slapping Dad in the face with that ultimate insult, shedding his name. Roy remembered silently envying his sister the freedom she must feel, even as he had condemned her aloud. Well, better late than never. Soon he would feel that same exhilaration. No longer be merely Royal Akers's "Junior," but his own man.

He was ready. In his mind. In his spirit. And in a little brass-plate bank in George Town. There, on Grand Cayman Island, his ace in the hole awaited him. A signature account set up years ago. Around six hundred thousand now. Small change compared to what he was leaving, but definitely worth claiming.

For sure, this was the way to look at it, down the straight and narrow path to a future so exciting, it made him catch his breath. A life of adventures. A glamorous life. A life where, at long last, he came into his own. Dazzling images swarmed in his brain.

Then one of a little boy with his turquoise eyes and without a father.

EMERGENCY ROOM,
NORTHWESTERN MEMORIAL HOSPITAL
12:35 P.M.

The diagnosis, based on patient observation plus EKG readings and preliminary cardiac enzyme data, was acute myocardial infarction—heart attack. The decision was made to move

Mr. Johnson immediately to the intensive care unit on seven.

Al Valladares, the ER orderly who was guiding the gurney into the first-floor hallway, saw the patient open his eyes again. Hang in there, Mr. J., Valladares thought. Johnson had been in and out of consciousness since being brought in. Then, all of a sudden, the patient's eyes went wide and panicky.

¡Dios mío! The poor guy was having another one!

Valladares braced to do a one eighty back to the ER. Then he noticed Johnson was pointing at something—a bank of pay phones across the corridor.

"What is it, Mr. J? You need to make a call?"

"Yes." It was a croak, but Johnson had definitely spoken— the first word Valladares had heard him utter. Now the patient was nodding "yes." Nodding frantically.

"Okay, calm down, my friend. You'll get your call. But first we gotta get you upstairs, get you stabilized, okay? Now here comes your elevator."

Johnson shook his head, just as frenziedly as he'd nodded a moment earlier. His eyes were shifting from side to side, searching. He forced more words out: "Life . . . or death."

"You're right, it *is* life or death. And that's why—"

"No . . . *phone* . . . life or death."

"I hear you, Mr. J. But you gotta talk to the folks upstairs about that, okay?" The elevator doors wheezed open.

As Valladares waited for the people to get off, Johnson's right arm snaked out—uprooting his IV needle—and snatched a cell-phone from the hand of an intern standing beside the gurney.

"Hey!" the intern yelled. Valladares was shocked. Mr. Johnson was shielding the phone with his shoulder while poking feebly at its keypad.

"Don't worry, Doc," Valladares said, bent on hiding how impressed he was by his charge's determination. "I'll get it back for you."

The intern waved him off. "Let the poor man make his

bloody call. As long as it isn't overseas."

TRAVELING SOUTH ON THE DAN RYAN EXPRESSWAY, CHICAGO
12:40 P.M.

"Chevy? I can hardly hear you. What's happened?" Jane's anger at him was instantly displaced by urgent concern. His voice was so faint, it was barely recognizable.

"Don't . . . fly," came the fragile response. "Grooms . . . sabo . . . tage . . . "

"Oh, God, no! Roy!"

She turned to Wade, who was driving their rental Malibu. "Chevy thinks Grooms may have sabotaged our plane. Call Meigs right away—they've got to keep trying to get through to Roy, bring him down." She went back to Chevy. "There's no reason to worry about us, Chevy, we're on the ground. But my brother took the Tiger up. Where are *you*?"

"Hos . . . pi . . . tal . . . Heart . . . "

"*Which* hospital?"

"A . . . ttack . . . North . . . west. Stop . . . Grooms . . . . "

"The bastard already got away, Chevy, but we'll catch him somehow. Northwestern Memorial?"

"Yeah."

Jane gestured at Wade to turn around and head back downtown. But Chevy was still talking, or trying to.

"Ed . . . can . . . track . . . G . . . P . . . S . . . my . . . unit . . . " The voice faded out.

Someone else came on the line, speaking rapid, Hispanic-inflected English, asking who she was. Jane gave her name, asked about Chevy's condition, then said, "Listen carefully, this is urgent! Your patient's heart attack was probably triggered by an injection of potassium chloride. Tell the doctors that immediately. You must do that! Promise me!"

"Yes, I promise."

"Thank you. We're on our way there now."

She hung up, turned again to Wade, who was putting away his own cell-phone while exiting the expressway. "Northwestern Memorial, Wade. It's on—"

"I know where it is."

"What about Roy?"

"They've been trying to reach him, and so has Air Traffic Control in Aurora, but his radio is still turned off."

"Shit!" Jane clamped her head in her hands, trying to think of something else she could do. "You heard all that about Chevy?"

"Some of it. He's had a heart attack. Courtesy of Grooms, I take it."

"Yeah, they're taking him to the ICU. He's still alive, thank God, with at least a fighting chance, according to the orderly. Maybe Grooms didn't give him the full dose."

"Grooms must have been playing that photographer Reggie mentioned. The guy's gotta be stopped."

"I almost forgot! Chevy said something about that."

"About what?"

"Tracking Grooms. His voice was real weak, but it sounded like, 'Ed can track him.' Then three letters."

"GPS?"

"Yeah, thanks! That makes sense, right—global positioning satellite navigation? Then something about Chevy's GPS unit. But who's Ed?"

"Ed Basinski, our transportation manager. It makes perfect sense. Ed has all kinds of GPS units on board the trucks, some portable units, handhelds, all tied in to his computers, which run vehicle-location software. Is it possible that your flyboy friend somehow managed to plant one of those portable units on Grooms's van? Let's give Ed a call."

"Go ahead, Wade. I'm going to try Meigs again and see if they've reached Roy."

OVER LAKE MICHIGAN
12:43 P.M.

Roy shook his head violently, banishing the vision of his little boy. Better for everyone, Tyler included, that Roy just disappear. He'd made a mess of fatherhood anyway, just as *his* father had. Even now, Roy could not conjure up Roy Sr. without feeling belittled by how he was continually compared to his younger siblings. Their achievements were always superior to his. They outdid him at every turn. At least in his father's eyes—an opinion it never occurred to the old man to keep to himself.

The final insult was his father's choice of Don Landsman—a black guy, for God's sake, who'd given less to the company than he had. And, of course, been asked to take a whole lot less grief from Dear Royal Dad.

Now, thanks to Jane the Brain, Don was back from exile, running the company in tandem with her, bucking all odds, both of them, succeeding where Roy had failed—and thought no one could succeed. Incredibly, the unlikely pair had even overcome the ongoing inventory-replenishment sabotage and resuscitated employee morale.

So let them have the goddamned company, if they wanted it so much.

A dark arrow of geese angled below the plane, and a white arrow furled below that, on the blue water, carved by a powerboat. But Roy was staring farther down, into the deep water of the past. Should he have just absorbed the insult of his own father picking Landsman over his name-sake? Should he have walked away with all his stock? But, dammit, he'd paid his dues to the company for twenty years, while Jane and Alex went AWOL. He was owed! *By rights* he was next in line.

But how had events spiraled so quickly out of control? Roy shook his head, not because he didn't know the answer to that, but because he remembered the exact moment everything turned.

He was sitting in Robert Grooms's high-tech van. They were parked behind Roy's property, checking out the new home-security system Grooms had installed—perimeter fencing, surveillance cameras, motion detectors. A lot more than Roy needed, but selling Roy on his sideline, Grooms was as persuasive and relentless as any mattress salesman.

Roy had brought along a bottle of Courvoisier XO, and Grooms produced smuggled Cuban Montecristos. The ex-Agency spook had been telling wild tales out of school, full of unauthorized exploits. Selling illegal detonators to Colonel Qaddafi and weapons to both sides in the Angolan civil war. Trading in Soviet war-surplus items, including unexploded cluster bombs. Taking out a top-ten terrorist in Pakistan by cleverly sabotaging his motorbike, then a Bahamian drug runner by detonating his Beechcraft in midair—a plane, Grooms had remarked idly, almost identical to the one Roy's father liked to fly.

As the two men sat there, Roy could see an indent of Porter Lake through the willows and Landsman's property on the far side. He was adding on a new deck over there. Roy had no recollection of asking for further details about the plane sabotage, but suddenly Grooms was rummaging in the back of his van and returning with a small green plastic box in the palm of his hand.

"What's that?" Roy had asked.

"Oh, a nifty little item I picked up on the streets of Karachi a few years ago—though it was actually made in Prague." Grooms reversed the box, exposing a setscrew, a square metal tongue, and four corner magnets. Grooms let the device clamp on to the metal dashboard, then pried it off. "This strip comes off and can tighten this screw, which seals a pressure changer inside, creating a small aneroid barometer—for measuring altitude. At a specified atmospheric pressure, the altimeter-triggering device detonates the main explosive charge. It came equipped with three ounces of Semtex—a helluva bang for the buck, I assure you—but any heavy explosive will work."

Roy was horrified, even after Grooms assured him the device wasn't armed. "Put that fucking thing away—now!"

"Of course," the security man said. "I've obviously misjudged the extent of your curiosity about aircraft sabotage. There are many lesser ways to bring down a plane, by the way. More cognac?"

Roy took a searing throatful, shaking his head as he stared out at the lake. He was recalling some mutinous grumblings about Royal he'd shared with Grooms earlier in the afternoon.

"Bobby," he said finally, "I'm gonna pretend I never saw that damned thing, and we'll never mention it again, or what you were just speaking about."

"Sure thing . . . "

But Grooms had obviously *not* forgotten that little chat by the lake. Christ, Roy should have fired the man on the spot. Instead he'd let Grooms smooth-talk him into a promotion as Akers's security chief.

It hadn't happened overnight, but he could see now that the man had planned it. How, with an office close by Roy's in the executive wing, Grooms had made himself an even handier confidant—and after-hours drinking buddy. Roy had failed to heed all the telltale signs of the man's psychotic personality. The way Grooms had fondled that little bomb. His inappropriate mirth over grisly war stories. His frequent hints at even darker deeds in his past—his pride in how, even in his Agency days, he'd never restricted his activities to the original mission.

Then came Royal's fatal plane crash.

Roy had dismissed his instant suspicion that Grooms might have had a hand in it. That made absolutely no sense. And the

NTSB's preliminary report quickly confirmed the Beech Bonanza crash as an accident. But Roy's suspicion surfaced again when Elie Grey trucked the wreckage over to his farm and hired his own crash investigator.

And finally, today, in Soverel's conference room, the truth had come crashing down on Roy.

When Zach Grey pointed to Grooms as the man who had killed his grandfather, Roy had instantly grasped the full hideous extent of Grooms's crimes. The madman had murdered not just Elie, but Royal as well—by sabotaging the plane. With the realization had come corrosive guilt, the conviction that he himself was somehow complicit in those murders.

Not by commission, but omission—by not stopping Grooms. And what of the ongoing destruction of the Akers empire? Surely Grooms had to be involved in the crippling supply-chain sabotage as well? But Roy had a share of guilt there, too. As soon as he had understood how crippling the computer damage was, and sensed his inability to deal effectively with it, he had gone to Soverel and struck a deal for a lucrative sellout.

With Grooms on the run now, his twisted motives might never be revealed. Or, if he *were* caught, the ex-spook might even implicate Roy.

And who would believe him innocent? Not the employees of Akers. Not his family, except for Grace, perhaps. And surely not his beloved sister.

When you came right down to it, he didn't believe he was innocent, either.

Sooner or later they'd know enough, or figure out enough, to come see Roy.

"Mind if we ask you a few questions, Mr. Akers?"

That scene was never going to take place. Roy smiled grimly as he held the plane steady. He was free, or about to be—utterly untethered, his eyes on the northern horizon, his wings stretched wide.

He glanced at his instruments. The altimeter needle was inching upward. He'd been creeping vertically, all the way to 2,400 feet. He stared up through the canopy at a region infinitely blue. He'd better keep the Tiger below 3,000, well out of any commercial traffic and, he hoped, low enough so that Air Traffic Control would leave him alone.

TRANSPORTATION MANAGER'S OFFICE, AKERS HQ
1:04 P.M.

Ed Basinski's office hadn't been so crowded since his sixtieth birthday, when a whole parade of drivers had marched in, singing and carrying a blazing cake. Assembled behind his chair and around his oversize monitors were a trio of FBI men, the county sheriff, a couple of deputies, an ex-FAA investigator, and an odd assortment of Akers people, including Don Landsman, an Arabian guy dressed for a cocktail party, and some Indian computer consultant with a bandage that was wound like a turban around his head.

All here to see a show.

And a damned good show it was, Basinski thought. It had started about ten minutes ago.

"Let's just give it a try," he had told the head FBI man, a pudgy guy name of Macomber, with a fringe of sandy hair. Basinski had plopped down on his swivel chair and brought up a database screen, highlighting the number of the portable GPS unit he had loaned to Chevy Johnson. Right beside the unit number appeared a phone number.

"You're going to call that unit now?" asked Petrossian, another FBI guy. "And Grooms won't hear it?"

"Right on both counts," Basinski said. Everyone in the room could hear the amplified electronic arpeggio of a modem dialing out. "'Course, we don't know for sure that Chevy actually planted the unit on Grooms's van. But, *if* he did, this call will activate its microprocessor and GPS receiver, which will start triangulating the van's position and calculating current speed and direction of travel, then transmit that data to us, right here. What you'll see then is—Wait a minute, we just locked on!"

There was a murmur of excitement, but the spectators heeded their instructions to leave the talking to Basinski and the law officers. The Akers fleet manager toggled the screen display to a digitized state map. "I'm patching that transmission into our AVL software," he explained. "That's automatic vehicle location—"

"We're familiar with the technology," Macomber said.

"—same as I use for monitoring our fleet. Minus, of course, the messaging I use with my drivers."

"Is that him?" Petrossian said, pointing at a blinking icon near the southern curve of Lake Michigan.

"Can't tell you *who* it is," Basinski said. "But that's the unit. And, yep, it's definitely moving! We got *somebody!*"

A cheer went up. Macomber whipped out a cell-phone, punched up a number. Petrossian, beside him, pointed again. "Can you enlarge this?"

"How's that?" Basinski produced a full-screen map of Chicago and environs. The blinking vehicle icon now showed west of the lakeshore. An on-screen window provided the GPS unit number, velocity, heading, latitude, and longitude. "And we can do better yet."

He used a mouse to trace a box around the icon, then double clicked. The map display jump-zoomed into a close-up with street names, numbers, and landmarks. And now the icon could actually be seen to move.

"We got him!" Petrossian said.

Macomber was leaning forward, squinting at the display. "What is that, the Kennedy?"

"Yep," said Basinski. "He's on the Kennedy Expressway, heading northwest at fifty-four, fifty-three miles an hour. Just crossing Irving Park Road."

Macomber relayed the information into his cell-phone.

"He's gotta be heading for O'Hare," Petrossian said.

Basinski was as excited as anyone now, finding himself at the nerve center of a crucial manhunt. There'd been an all-points bulletin out on the van for a half hour, he knew, and radio cars cruising all over the city—state police, Chicago PD, FBI—with no sightings. But now, thanks to the team of Johnson and Basinski, the authorities knew exactly where in the haystack to find a moving needle. He heard Macomber behind him directing any unmarked vehicles in the vicinity to close in.

"You know what would be a really cool enhancement of this system?" piped a voice somewhere in the rear. Basinski turned in annoyance; it was the Indian computer guy's ponytailed side-kick. "A joystick with an autofire button, so you could like target the icon, then vaporize the asshole right out of the galaxy."

TRAVELING WEST ON THE KENNEDY EXPRESSWAY
1:16 P.M.

The excrement was definitely hitting the oscillator, but that was nothing new to Robert Grooms. He'd been in much tougher scrapes than this one—in Vietnam and Afghanistan, and he wouldn't want to forget Nicaragua or Angola. The truth was, he *needed* a periodic dust-up to get all his juices flowing again. Nothing could make a man feel more alive than to slip out the postern gate as the barbarians swarmed over the drawbridge, their mouths foaming in anticipation of sacking and burning the palace and raping all the damsels.

In this instance, he had managed to do a good bit of palace sacking himself—getting damned good money for his take-down of a corporate dragon. And he'd done it all with a mini-mum of fuss—just some judicious arm twisting, metal filing, geezer needling, all culminating in this afternoon's Meigs flight to nowhere. Someday, in a bar somewhere across the world, he'd regale the right party with how he'd single-handedly punctured the heart of the Akers empire. (Well, he'd had a *little* help.) Grooms smiled at the prospect of sharing his wit.

The ironic thing was, the original corporate target, Tom Soverel, was ending up the beneficiary of Grooms's piecemeal dismantling of rival Akers. But Grooms's stint at Soverel's company had yielded only a series of dead ends—except for the serendipitous fact that it had eventually turned up Roy Jr., a textbook candidate for executive manipulation if ever there was one. Who could resist such a rich opportunity? Not Robert Devereux Grooms.

As it turned out, Akers's corporate links to South Korea were far more extensive than Soverel's. No amount of havoc wrought upon Big Tom could have created the political chaos now merrily under way in the streets and suites of Seoul. Mighty Hyon had fallen. The South Korean government could very well be next.

It was bonus time for Grooms, surely.

In any case, his employers in the North were exceedingly pleased. A tangible token of their gratitude awaited him momentarily.

But there were still a few yards to the goal line. Grooms kept the speedometer sedately under the speed limit, obeying every law in the vehicle code, while scanning continuously for cop cars of various denominations. They'd be looking for the van, but it was a popular model and color, and he'd switched the plates.

He ought to have made sure of Johnson, though. The big dude clearly hadn't taken a lethal dose. Probably his ticker had only skipped a few beats. Truly embarrassing.

Of course, if Grooms had driven straight to O'Hare and dumped the body in a parking lot, as he'd planned, it would have been okay. When the guy came to in the van, Grooms could have finished him off. The real mistake was stopping off at the Soverel Building to silence Roy. He did like to tie up loose ends, but it had already been too late for that.

Fortunately he kept his escape kit in the van. Grooms now had a new set of identity papers in his back pocket, including a passport with photo matching his new look—curly gray toupèe and horn rims, with colored contacts to make his eyes ordinary. He also had a ticket to El Paso. Tonight he'd stroll across the bridge to Juarez. By Thursday he'd be in Mexico City, sipping a cold one and spinning the globe in search of his next adventure.

Unless Pyongyang wanted to bid for his services again.

It was the way life *should* be lived, he thought, in serial installments, each with a cliff-hanger ending.

He saw the sign ahead for O'Hare economy parking and moved into the exit lane. The cops would tumble to the van in a day or two and have lots of fun inventorying his high-tech toys. Let them. It would only add to his legend.

The one who got away.

TRANSPORTATION MANAGER'S OFFICE
1:19 P.M.

"It's definitely O'Hare," Macomber said into his cell-phone, acting now as the command post in a tactical net. "He could be heading for lot D by terminal 5, or economy lots E and F. Stay on him, and we'll keep you updated."

They had their people in place all around O'Hare now. Several unmarked cars were cruising the parking garage near the main terminals; more were positioned in lot D by the international terminal—a likely escape point, Macomber thought. Finally, a team of FBI agents was cruising economy lot E, while plainclothes police detectives loitered about the people-mover station in lot F.

Macomber watched the blinking icon, waiting for the mouse to choose its trap. When it happened, Macomber provided play-by-play. "Okay, turning right now on Bessie Coleman Drive, heading for E and F. Let's close in."

"This is so beautiful!" Singh said under his breath.

"Yeah," Mimi said wryly. "Lenny would love it."

INTERNATIONAL TERMINAL FIVE, O'HARE AIRPORT
1:20 P.M.

The short-statured man leaned against the rear fender of the dark green Lincoln Town Car. Though the uniformed man's features were concealed by a peaked cap and mirrored sunglasses, there could be no doubt about what he was— a bored limo driver, killing time until his next pickup. He puffed on a cigarette, glanced at his watch and occasionally ahead at the gleaming line of luxury vehicles in the terminal's limo lane.

Jong Ye Nam, not a chauffeur, but a colonel in the Reconnaissance Bureau of the North Korean People's Army, smiled. He rather enjoyed his disguise, because it involved more than wearing a carefully assembled civilian uniform. His casual posture and behavior were quite as important. What he perhaps enjoyed most was his certitude that, beneath his casual air, he remained an experienced soldier.

Jong was fully alert. Especially for the last two minutes.

Ten feet behind him, across the median divider, a Chicago police car was idling, radio squawking. A female officer was out on the sidewalk, beckoning furiously to her chubby male partner, heavy-footing it across the road from the terminal building while swinging a fast-food bag.

"Move it, Abe!" she shouted. "Looks like they got that Grooms dude cornered! His van's heading into lot E or F. We gotta cruise the perimeter, ready to tail him if he tries to duck out fast."

"Sorry, Val," Abe gasped. "Bozo at the burger counter said . . .

all out of action figures . . . My kid collects 'em . . . . Hadda call the damn manager."

Moments later the police car rocketed away. Without seeming to rush, Colonel Jong followed quickly, accelerating past the limo queue, then squeezing expertly into the main airport exit lanes ahead of a hotel courtesy bus. He had the dash radio on, scanning for police channels.

Economy lots D and E were close by. The colonel wouldn't be the first one to arrive. But he wouldn't be far behind.

TRAVELING EAST ON BESSIE COLEMAN DRIVE, VICINITY OF O'HARE AIRPORT
1:21 P.M.

Special Agent Gary Wilburn jockeyed his unmarked, unwashed Saturn coupe directly behind the boxy blue van, with ladder and spare tire on the rear door. Out of sight of the driver's mirror, Wilburn dictated the license number into his radio.

The answer came back while he was stopped behind Grooms at a red light. The number matched a Geo Metro registered to some guy who lived on Archer Avenue over on the South Side. Grooms had been playing tricks.

The light turned green. "Moving again," Wilburn said. Then, half a minute later, "Okay, he's got his blinker on, he's turning into lot E."

"You keep straight on!" came the squawk-box command of Wilburn's boss, Ernie Macomber, special agent in charge of the Chicago field office. "Don't worry, we got a little reception committee waiting."

ECONOMY PARKING LOT E, O'HARE AIRPORT
1:24 P.M.

Grooms nosed the van into a slot, closed the windows, and killed the engine. "*Adios*, good and faithful servant," he said, patting the dash. Then he hefted his well-traveled canvas kit bag—his entire golden parachute, not counting the money belt girdling his gut and the hefty payoff he was about to get from his affluent friends in North Korea. Four steps away from the van, he triggered the electronic door locks, then pocketed the keys—out of habit, since he was never coming back.

After another four or five paces Grooms turned for a final, sentimental look and offered the van a salute. Funny, it seemed like there was an extra dimple along the van's high roofline, one he couldn't recall. Probably the sunlight playing optical tricks with his unfamiliar glasses and new contacts.

He headed for the nearest shuttle bus stop, one aisle over, and immediately spotted a shuttle coming his way. He watched it stop, let off a passenger, start up again. Another minute and he'd be on his merry way.

But his luck was even better. Suddenly, swinging out of a nearby aisle directly toward him, came another shuttle bus with maybe half a dozen passengers.

Grooms waved as the bus pulled alongside, and the front and side doors opened. He went up the front step, glancing at the other passengers—a scattering of men, blue-collar and white-. Grooms slid into the first empty seat behind the driver as the door slapped shut.

Okay, finally time to relax a little, he thought, setting the bag between his feet. The strain of the last few hours had definitely taken its toll. He shut his eyes a moment, inhaling deeply. But something was wrong.

The bus hadn't moved.

Grooms opened his eyes and turned. The driver was standing in the aisle, legs spread, both hands wrapped around a revolver pointed directly at Grooms.

Grooms made a fast peripheral scan. A half dozen weapons were trained on him. Every man on the frigging bus had a gun, each carefully positioned with a separate line of fire.

"End of the line, Grooms," the driver said.

Robert Grooms, stripped now of all weapons, his wrists cuffed tightly at his back, didn't twitch, didn't even blink. In rigid fury, he stared straight ahead.

A moment later the shuttle bus swung into another aisle. Dead ahead, through the wraparound front windshield, Grooms saw an unmarked white Blazer blocking the lane. A Blazer with dark windows and a telltale whip antenna—and several men clustered around it.

All waiting for him, Grooms realized. He was about to vanish into the FBI gulag. But how the hell had they known where he was heading? Only Colonel Jong knew today's rendezvous point. But why would that ingenious little agent, or his Politburo bosses in Pyongyang, tip off the feds? Did they *want* him to spill his guts? Did they actually *want* to take credit for the clandestine op and the resulting chaos in the South?

Made no sense.

Grooms's torturous thoughts were sheared off by a shriek of rubber, the crumpling thud of vehicular impact. The bus veered right, rocked to a stop. Grooms, unable to brace himself with his hands, spilled to the floor. The other passengers sprang to their feet, guns out.

From the floor, Grooms heard the doors wheeze open, agents clattering off the bus. A voice close by growled, "Stay the fuck on the floor, Grooms. Don't move a muscle."

He obeyed but was fully alert now. *Something was happening.* From outside arose a strident medley of voices. One particular voice, much louder than the others, galvanized Grooms. Without knowing the language, he could identify the guttural cadences of rapid-fire Korean.

Colonel Jong? Had the gutsy little bastard staged a crash with a busload of federal cops? If so, bravo! What a prelude to a rescue! And Grooms would play his part, react with lightning speed when the instant came.

Several agents scrambled back aboard, one saying, "Driver can't speak a freakin' word of English, obviously shouldn't be driving anything bigger than a bicycle, acts like it's *our* goddamned fault. Get Grooms out of here and into the Blazer."

He was manhandled to his feet, gang-marched down the front steps of the bus. Outside he got a peripheral glimpse of a black luxury sedan, still idling, driver's door open, dark-uniformed chauffeur standing there, waving his arms, flanked by dark-suited feds.

It was Jong all right! What magnificent *cojones!* Grooms felt a surge of adrenaline as he was jostled toward the Blazer. The chauffeur was gesturing indignantly now at the damage to his right rear fender. Then he yanked open the right rear door and pointed at the door panel.

That open door was only four strides away from Grooms. Now three—

He wrenched sideways, sprinting toward the waiting door, the men around him yelling. There was just enough time, Grooms estimated, for Jong to leap behind that door, then into the front seat and behind the wheel, while Grooms flung himself into the back. It'd be a hard fall with cuffs on, but so what? Assuming the big sedan was bullet proofed, they could burn rubber and disappear into the maze of streets and expressways circling O'Hare, all heavily trafficked by look-alike limos and luxury sedans.

Grooms launched his desperation dive—headfirst and handcuffed—toward Colonel Jong and the sedan's open door beyond.

But Jong didn't move out of the way. Instead he screamed as if in terror, then snatched from his jacket a small, snub-nosed pistol. But to Grooms it looked like a cannon's mouth as he flew helplessly toward it.

Fucking yellow-skinned assassin!
It was his last thought.

**INTENSIVE CARE UNIT,
NORTHWESTERN MEMORIAL HOSPITAL
1:37 P.M.**

Jane got the call as she was getting ready to be let into the ICU. She expected it to be bad news about her brother. But it was Singh, calling from the Akers transportation office with the good news about Grooms.

Having washed her hands, Jane was allowed in—one visitor at a time, the ICU nurse said sternly. Wade shrugged his understanding. Chevy's bed was facing the door, his face obscured by an oxygen mask, feeding tube, monitoring wires. The wall above his headboard bristled with gadgetry.

Please, God, help him get out of here and back in a jet or a sailboat—or wherever he wants to be.

"He's coming along nicely," the nurse whispered. "But he seemed a little too excited when I gave him your name. So, please, make it *very* brief."

She nodded, then smiled as she approached, covering her distress at seeing him so enfeebled. From the instant she'd first glimpsed him in the inferno glow of that hellish night in Dhahran, Chevy Johnson had been cast firmly in heroic, chin-thrusting mold. Now there was an almost apologetic mist in his blinking eyes, a quiver at the corners of the mouth that shaped a silent hello. Chevy's skin looked waxen, the back of his big right paw livid where the IV was inserted.

And yet, and yet, this new vulnerability made him somehow more appealing to her than ever. A puzzle to be worked on another day.

The nurse had said he'd be okay, so there *would* be time to work it out.

She'd do her part. Make the time. Definitely.

Jane laid a hand gently on Chevy's corded forearm. "You saved the day, Major. Thanks to you, the FBI tracked Grooms's van to O'Hare. They just arrested him in a parking lot."

Chevy was already smiling, or trying to, so Jane missed his reaction at first. It took her a second to notice his hands.

Both his thumbs were sticking up.

OVER BATCHAWANA BAY, ONTARIO, CANADA 4:19 P.M., E.T.

Five minutes earlier, radio off and head throbbing, Roy Akers Jr. had crossed the border just east of Sault Sainte Marie, where Lake Huron meets Lake Superior. He'd deliberately skirted the twin cities and the airport of entry on the Canadian side. But he'd been near enough so that, in the afternoon haze beyond his portside wingtip, he could see the International Bridge linking the two countries.

Air Traffic Control was probably eyeing him down there, trying to contact him. There'd been no signs of pursuit, but an unidentified private plane didn't exactly warrant interceptor jets.

Worry about this plane. And, indeed, according to his fuel gauge, he needed to find a safe, inconspicuous spot to set this contraption down—damn soon. Not *too* far into the interior, though; he was going to have to make his way back to a populated area. Which was why, with Sault Sainte Marie behind him, he'd headed west, back toward the iridescent shimmer of Lake Superior.

Inland from the scalloped, forested shoreline was a scattering of small lakes. Roy took a close look at several before losing his nerve. Lacking floats, he didn't have the guts to belly flop on water. What he wanted was a road out in the boondocks, preferably paved but definitely unpatroled.

His maps told him he was about twenty miles northwest of the border, following Highway 17 along the indent of

Batchawana Bay. He swung back east, losing altitude.

The beauty of the pines and cedars that now came rolling under his wings in endless green waves was lost on him. The euphoria that had fueled his escape was long gone. A pounding head and stomach-churning nausea were his only cockpit companions now. But the pain blunted memory and thought, and that was good. Because there could be no going back, no undoing what was done. Better to look ahead, gather his wits for the next few critical moments.

Then he saw it: a small paved road leading off from Highway 17. Untrafficked. Maybe a little-used logging road. But, dammit, there was too much flanking foliage to afford the necessary wing clearance. Still, he couldn't see any other option. He decided to check it out. Maybe he'd find a treeless patch.

After only a few miles more, he got his wish. The underbrush thickened, but the conifers began to thin out. There was maybe a mile of straight blacktop with scrub shoulder, wide enough—barely—for the Tiger's thirty-foot wingspan.

Okay, this is it! His hands went through the necessary motions, lining up for a straight shot, but in his belly Roy felt as though what happened between now and the ground was out of his hands. The Tiger's landing gear was already down, permanently locked. He shed altitude and airspeed, put in ten degrees of flaps, then twenty, held seventy knots on final approach, with asphalt and greenery rising fast to meet him.

Dammit! The pavement was cratered with potholes!

He could still abort! Except that he was too exhausted to pull out. One way or the other, let the whole damned thing be over. He flared the wings.

With sickening impact, the wheels slammed down, and something snapped—main gear legs, probably, because the undercarriage collapsed, the aluminum belly tearing open with a banshee shriek. Roy tried to control the skidding fuselage but discovered he was strictly a passenger, buckled inside a careening hunk of metal and relieved of making further decisions. He offered up a small, guilty prayer and waited to find out how it would end for him.

Now I know how Dad felt.

The Tiger leapt up a final time, as if attempting takeoff, then slewed wildly off the pavement into the bordering underbrush. The propeller and wings attacked the thick growth, plowing and slicing a desperate swath. But the thick vegetation slowed the onslaught, rank upon rank of black branches battering and slashing, choking the propeller, and, finally, shattering the cockpit canopy.

There was a strangulated scream—then . . . nothing.

It wasn't the end. Roy could hardly believe his good fortune as he climbed tentatively out of the wreckage—apparently uninjured. As he stepped from wing to ground, he twisted his left ankle—and began to laugh. Even that minor mishap took place after he'd landed! To add to his amazement, his headache was diminished. He felt more numb than anything else, hardly noticing that blood trickled from his scalp.

He staggered through mangled brush, reached asphalt, glanced back at the wreckage. A spacious silence seemed to mock the twisted remains of the plane. A jay squawked nearby.

Okay, you're still alive. But that's the only surprise. Nothing else has changed—you're still on the run. So get away from the plane. You need water, food, a place to rest. Keep moving—the ankle isn't that bad.

Pieces of the plane were strewn along the asphalt. A large aluminum panel. Dozens of nuts and bolts. A jagged fragment of fiberglass wheel fairing. Roy walked on past the debris.

Until, on the sandy shoulder, he saw something that made him stop.

He shuffled over, picked it up. Shiny green plastic case, about the size of a cigarette box. He turned it over, saw the corner magnets, felt fear pinch his heart. The hand that held the neat little contraption began to shake.

Because what it held was the Czech bomb, the one he'd last seen in Grooms's palm! The only difference was that the metal strip was gone—the one used to turn the arming screw. Dear God! The bomb must have fallen from the Tiger's undercarriage, dislodged during the bang-down landing. Which meant Grooms must have put it there to blow up his plane!

Roy crouched, his legs trembling as badly as his hands now, and set the plastic case on the ground as gently as he could. His eyes fixed on it, he stood, then backed away. After a few steps he whirled, walking too fast for his turned ankle—then breaking into a run. He hit a pothole and went sprawling.

He scrambled up instantly, ignoring the searing pain in his ankle, and lurched on down the road. Several panicky minutes later he tumbled, exhausted, alongside the asphalt. And this time he did not get up.

Think.

The bomb wasn't going to detonate now; his running had been for nothing. Grooms would have set it to go off at a certain altitude. Roy had saved himself by flying low. Still, eventually, as the land rose, he'd climbed above 2,000, even flirted with 2,500. Another quiver of the altimeter needle and he might have been blasted into atoms along with the plane.

Except Grooms hadn't known Roy would be taking the plane.

That bomb wasn't meant for him, it was intended for *Jane!* And maybe for Wade. Johnson, the new pilot, would have been a throw-in. Roy rolled sideways and vomited into the dust. *God forgive me!*

First Royal, then Elie Grey. Now his sister! Who was next?

But Grooms would come after Jane again, the instant he learned about the plane switch.

Roy had to go back. His own life was now scattered debris, just like the plane, but suddenly that was beside the point.

Whatever the consequences—of underestimating Grooms's lust for evil, of running away, of deserting his loved ones—he had to warn Jane and stop Grooms, this monster Roy had unwittingly unleashed on his own family.

Focusing on this single goal—and on the terror that would take place if he failed—Roy picked himself up and began to shuffle quickly along the potholed road, banishing the throb in his ankle, his thirst, and his hunger. He had to make his way back to Highway 17, flag down a vehicle, get to a telephone . . . .

Roy was still a distance from the main road when, twenty-five minutes later, a white sedan with light bar and official logo came thumping over the potholed surface toward him. Even as Roy raised his arms to wave it down, the car pulled to a stop and the driver's door swung open, revealing the logo of the Ontario Provincial Police.

A tall man in a trooper's hat and creased khakis climbed out, hand on his holster.

"Thank God you're here," Roy said.

"Are you Royal Akers Jr.?"

"Yes, sir, I am. And I need to—"

"Hold on there." The trooper's voice held command.

"I'm not armed," Roy said. "I swear."

"I can see that. But I can also see you're hurt, so stand still, will you?" The trooper moved forward, his hard squint softening slightly. He took Roy firmly under one arm. "Let me help you."

"It's nothing, Officer. The help I need is to call my sister. She's in terrible danger!"

AKERS PARKING LOT
4:32 P.M.

As Wade and Jane turned in to the main gate, a two-person camera crew jumped out of a media van parked nearby and came hustling forward. Jane, seated on the passenger side, realized she was in the direct line of fire of the approaching shoulder-mounted videocam.

"Wade, get us inside—now!"

Behind the wheel, Wade was waving his ID and calling to the guard, but the barrier arm remained down. Suddenly three men in business suits converged on their car—all on Jane's side, effectively blocking the cameraman. A moment later Wade toed the gas and they were scooting under the lifting gate arm.

"Thanks for the rescue," Jane said as the three men caught up to them in the parking lot.

"I meant to head it off altogether," said a man with a clownish fringe of hair and dead serious expression. "Ernie Macomber."

"Finally we meet." Jane took his hand. In the last critical hours she'd had several cell-phone conversations with the head of the Chicago FBI office.

"There's actually two news crews staked out on the street," Macomber added as they all set off toward the general offices. "Both from Champaign stations. These guys monitor police frequencies and obviously picked up on us chasing Cristofaro. When word gets out about everything else that's happened today, it's gonna get a helluva lot worse around here."

Jane nodded, but an impending media siege wasn't her foremost concern at the moment. "What about my brother? Is he—has he landed?"

"He's down and okay."

"Thank God! Was there a bomb?"

Macomber nodded. "We just got a call from the Ontario Provincial Police. Their explosive disposal unit has the device now—your brother found it. Lucky guy—it hadn't detonated."

"Ontario? Roy's in Canada?"

"He crash-landed his plane on a logging road in the bush about twenty miles northwest of Sault Sainte Marie. Our Air Traffic Control had alerted Canadian ATC, so when the plane disappeared from their radar screens, they sent troopers out looking. One of 'em picked up your brother hiking out from the crash site."

"He's really okay?"

"Well, he's pretty banged up—nasty scalp laceration, probable concussion—but, like I said, obviously lucky to be alive, especially considering the bomb. He's in a hospital up there." Macomber drew a slip of paper from his suit pocket. "Plummer and General in Sault Sainte Marie. He's expecting your call."

Macomber opened the door to the general offices, motioning for Jane to precede him, but she came to a full stop. "What did Roy say? Is he in custody?"

"You might say that. He's definitely being held for questioning. Both the FAA and the Canadian Aviation Safety Board will likely be conducting inquiries. There are federal regulations against careless and reckless flying—deviating from a flight plan, staying out of contact with ATC, entering Canadian airspace without clearance. I'd be kinda surprised if your brother didn't lose his pilot's license and have to pay some kind of civil penalty. The FAA itself can't initiate a prosecution, but individual states can, and in this case that'll be up to Illinois, Wisconsin, and Michigan.

"And, of course, *we* want to ask your brother a whole lot of questions about his connection to Grooms and Cristofaro, and about why he was trying to flee the country."

"But Roy did ask to speak with Jane, Mr. Macomber?" Wade put in. "That's a hopeful sign, isn't it?"

"Yes, apparently the first thing he told the trooper was that he had to warn his sister about Grooms."

Jane led the others inside, eager now to talk to Roy herself, to try to fathom what had happened—and to help him, if she could. A few strides down the corridor, Macomber caught up to her. "You've heard about Grooms, Ms. Akers?"

"That you caught him at O'Hare. Singh phoned me."

"Old news. Robert Grooms is dead. I think we'd better talk

before you make that call to your brother."

In Jane's office, Macomber introduced his two colleagues—Pete Petrossian of the FBI's Computer Crimes Division, and Ben DiLucca, a boyish-looking Justice Department prosecutor for computer-related crimes—and then suggested to both men that they attend Singh's system meeting down the hall. As they were leaving, Don Landsman and Lynford Harlow arrived to sit in on Macomber's briefing, for which Wade also stayed.

Jane accepted the delay before she could talk to Roy. Just knowing that he was alive was enough—for the moment. Her whirling gut understood that before long she'd have to learn more, and that it was unlikely to be good. But the truth was, for all their fierce clashes in recent days, she desperately wanted to believe the best of Roy—as long as she could. Because she remembered how she'd felt as she'd watched his tiny white plane fading into the blue, sensing that it was a doomed flight even before she'd learned that Robert Grooms had engineered a death flight for her and for Wade. What she'd felt, mostly—and unexpectedly—was love for her brother.

"Before we get started," Macomber said when they were all seated, "I want to acknowledge our debt of gratitude to your company pilot, Mr. Cheval Johnson, who really deserves the credit for our capturing Grooms. I understand Mr. Johnson is doing fairly well?"

Jane nodded vigorously. "He's got what it takes to make it."

"Unfortunately, we didn't hold on to Grooms very long." Macomber gave a brief account of the shooting in the airport parking lot. "The limo driver is claiming self-defense, saying he got scared when a total stranger in handcuffs rushed him, and that's why he went for his gun." Macomber paused, his expression dubious. "And somehow shot him right between the eyes?"

"Sounds more like a professional hit than someone acting in self-defense," Don Landsman said.

"Doesn't it?" Macomber said. "It sounds, in fact, like a textbook terrorist hit—every detail, from the way the driver staged the crash to the point-blank fusillade. Grooms clearly knew this guy—he was running toward the open back door of the limo. Must've figured it was a rescue. The driver's documentation says he's Yoo Ho Shim, a Canadian citizen of Korean descent, born and raised in Japan, who came to the States in 1995 on a temporary work visa. In addition to being employed by Sunrise Limousine Services of Chicago, Yoo does occasional security work and has a license to carry a gun.

"I wasn't convinced. So we faxed Mr. Yoo's photo to South Korea's National Security Planning Agency, and they've just faxed back. They ID him as Jong Ye Nam, a colonel in the North Korean People's Army, a guy they lost track of a few years back. Whoever he is, he's being held for further questioning, but we may have to just ship the SOB back to Canada after a closed deportation hearing based on classified information."

Macomber glanced at Jane. "Sort of intriguing, isn't it, a possible North Korean connection to Grooms, what with you folks being accused of damaging the South Korean economy and government because of all your computer viruses?"

"*Very* intriguing," Jane said.

"Maybe now we know who's been making it happen," Don Landsman said. "And why."

"Unfortunately," Macomber said, "Colonel Jong or Mr. Yoo or whoever he is isn't answering any questions. And we can't talk to Grooms. Which leaves—"

"What about Lenny?" Wade asked.

"Ah, Cristofaro, yes. He apparently told Sheriff Bevins he was totally innocent, but Ben DiLucca, whom you just met, had a few minutes to talk with Cristofaro in the presence of a public defender. He explained we were investigating a conspiracy involving not only computer sabotage, but several homicides as well. Once Cristofaro heard the word 'homicide,' according to Ben, the public defender had to basically shut him up. Cristofaro implicated Grooms right away, but said nothing about any Koreans—or, for that matter, Roy Akers."

Hope rising in her, Jane said, "Mr. Macomber, can you give me a number to reach my brother in the hospital?"

"Right. But I don't want you volunteering any information about Grooms—even that he was killed—to your brother, not till we question him." His eyes told Jane he wasn't making a request. "Do you understand?"

Jane nodded. He gave her the number and inclined his head toward the phone. It was clear he would stay while she called. She assumed the call would be recorded.

She dialed.

When he answered, for a long second her voice jammed in her throat. Then she managed, "Roy? It's Jane."

"Jane! Thank God you're okay! You got the message about Grooms? He's got—"

"Yes, Wade and I are both okay, don't worry."

"Jane, that man has to be stopped. And the family, have you talked to everyone? Are you still in Chicago? Is Grace there?"

"We're all okay. I haven't talked to Grace, but I know she's okay. I'm in my office at Akers. The . . . the FBI are here. Everyone's being protected."

"But what about Grooms? The man's a maniac, Jane. Did they get him?"

"All I know, like I said, is the FBI is on the case, and . . . all appropriate steps are being taken. What about you?" Jane stopped short, leaving the question vague and open-ended.

"Me? Oh, I'll survive." There was a long pause. "I was running away, Jane. I know, it's despicable. I'm hoping that somehow Tyler won't ever have to know about that. But I panicked, thought I couldn't face . . . everything—everyone. I have to, though, I realize that now. I didn't mean to do anything that could hurt anyone. I'm not saying I'm innocent. God, no. I tried to take advantage of the company's troubles for a sale to Soverel. But I'm no saboteur. I didn't know what Grooms was planning. Until this morning, when that Down's boy pointed to him, I never dreamed Grooms had anything to do with Dad's crash, or Elie's death. Jane, you have to believe that!"

"I'll try, Roy. I'll really try."

"That's all that matters to me now—the family. I'm finished with the company, I know that. But what you and Alex and Grace think of me, and Louise and Anne—that matters more than I ever understood till now. I pray to God that you can all forgive me. And I hope it's not too late for me to have another chance at being a father. You can tell the FBI I'll tell them anything they want to know. I have a lot to answer for, but I didn't mean to hurt anyone. You do believe me, don't you, Jane?"

She searched herself. "I believe you, Roy." She glanced at Macomber. "And I'll tell them what you said."

Jane hung up feeling some relief. Five minutes ago she had been afraid that her brother was involved in terrible crimes. If he hadn't taken part in *those* things, well, she and the family could face the rest. Whatever it turned out to be.

Not that it would be easy, of course. There was sure to be a mountain of rocks thrown in the unavoidable scandal. The story would doubtless fill several spreads in every racy tabloid and a number of subtler but even more injurious columns in business journals. God only knew how tonight's TV news would treat Roy's flight—and his crash landing in Canada—especially with the inevitable links to several homicides. How could the always thirsty fourth estate *not* milk all of it?

If only they could just magically pull the TV plug for seventy-two hours! What they could do—had to do—was to monitor everything and react quickly, get their PR people out there on damage control . . . . She needed to schedule a crisis meeting as soon as possible.

Her mental checklist was interrupted by Macomber, who

clearly wasn't finished here.

His sole concern, of course, remained law enforcement. "The NTSB is sending its investigator back from their West Chicago office to examine Mr. Harlow's evidence in the matter of the plane crash," Macomber was saying. "All jurisdictions, including the Illinois State Police, are cooperating fully. Even the CIA is helping us in ferreting out Mr. Grooms's past."

The man clearly had experience in all kinds of dirty work, Macomber told them. Grooms's van, recovered from the O'Hare lot, was being examined inch by inch, as was the duplex apartment he'd rented in Ocean Plains. In both van and apartment they'd quickly recovered weapons and electronic surveillance gear. "Plus syringes and potassium chloride," Macomber added. "A little injectable death kit."

Which backed up the testimony of Zach Grey and the preliminary autopsy finding of lethal doses of potassium in Elie Grey's bloodstream.

"Could any of that surveillance gear in Grooms's van be used for monitoring cellular calls?" Wade asked.

"Sure could," Macomber said. Grooms had been driving around with a sophisticated radio direction-finding unit, it turned out, complete with a five-channel receiver for monitoring both sides of a conversation simultaneously. Further, at Lynford Harlow's suggestion, the FBI had searched for and located a direct tap on the phone in Elie's Grey's farmhouse.

"Which means Grooms was able to keep track of what we were doing on the farm," Harlow said. "He went into action the instant I told Elie about finding evidence of sabotage."

"I think he waited until Elie got hold of us in Chicago," Jane said. "Hearing Elie tell us to come back at once—that probably triggered it. Grooms must have immediately phoned the wrecking yard to come and pick up the plane. Then he attacked Elie, injecting him with a lethal dose of potassium chloride. That's how I'd read the sequence of events, anyway."

"We do know for sure Grooms was the man who met the wrecking truck at the Grey farm," Harlow put in. "Their truck driver was able to positively ID him from a photo this morning."

"That's two probable homicides," Landsman said. "Elie and Jane's father—and what about the attempt on her brother? That Grooms was a certifiable psychopath."

"I don't know about certifiable," Macomber said. "But from his CIA dossier, these wouldn't be the first murders he's committed, Mr. Landsman. Grooms was let go on suspicion of siphoning off Agency funds, but he was also suspected of killing some of his own agents in Pakistan. Working with North Korea would be right up his alley."

"Thank God he didn't go after you, Mr. Harlow," Jane said. "I mean, you were a real threat to him."

"Someone definitely tried," Macomber said. "We removed a plastic explosive device wired to the ignition of Mr. Harlow's car— could be the same type of device used to blow up the plane."

Jane turned to Harlow. "But you've been driving your car!"

"Not that car, I haven't. When I learned of Elie's death, I took a few precautions. Stayed away from my apartment and my car, checked into a motel in Red Knoll, drove a rental."

"Smart moves," Macomber said, nodding affirmation. "As furious as I am that we let someone get to Grooms while he was in our custody, and that we don't have him around to interrogate and prosecute, I have to admit it's a relief to know the bastard has definitely left the planet."

It took several calls to locate Alex at the Drake. He was down in the Oak Terrace restaurant, having a late lunch with Anne. Grace Akers had been given five milligrams of Valium, he reported, and was sleeping soundly, with Aunt Lou doing her needlepoint at Grace's bedside. Jane gave him an extensive update.

"Alex, we just don't know yet how involved Roy was in everything that's happened. He's filled with remorse about *something.* But it seems to me that if we all pull together, we'll get through it . . . together. This isn't the way any of us would have chosen, but maybe we can become more of a family than we've been for a long, long time.

"In the meantime," she said firmly, "you've got to take charge. Grace and Louise and, especially, Tyler are going to need you now. Desperately. I'll do whatever I can to help, but you're closer to them, you're the one they'll be able to turn to."

"I'll do it, Jane. You wouldn't know this, no one does. But I've always had a secret longing to take care of *somebody.* You're right, you know, this *is* a rotten way to get my chance—but I'm going to grab it with both hands."

"I just bet you will."

"Thanks. Do you want me to tell Mother?"

It took Jane only a second to decide: "*Would* you?"

45 MAPLE STREET
9:02 P.M.

Jane was slumped in her father's big La-Z-Boy. At the other end of the family room, Alex got up and switched on the TV—then, glancing over at Jane, immediately turned it off.

"I'm going upstairs to check on Tyler," he announced.

The boy's mother was also up there now, still being watched by Louise. After being brought back from Chicago by Frank Kilfoyle, Grace had spoken at length to Roy, who was still in Canada, and she was apparently doing well enough now to be off the sedatives prescribed by Dr. Ling of the Ocean Plains Clinic.

The clear priority, Jane thought, was family first, company second. Although, in an important sense, all Akers *was* family. Plainly, she'd just have to get through the days one at a time. That's what they all had to do. So she had deliberately unplugged from the law enforcement investigations for the duration.

"Jane?"

"Yes, Anne." Jane looked across at her stepmother. The older woman's face was in shadow. Her posture on the sofa was, as always, perfect, though she had to be utterly exhausted.

"You mustn't blame yourself, not in any way." Anne was responding to a remark Jane had made—what? an hour earlier—about perhaps having pushed Roy too hard. "Roy's alive, Jane. He's coming back. That's a blessing—if not a miracle. As for whatever happened, it happened around you, not through you."

"I bet Aunt Lou doesn't feel that way. And Grace—I'm even afraid to see her, Anne. Why should I feel that way, if there's no reason for me to feel guilty?"

"Because it wouldn't be human not to."

Jane tried to take that in, to believe that was the only answer. The silence in the room was italicized by the ticking of the hall clock. After a few minutes Jane heard footsteps coming downstairs.

In the doorway Alex said, "Ty's asleep. Either of you want anything from the kitchen?"

Neither woman did, and Alex left. After a moment they heard the refrigerator door open and, moments later, close.

Anne said, "You remember what you asked me—oh, it seems forever ago, but probably it's only three or four days—when we had our little talk in the breakfast nook? You asked me how I won over the family, something like that."

"Of course I remember. I remember what you said, too."

"Well, I don't. But here's the answer I'd give today. It took a long time. I didn't *try* to do it all at once—you know, show up on the heels of Joycelyn's departure and just charm everybody, do handstands so they'd accept me. The truth is, I had no plan, not even a specific goal. I just knew I had to take things day by day, offer my best to each person, ask myself, How can I help him or her? What needs to be done, and what part of that can I attempt, without offending? How can I reach out to this person? You see? Now why should I have taken such elaborate pains when I wasn't really guilty of anything, either, at least in my eyes?"

"You're saying—?"

"I'm asking, Jane. Do you want to be part of this family? Or . . . "

"If you're asking if I'm planning to stay . . . I don't know if that's possible, for a number of reasons. But I can tell you that, here or back in San Francisco, I very much want to be a part of this family—what's left of it."

"I'll settle for that, for now," Anne said. "I want you to do what's best for you, but do think about this. We all need each other a little more now. Especially Tyler. That little boy could definitely use a full-time aunt."

A full-time aunt. Not part of her vitae, but an addition she'd very much like to make. Only it would be a title she'd have to earn. Roy might actually welcome it, she thought, judging from the contrition she'd heard in his voice this afternoon. Grace would probably be skeptical and protective of the boy at first, as she should be. But with attentiveness and persistence . . . and love . . . the title could be won.

"Speaking of Tyler and his needs," Anne went on, "one small miracle I'd like to see you pull off is talking Alex into settling down here."

Jane actually chuckled. "Anne, getting Alex to *settle* anywhere will be the tough part. But you're right. It would be wonderful for Tyler. Maybe for Alex, too. Because he's needed by more than Tyler. He'd make a great addition to Akers. The fact is, he's a living, breathing PR magician. The hard part will be persuading him of that. But maybe God and I *can* manage one more miracle."

DAY 9

EXECUTIVE CONFERENCE ROOM, AKERS HQ
10:00 A.M.

The emergency board meeting was held in a charged atmosphere. Although they had been notified of Roy's flight and crash landing before the Monday evening news, Akers's directors were still trying to take it in. To that shock were added allegations of criminality upon which the media had pounced with ferocious appetite. As Jane looked around the table, her eyes came to rest on several vacant chairs—Roy's, Elie's, and Wally Conover's.

In a late evening conference call, she and Don had asked Wally Conover to convene this meeting, and he had agreed to do that. "But," he said, "I won't be there. You've both deliberately kept me out of the loop. You can't really expect me to come in there and act as if I know what's going on—when I have no idea!" He hadn't wished them luck.

Madeleine Archer, the board's first female member, a long-time director and, since Royal's death, acting chairperson, tapped on her water glass, waited for the murmuring to stop, and began the meeting.

After only a very few soft-spoken introductory words, she yielded the floor to Jane, who took it upon herself to clarify for the board members what was known so far. She did not try to gloss over the possibility that there might be more calamitous news to come, admitting that there was no way to know yet what the final toll on the family and the company would be.

"Obviously we're going to be cooperating fully with an ongoing criminal investigation here," she said. "And I know that all of us will do whatever we can to assist in that. The media will be dogging our steps at every turn. We can't stop them, but we don't have to get sucked in by them, either. We'll be addressing how best to deal with the media in a separate PR meeting later today." She explained that the rumored takeover deal by Soverel

Corporation was defunct and that Tom Soverel was distancing himself from their company and its woes.

"Which is *not* bad news. But it does mean that, in the midst of everything, we have to go full speed ahead to try to save this company, for all of us and each of us. It's my opinion that putting the company back on track and running smoothly is also the key to restoring cohesiveness to the Akers family and honor to the Akers name.

"Permit me to make absolutely clear that when I speak here of the Akers family, I don't mean only my immediate family, although I mean them—us—too. But when my father built Akers, he extended his own notion of family to cover everyone in the company. True, his wife and his children and grandson are his heirs. But it seems to me that, given the way my father thought and operated and *lived*, his real heirs may be not his immediate family, but the extended Akers family—the thousands of men and women across this country and the world who have invested their working lives in it. I think *that* family cares as much as Dad's blood relatives do about Royal Akers's goal of giving the customer the best product at the best price. Thousands of Akers people give *their* best each day to that effort. And their doing that is the best reason for fighting to save the company. Not to the end—till we *win*."

She paused. They had lost their look of distraction, and even the directors whose expressions were more skeptical than convinced were paying close attention.

But she had to get the support of them *all*.

"Yesterday," she said, "we requested that markets suspend trading in our stock until the facts about our news—both the tragic and the hopeful—can be assimilated and, we hope, interpreted properly by the Wall Street gurus. Our request was granted. But, as you know, the share price has already fallen through the floor. If there's any good news, marketwise, it has to be that there's nowhere left to go but up."

As Jane glanced around the table, it was clear that her words hadn't persuaded the naysayers. But it was part of the whole picture, and the best thing was to get it out in the open.

She took a breath and managed a small smile. "On other fronts, I am pleased to be able to report, there is very good news indeed. R. K. Singh, who has headed our internal investigation of the sabotage of our computer system, has informed Don Landsman, Wade Crain, and me that he has full confidence he can bring up the integrated logistics system and enable EDI. As *we* have full confidence in *him*, that is good news indeed. After the most thorough possible investigation, Mr. Singh discovered nothing wrong with the hardware or the software."

She waited a beat, then said very slowly, "Ladies and gentlemen, the problem with our computer system was that its prime caretaker was also its prime saboteur, having undertaken his filthy task, as some of you are now aware, at the behest and coercion of our security chief, Mr. Grooms."

As several directors *hadn't* yet heard these astounding revelations, it took several minutes before Jane could quell the hubbub and get their attention back. "By tomorrow morning," she said then, "Mr. Singh has assured us, we'll be ready to boot up all systems, virus-free, and inoculated against further infection. He is working with various department heads on a timetable for bringing our DCs and our stores on-line thereafter . . . .

"Please," she said, holding up her hand to stop the scattered applause. "There's something else I want to tell you. I'm not sure every cloud has a silver lining, especially not when the cloud's the size of the one that's been hanging over Akers. But the treachery of Lenny Cristofaro and Robert Grooms offered us two golden opportunities. First, it made Akers people pull together in a way that lots of folks don't think happens anymore—*but it happened here.* And none of them—or us—will ever forget that.

"Second, the crippling of our computer system gave us a chance to prove that even in this age of technological wonders, *they* can fail, but the human mind can keep on working without them. We showed here that while logistics *uses* technology, it *depends* on training and skill and experience and good judgment and, perhaps most of all, the willingness of a logistician to trust that his or her intuition—that blessed educated gut we develop—will be able to figure out a way to get the job done.

"I'm pretty proud of what we've made of both opportunities, in spite of my contempt for the people who inadvertently handed them to us."

She swallowed. Twice. "And I'd like to think that my father, wherever he is, is a little proud right now of what all of us have done together."

There was silence as Jane sat down. Had she talked too long, keeping them from their unfinished business?

Because the remaining issue to be put before the board had been hanging in the tension-thick air of the boardroom since the meeting began. The vacant chairs at the conference table kept it alive throughout the meeting.

Who was going to run things on an interim basis?

Madeleine Archer was the first to speak up. "Jane—may I call you Jane . . . "

"Of course. Haven't I suggested we're all family here? Please, go on."

"Well, you've been totally involved in returning the company to working order. I think you've earned the right—and perhaps you also have the obligation—to tell this board who you would recommend to carry things forward. You yourself have said there's still plenty to be done. Well, who do you think is best qualified to lead us—at least for the time being, so we keep moving forward and can finish the job with deliberate speed—the way you got us to this point?"

Jane was ready. "I won't pretend I haven't thought about

this—in my fifteen or twenty free seconds each of the last couple of days. But I have to admit it didn't require a lot of thought. I think you ought to see if you can persuade Don Landsman to be acting CEO—maybe you can even talk him into taking the job permanently. I think you may want to ask him as one of his first tasks to put together a search committee for a new president. If Wally Conover wants to stay on, which I somehow doubt, I'd suggest moving him one step down to executive vice president."

Jane waited until the murmuring died down before she said, "After working with him the past ten days, I can't imagine anyone more qualified to be director of distribution than Wade Crain. I also want to recommend to you my brother Alex, who was so instrumental in rebuilding goodwill between us and the Hyon Group, when that appeared nearly impossible to accomplish. I haven't had a chance, given the circumstances, to broach the subject with Alex, but I dearly hope he will want to take an active role in exactly that kind of thing, improving relations with our major suppliers, and with community outreach programs in Akers towns."

"What about you?" Sam Spencer asked. Sam had been on the board ever since he retired, after putting in twenty years running the Pacific Northwest region for Akers.

Now Jane did hesitate. Both Don and Wade had been pressing her to come back to the company on a full-time basis. In fact, Don told her that even his staying on temporarily would be contingent on her doing the same. A permanent commitment to Akers would of course mean the end of her own shop, which she had fought so hard to create and which, with the TranSonic Airex deal now settled, was finally beginning to show results. And what about Mark, whom she had lured away from academia—and on whose shoulders she had placed so much during the past ten days? Not to mention Bridey, who was due back next Monday after suffering her own family tragedy. There was no chance she'd leave Bridey in the lurch at such a time.

As she stood there, unsure of how to answer Sam Spencer's question, she remembered once telling Chevy Johnson that it almost didn't matter where in the private sector she found her challenges. But suddenly that seemed facile. It did matter. As much energy and passion as she had poured into Malcolm and Associates, Akers had come first in her life. And Akers still came first. This was where she belonged. Maybe Mark and Bridey could be the first members of her new team.

That's my girl.

Guess I am, Dad. Besides, if I left here, how do I know you'd follow me back to San Francisco and keep giving me advice?

I'd come.

All the more reason for me to stay, then.

"Thank you all for being patient. Frankly, I'm used to thinking fast on my feet, but not when my answer has to do with the whole rest of my life."

"If you need time to decide, we darn well ought to give it to you, is what I think," Madeleine Archer interjected.

"Hear, hear," came a few voices.

"That's very generous, and I appreciate the offer. But I have the answer now. I would like to stay on at Akers. Very much. As vice president, corporate logistics. That's the job I want," she said.

"We never had one of those," someone muttered.

"We never had anyone else aboard with Royal's smarts, either," Madeleine Archer said. "I say we give this young woman the job she wants, and thank our lucky stars she's going to be here."

A whole chorus of amens followed that.

Vice president, corporate logistics. It wasn't the same as being president of Malcolm and Associates, but it felt right. Good, too.

One thing she had to do before she felt *all* right was to call

Mark and Bridey, too, and discuss their job futures. Both calls would definitely include an offer to join Akers—in responsible jobs, with meaningful titles. Like her new one.

Suddenly she remembered that vice president, corporate logistics, was the same title Tom Soverel had dangled in front of her. How could that have slipped her mind? Now to beat the pants off him and 4-Mor Shopping. She smiled to herself at the prospect.

Jane. You really forgot?

Honest, Dad. Got a lot on my mind here.

A sigh.

Then: *I kind of hoped I could go back to spending my time here learning how to loaf. Play some chess with Elie, maybe do a little fishing. But if you're forgetting things already, maybe I better hang around and keep an eye on you.*

Dad, there's fishing there? Where are you, anyway?

Certainly not where I expected to be this soon.

That's an answer?

It's all the answer you're getting, daughter. Now, seems to me you've got a fine job offer there. There's work that needs doing. Let's do it!

Yes, sir!